Formation, Exploration and Development of Natural Gas Hydrate

Formation, Exploration and Development of Natural Gas Hydrate

Topic Editors

Hongsheng Dong
Lunxiang Zhang

MDPI

Basel • Beijing • Wuhan • Barcelona • Belgrade • Novi Sad • Cluj • Manchester

Topic Editors

Hongsheng Dong
School of Marine Science and Technology
Northwestern Polytechnical University
Xi'an
China

Lunxiang Zhang
School of Energy and Power Engineering
Dalian University of Technology
Dalian
China

Editorial Office
MDPI AG
Grosspeteranlage 5
4052 Basel, Switzerland

This is a reprint of the Topic, published open access by the journals *Energies* (ISSN 1996-1073) and *Journal of Marine Science and Engineering* (ISSN 2077-1312), freely accessible at: https://www.mdpi.com/topics/32C29LI6Y7.

For citation purposes, cite each article independently as indicated on the article page online and as indicated below:

Lastname, A.A.; Lastname, B.B. Article Title. *Journal Name* **Year**, *Volume Number*, Page Range.

ISBN 978-3-7258-3713-7 (Hbk)
ISBN 978-3-7258-3714-4 (PDF)
https://doi.org/10.3390/books978-3-7258-3714-4

Contents

About the Editors

Hongsheng Dong

Dr. Hongsheng Dong is currently a Professor at Northwestern Polytechnical University. He earned his B.S., M.S., and Ph.D. degrees consecutively from Dalian University of Technology, complemented by academic visits to the University of Edinburgh and postdoctoral training at the Dalian Institute of Chemical Physics (CAS) and Xi'an Jiaotong University. Currently serving as a doctoral supervisor, Prof. Dong leads cutting-edge research in underwater vehicle thermal management systems and marine energy conversion technologies at the intersection between naval architecture, power engineering, weapons science, and advanced materials. His work pioneers a comprehensive "material–device–system–product" research framework for marine applications. Notably, he contributed as a core member of the Autonomous Underwater Vehicle Innovation Team developing next-generation solutions for underwater energy systems. With dual excellence in academia and industry, Prof. Dong served as Senior Engineer and R&D Project Manager at Fortune 500 corporate Midea Group, where he bridged theoretical research with commercial manufacturing. His scholarly credentials include leading several national and provincial research projects, authoring over 70 papers, and securing many patents.

Lunxiang Zhang

Dr. Zhang Lunxiang is currently an Associate Professor at Dalian University of Technology, where he completed his B.S., M.S., and Ph.D. degrees. He serves as Deputy Director of both the Institute of Energy and Environmental Engineering and the Liaoning Provincial Key Laboratory of Hydrate. His scientific interests center on phase transition theory and control technologies for natural gas hydrates, with applications in flow safety assurance and multi-energy storage/utilization. Dr. Zhang has led over 10 significant research initiatives including sub-projects and projects of the National Key R&D Program and the National Natural Science Foundation (General, Youth, and Key Program categories). Some of his recognitions include the CAST Youth Talent Support Program. His honors encompass first prizes in Natural Science from the Ministry of Education and Chinese Society of Engineering Thermophysics, first prize in Technical Invention from the China Petroleum and Chemical Automation Association, as well as provincial awards for scientific advancement and water conservancy technology. In the past five years, he has published over 60 first/corresponding-author papers in top-tier journals including *Nature Communications*, *ACS Nano*, *NPJ Clean Water*, *Engineering*, and *The Innovation Energy*.

energies

Editorial

Formation, Exploration, and Development of Natural Gas Hydrates

Hongsheng Dong [1], Lunxiang Zhang [2,*] and Jiaqi Wang [3,*]

1 Dalian National Laboratory for Clean Energy, Dalian Institute of Chemical Physics, Chinese Academy of Sciences, Dalian 116023, China
2 Key Laboratory of Ocean Energy Utilization and Energy Conservation of the Ministry of Education, School of Energy and Power Engineering, Dalian University of Technology, Dalian 116024, China
3 College of Power and Energy Engineering, Harbin Engineering University, Harbin 150001, China
* Correspondence: lunxiangzhang@dlut.edu.cn (L.Z.); jiaqiwang@hrbeu.edu.cn (J.W.)

1. Introduction

Currently, natural gas hydrates (NGHs) have been proposed as promising and environmentally friendly carbon-based energy sources that are beneficial for mitigating the traditional energy crises [1]. NGH is an ice-like crystal composed of methane molecules enclosed in a water lattice under low temperature and high pressure, and it is mainly enriched in deep marine reservoirs or permafrost zones [2]. The energy density of NGH is extremely high because 1 m^3 NGH can disassociate into 164 m^3 methane at standard temperature and pressure. In addition, the amount of organic carbon contained in NGH is twice that of other traditional hydrocarbon reservoirs, according to the resource's assessment [3]. Therefore, studying the formation, exploration, and development of NGH at a greater depth is imperative for promoting the commercial production and utilization of NGH.

The exploration, exploitation, and further application of NGH are interrelated. Understanding the NGH's formation and dissociation is the foundation of its exploration, development, and applications. The exploration serves as a connecting link between formation and development. The exploration is highly dependent on understanding the formation characteristics and provides references for selecting development sites and methods. The development directly tests the theories and practices of the formation and exploration and advances NGH to commercial production. Although some pilot productions of NGH have been carried out in some countries and regions, some unforeseen problems still remain, which limit its commercial exploitation [4,5]. The comprehensive research on the properties and formation of NGH is paramount for guaranteeing its efficient and effective exploration and development. An in-depth study of the fundamental properties of NGH formation and dissociation can provide significant guidance for its exploration and development [6]. The practices on the exploration and development, in turn, can also provide helpful insights into its formation and dissociation. The Special Issue of *Energies* on the subject area of "Formation, Exploration and Development of Natural Gas Hydrate" aims to collect the latest research outputs on the theory and practice in basic properties, novel exploration technology, and highly efficient development process. We hoep that this Special Issue can spur advancements and provide a more effective strategy for the field production of NGH.

2. Formation and Exploration of Natural Gas Hydrate

Distinct from conventional oil and natural gas deposits, marine NGH is widely enriched in argillaceous low-permeability sediment reservoirs, which are characterized by non-diagenesis and weak cementation. The basic physical characteristics of marine hydrate reservoirs and their spatial-temporal evolution are the decisive factors for understanding

Citation: Dong, H.; Zhang, L.; Wang, J. Formation, Exploration, and Development of Natural Gas Hydrates. *Energies* **2022**, *15*, 5951. https://doi.org/10.3390/en15165951

Received: 28 July 2022
Accepted: 16 August 2022
Published: 17 August 2022

the reservoirs, which transform and realize the development of hydrate resources. Although the phase equilibrium [7], structural properties [8], mass transfer mechanism [9], heat transfer analysis [10], growth kinetics [11,12], and phase change parameters [13] of NGH have been well-studied in the past, a need for a fundamental and thorough understanding of NGH formation and occurrence still remains. In particular, developing high-precision, multi-dimensional, and comprehensive exploration technologies are the future direction of deep-sea NGH exploration [14]. The advancement in this field will be conducive to ascertaining the geological condition, revealing the reservoir-forming mechanism, proving the NGH reserve, and disclosing the NGH distribution. This Special Issue plans to collect the latest advancement in the formation mechanism and exploration technologies of NGH and provides a basis for screening NGH's sweet spots.

3. Dissociation and Exploitation of Natural Gas Hydrate

NGH is considered the most promising potential clean energy source for substituting traditional energy sources [15]. Many countries, including Russia, Canada, the USA, Japan, and China, have launched pilot field productions of NGH and have made rapid and important progress [16]. The development of NGH has moved from early field surveys and laboratory tests into field production, and it is at the transitory stage of commercial production [17]. Although some technologies, including depressurization [18], heat injection [19], CO_2 replacement [20], and other novel methods [21], have been widely studied, these technologies are still immature because of their low yields, high costs, and low efficiencies. In addition, the spatial-temporal variation and multiple-time-scale characteristics of heat, fluid, stress, structure, and distribution in NGH reservoirs are unknown, impeding the large-scale production of NGH. The joint development of hydrate and oil and gas with multi-type, different occurrence forms, and various depths will be the critical points of future deep-sea hydrate development strategies. This Special Issue will focus on the recent advancement of the highly efficient exploitation method of NGH and drive the large-scale and commercial development of NGHs.

4. Flow Assurance of Hydrate Blockage

Safe and efficient development has always been a central topic in the oil and gas industry. Since 2000, significant oil and gas discoveries have been made in deep waters around continental edges. Compared to onshore development, offshore development safety issues are more prominent because of (i) more expensive offshore facilities and (ii) marine ecological disasters caused by oil leakage and geological disasters by hydrate blockage [22]. Continuous accidents due to hydrate blockages were reported, such as Statoil Tommeliten-Gamma, Norway, Wyoming Werner-Bolley, America, Jinzhou 20-2, China, Roncador Field, Brazil, Marlim Field, Brazil, and others. Therefore, the flow assurance of hydrate blockage management attracts increasing investigations from both scientists and engineers. Extensive efforts have been made for accurate predictions as well as detection, effective prevention, and functional remediation [23]. This Special Issue provides accounts of recent concepts and technologies in flow assurance, and some novel methods such as interfacial modification and green inhibitors would help the management of hydrate blockage.

5. Further Applications of Gas Hydrate

In recent years, hydrate-based technologies have been considered as promising alternatives for solving numerous energy- and environment-related issues, such as H_2 storage and separation [24], CO_2 geological sequestration [25], refrigerant as suitable cold storage [26], heavy metal separation and desalination [27], and others [28]. Extensive efforts have been dedicated to optimizing the hydrate growth rate and the application efficiency across various scales. The latest research studies reported impressive experimental results and functions with great promise; however, this field is still far from industrial application and is limited by effective rapid formation methods and suitable adjustable systems. Thus, this

Special Issue also pays attention to the new discovery of well-controlled hydrate phase changes and development to promote the further application of gas hydrates.

Funding: This research received no external funding.

Conflicts of Interest: The authors declare no conflict of interest.

References

1. Aman, Z.M.; Koh, C.A. Interfacial phenomena in gas hydrate systems. *Chem. Soc. Rev.* **2016**, *45*, 1678–1690. [CrossRef] [PubMed]
2. Sun, L.; Wang, T.; Dong, B.; Li, M.; Yang, L.; Dong, H.; Zhang, L.; Zhao, J.; Song, Y. Pressure oscillation controlled CH_4/CO_2 replacement in methane hydrates: CH_4 recovery, CO_2 storage, and their characteristics. *Chem. Eng. J.* **2021**, *425*, 129709. [CrossRef]
3. Yu, Y.; Zhang, X.-W.; Liu, J.-W.; Lee, Y.; Li, X. Natural gas hydrate resources and hydrate technologies: A review and analysis of the associated energy and global warming challenges. *Energy Environ. Sci.* **2021**, *14*, 5611–5668. [CrossRef]
4. Yan, C.; Ren, X.; Cheng, Y.; Song, B.; Li, Y.; Tian, W. Geomechanical issues in the exploitation of natural gas hydrate. *Gondwana Res.* **2020**, *81*, 403–422. [CrossRef]
5. Wei, W.-N.; Li, B.; Gan, Q.; Li, Y.-L. Research progress of natural gas hydrate exploitation with CO_2 replacement: A review. *Fuel* **2022**, *312*, 122873. [CrossRef]
6. Wallmann, K.; Pinero, E.; Burwicz, E.; Haeckel, M.; Hensen, C.; Dale, A.; Ruepke, L. The global inventory of methane hydrate in marine sediments: A theoretical approach. *Energies* **2012**, *5*, 2449–2498. [CrossRef]
7. Khan, M.N.; Warrier, P.; Peters, C.J.; Koh, C.A. Advancements in hydrate phase equilibria and modeling of gas hydrates systems. *Fluid Phase Equilibria* **2018**, *463*, 48–61. [CrossRef]
8. Fandino, O.; Ruffine, L. Methane hydrate nucleation and growth from the bulk phase: Further insights into their mechanisms. *Fuel* **2014**, *117*, 442–449. [CrossRef]
9. Liang, H.; Guan, D.; Shi, K.; Yang, L.; Zhang, L.; Zhao, J.; Song, Y. Characterizing mass transfer mechanism during gas hydrate formation from water droplets. *Chem. Eng. J.* **2022**, *428*, 132626. [CrossRef]
10. Wan, Q.-C.; Si, H.; Li, B.; Li, G. Heat transfer analysis of methane hydrate dissociation by depressurization and thermal stimulation. *Int. J. Heat Mass Transf.* **2018**, *127*, 206–217. [CrossRef]
11. Yin, Z.; Khurana, M.; Tan, H.K.; Linga, P. A review of gas hydrate growth kinetic models. *Chem. Eng. J.* **2018**, *342*, 9–29. [CrossRef]
12. Li, X.; Wang, C.; Li, Q.; Fan, Q.; Chen, C.; Sun, C. Study on the growth kinetics and morphology of methane hydrate film in a porous glass microfluidic device. *Energies* **2021**, *14*, 6814. [CrossRef]
13. Tsimpanogiannis, I.N.; Michalis, V.K.; Economou, I.G. Enthalpy of dissociation of methane hydrates at a wide pressure and temperature range. *Fluid Phase Equilibria* **2019**, *489*, 30–40. [CrossRef]
14. Li, X.; Liu, Y.; Zhang, H.; Xiao, B.; Lv, X.; Yao, H.; Pang, W.; Li, Q.; Yang, L.; Song, Y.; et al. Non-embedded ultrasonic detection for pressure cores of natural methane hydrate-bearing sediments. *Energies* **2019**, *12*, 1997. [CrossRef]
15. Chatti, I.; Delahaye, A.; Fournaison, L.; Petitet, J. P. Benefits and drawbacks of clathrate hydrates: A review of their areas of interest. *Energy Convers. Manag.* **2005**, *46*, 1333–1343. [CrossRef]
16. Yang, M.; Zheng, J.-N.; Gao, Y.; Ma, Z.; Lv, X.; Song, Y. Dissociation characteristics of methane hydrates in South China Sea sediments by depressurization. *Appl. Energy* **2019**, *243*, 266–273. [CrossRef]
17. Qin, X.; Liang, Q.; Ye, J.; Yang, L.; Qiu, H.; Xie, W.; Liang, J.; Lu, J.A.; Lu, C.; Lu, H.; et al. The response of temperature and pressure of hydrate reservoirs in the first gas hydrate production test in South China Sea. *Appl. Energy* **2020**, *278*, 115649. [CrossRef]
18. Boswell, R.; Myshakin, E.; Moridis, G.; Konno, Y.; Collett, T.S.; Reagan, M.; Ajayi, T.; Seol, Y. India National Gas Hydrate Program Expedition 02 summary of scientific results: Numerical simulation of reservoir response to depressurization. *Mar. Pet. Geol.* **2019**, *108*, 154–166. [CrossRef]
19. Wang, B.; Dong, H.; Fan, Z.; Zhao, J.; Song, Y. Gas production from methane hydrate deposits induced by depressurization in conjunction with thermal stimulation. *Energy Procedia* **2017**, *105*, 4713–4717. [CrossRef]
20. Horvat, K.; Kerkar, P.; Jones, K.; Mahajan, D. Kinetics of the formation and dissociation of gas hydrates from CO_2-CH_4 mixtures. *Energies* **2012**, *5*, 2248–2262. [CrossRef]
21. Deusner, C.; Bigalke, N.; Kossel, E.; Haeckel, M. Methane production from gas hydrate deposits through injection of supercritical CO_2. *Energies* **2012**, *5*, 2112–2140. [CrossRef]
22. Olajire, A.A. Flow assurance issues in deep-water gas well testing and mitigation strategies with respect to gas hydrates deposition in flowlines—A review. *J. Mol. Liq.* **2020**, *318*, 114203. [CrossRef]
23. Wei, N.; Sun, W.; Meng, Y.; Zhao, J.; Kvamme, B.; Zhou, S.; Zhang, L.; Li, Q.; Zhang, Y.; Jiang, L.; et al. Hydrate formation and decomposition regularities in offshore gas reservoir production pipelines. *Energies* **2020**, *13*, 248. [CrossRef]
24. Weissman, J.T.; Masutani, S.M. Hydrogen storage capacity of tetrahydrofuran and tetra-N-Butylammonium bromide hydrates under favorable thermodynamic conditions. *Energies* **2017**, *10*, 1225. [CrossRef]
25. Pandey, J.S.; Daas, Y.J.; Karcz, A.P.; von Solms, N. Enhanced hydrate-based geological CO_2 capture and sequestration as a mitigation strategy to address climate change. *Energies* **2020**, *13*, 5661. [CrossRef]
26. Cheng, C.; Wang, F.; Tian, Y.; Wu, X.; Zheng, J.; Zhang, J.; Li, L.; Yang, P.; Zhao, J. Review and prospects of hydrate cold storage technology. *Renew. Sustain. Energy Rev.* **2020**, *117*, 109492. [CrossRef]

27. Dong, H.; Zhang, L.; Ling, Z.; Zhao, J.; Song, Y. The Controlling Factors and Ion Exclusion Mechanism of Hydrate-Based Pollutant Removal. *ACS Sustain. Chem. Eng.* **2019**, *7*, 7932–7940. [CrossRef]
28. Zhang, L.; Sun, M.; Wang, T.; Yang, L.; Zhang, X.; Zhao, J.; Song, Y. An In-Situ MRI Method for Quantifying Temperature Changes during Crystal Hydrate Growths in Porous Medium. *J. Therm. Sci.* **2022**, 1–9. [CrossRef]

Short Biography of Authors

Dr. **Hongsheng Dong** is an assistant professor at the Dalian Institute of Chemical Physics, CAS. He received his Ph.D. degree in Engineering Thermophysics from Dalian University of Technology. His research interests include gas hydrates, phase change material, and thermal energy storage. He has published more than 50 peer-reviewed journal papers. As a visiting researcher, he spent one year at the University of Edinburgh.

Dr. **Lunxiang Zhang** is a lecturer at Dalian University of Technology. He received his Ph.D. degree in Engineering Thermophysics from Dalian University of Technology in 2019. His research areas focus on safe and high-efficiency exploitation of NGH, flow assurance of oil and gas, carbon dioxide capture and utilization, and green hydrate-based applications. He has published more than 60 peer-reviewed journal papers and has nearly 1000 citations.

Dr. **Jiaqi Wang** is an associate professor at Harbin Engineering University. She received her Ph.D. degree in Engineering Thermophysics from Dalian University of Technology. Her research areas focus on marine gas hydrate exploitation technology, hydrate-based technology application, and seepage and heat and mass transfer in porous media. She has authored or co-authored about 30 peer-reviewed journal papers and has nearly 600 citations.

energies

MDPI

Article

Principle and Feasibility Study of Proposed Hydrate-Based Cyclopentane Purification Technology

Xianbing Hu [1], Lingjie Sun [1], Chengyang Yuan [1], Man Li [1], Hongsheng Dong [1,2], Lunxiang Zhang [1,*], Lei Yang [1], Jiafei Zhao [1] and Yongchen Song [1,*]

1 Key Laboratory of Ocean Energy Utilization and Energy Conservation of Ministry of Education, School of Energy & Power Engineering, Dalian University of Technology, Dalian 116024, China
2 Dalian Institute of Chemical Physics, Chinese Academy of Sciences, Dalian 116023, China
* Correspondence: lunxiangzhang@dlut.edu.cn (L.Z.); songyc@dlut.edu.cn (Y.S.)

Abstract: The separation of azeotropic mixtures has conventionally been one of the most challenging tasks in industrial processes due to the fact that components in the mixture will undergo gas–liquid phase transition at the same time. We proposed a method for separating azeotropes using hydrate formation as a solid–liquid phase transition. The feasibility of hydrate-based separation is determined by analyzing the crystal structure and chemical bonds of hydrate. Taking the azeotrope cyclopentane and neohexane in petroleum as an example, cyclopentane (95%) was purified to 98.56% yield using the proposed hydrate-based cyclopentane purification technology. However, this is difficult to achieve using conventional distillation methods. The proposed method is simple in operation and yields a good separation effect. This study provides a new method for separating cyclopentane and neohexane.

Keywords: azeotrope; cyclopentane; neohexane; hydrate; phase transition

Citation: Hu, X.; Sun, L.; Yuan, C.; Li, M.; Dong, H.; Zhang, L.; Yang, L.; Zhao, J.; Song, Y. Principle and Feasibility Study of Proposed Hydrate-Based Cyclopentane Purification Technology. *Energies* **2023**, *16*, 4681. https://doi.org/10.3390/en16124681

Academic Editor: Mofazzal Hossain

Received: 11 March 2023
Revised: 16 April 2023
Accepted: 29 May 2023
Published: 13 June 2023

1. Introduction

In many areas of industry, the separation of liquid mixtures is an important technology [1]. Some components in these liquid mixtures have great economic value after purification, so the separation of these mixtures into their pure components is necessary. Of all known liquid separation techniques, distillation is the most widely used technique [2]. It has the advantages of simple operation and high controllability, and it is especially widely used in oil separation. However, when some liquid mixtures boil, the liquid and gas phases have the same composition. This condition is called azeotropism [1]. Due to the relative volatility ($\alpha = 1$), azeotropes cannot be separated by ordinary distillation [3]. Therefore, some other methods have been invented to separate azeotropes, such as azeotropic distillation, extractive distillation, pressure-swing distillation, liquid–liquid extraction, adsorption, membranes, etc. [4].

Cyclopentane (CP) and neohexane (22MB) are important industrial chemicals and typical azeotropic mixtures [5]. CP can be used as a foaming agent as well as to produce cyclopentanol and cyclopentanone [6]. When separating CP from a C5 hydrocarbon mixture, the purity of the product is limited by the coexistence of 22MB in the system, making the separation difficult [7]. For this near-boiling and azeotropic system, extractive distillation is commonly used in industry for separation [8]. Extractive distillation (ED) is an efficient technology for separating azeotropic and near-boiling-point systems [3]. Many researchers have studied extractants that separate CP and 22MB. For example, Lee [9] used mixed solvents as extractants for ED. Sun and Zhao [10] found that N, N-dimethylformamide (DMF), N-methylpyrrolidone (NMP) and cyclohexanol (CHOL) are also suitable organic solvents for separating the mixture of CP and 22MB. However, the ED process usually has drawbacks, such as smaller changes in relative volatility, large solvent usage and high energy consumption. So, we need to develop new technologies that are more energy efficient, environmentally benign and inherently safe.

Hydrate is a kind of nonstoichiometric cage-like crystal material formed by water and other small molecular objects, also known as cage hydrate [11]. When water and guest molecules have contact under proper conditions, the water molecules are connected by hydrogen bonds to form some polyhedral cages, in which guest molecules of appropriate size can be fixed to form solid hydrates. The hydrate former M can be described by the following hydration reaction equation:

$$M + xH_2O = M{\cdot}xH_2O \quad (1)$$

where x is the hydration number [12]. However, when environmental conditions change slightly, hydrates can be decomposed into water and guest molecules again. Hydrate generation in oil and gas pipelines will cause plugging [13], but it is undeniable that gas hydrate has the potential to be a new technology for the benefit of mankind. Because it is a reversible phase-change mechanism that works by controlling temperature and pressure, a series of hydrate-based technologies have been developed. The hydrate-based desalination process is the earliest hydrate utilization technology [14]. At present, hydrate utilization technology has been developed for gas separation, gas storage and transportation [15], solution concentration and separation, hydrate cold storage and replacement mining [16]. Among these technologies, hydrate-based gas separation, gas storage and transportation, and wastewater treatment are three of the most studied, mature and large-scale developments for commercial use [17]. At present, hydrate technology can be used to store methane under conditions close to normal pressure [18].

Hydrate-based gas-mixture separation technology is widely discussed because of its simple process, low cost, environmental friendliness and other characteristics [19]. Different gas hydrates have different phase equilibria. By controlling temperature and pressure, one kind of gas forms hydrate while the other gas remains unchanged, thus achieving gas separation [20]. Compared with the crystallization method and liquefaction method, the conditions for hydrate-based gas-mixture separation technology are milder, and the gas forming the hydrate more readily than the others will be concentrated in solid hydrate [21]. The pressure required for gas hydrate formation is often lower than the pressure for liquefaction, and using water as a promoter will not cause pollution. This means lower costs and environmental friendliness. Using CO_2 hydrate formation and dissociation to separate CO_2 from gas mixtures is perceived as a technically feasible method for carbon capture and storage [22]. Hassanpourouzband et al. [23] successfully realized that more than 40% of CO_2 can be separated in the form of hydrate by using gas hydrate formation and dissociation. This method can effectively reduce the emissions of CO_2 from fossil fuels and thus slow down global warming [24]. CH_4 separation from a gas mixture also has great significance and important research value [25]. Zhang et al. [26] recovered CH_4 from low-concentration coal-mine CH_4 by adding THF as the thermodynamic promoter. This is a good way to use the coal-mine CH_4. Hydrate-based gas separation is also very effective in separating many other mixed gases. For example, Ko et al. [27] proposed a hydrate–liquefaction combined method to separate SF6 from greenhouse gases. Kim et al. [28] analyzed the viability of CHF_3 separation using a gas-hydrate-based method. Wang et al. [29] separated H_2 from H_2 and CH_4 binary gas mixtures with various H_2 concentrations using hydrate-based technology, and the H_2 content could be enriched to up to 94%. These studies show that hydrate-based gas-mixture separation technology can achieve high gas-separation efficiency, and the method is especially suitable for the gases which easily can form gas hydrates [17]. However, at present, hydrate-based separation technology is mainly used to separate gas, and the separation of liquid azeotropes has not been confirmed.

This paper is the first to explore the application of hydrate-based separation technology in liquid azeotropes. CP and 22MB will form binary azeotropes and CP is an ideal guest molecule to form hydrate. CP can form hydrate under normal pressure [30]. 22MB cannot form hydrate under normal pressure and will be excluded from the crystal structure. At the same time, CP is insoluble in water, which makes it easy to separate the CP from water after the subsequent decomposition of the hydrate. At present, some researchers have verified the feasibility of separating CP and 22MB by hydrate phase change under normal pressure through molecular simulation [31].

In this study, we conducted an experimental exploration of the treatment of a CP and 22MB mixture using hydrate-based separation technology. The hydrate crystal structure and purification efficiency were studied. The results show that the hydrate phase-change separation method has broad prospects for treating the mixture of CP and 22MB, high purification efficiency and short treatment time.

2. Materials and Methods

2.1. Experimental Apparatus and Materials

Deionized water with a specific resistance of 18 MΩ was produced using Aquapro2S via reverse osmosis. CP and 22MB were purchased from Aladdin Industrial Corporation (Shanghai, China). An Oxford Instruments Low Field Nuclear Magnetic Resonance System was used (Figure S1, Abingdon, UK). The model of X-ray diffractometer used was a Bruker D8 Advance (Figure S2, Billerica, MA, USA) and the Raman spectrometer model was a LabRAM HR Evolutione (Figure S3, Horiba, Kyoto, Japan). The model of refrigeration constant-temperature circulating water bath was a Labtemp CC-4008E (Guangzhou, China). The model of the magnetic stirrer was an AS ONE CS-4 1-4609-25 (Osaka, Japan). The model of diaphragm vacuum pump was a GM-0.5A (Tianjin, China). The model of the high-speed freezing centrifuge was an H1850R (Changsha, China).

2.2. Experimental Procedure and Conditions

CP and 22MB were used to configure mixtures of different concentrations. The concentration of CP in the mixture was 70–95%. The experiment was divided into three parts. The experimental process is shown in Figure 1. The first part of the experiment was to verify whether hydrate can be formed. A sealed glass bottle was used as a reaction container. The 70% CP (20 mL) was mixed with water (80 mL) at 2 °C and the electromagnetic stirrer was set to 700 rpm. The reaction time was 12 h. It has been reported that it is very hard to form CP hydrate in the static state, while NMR measurements cannot be performed in the agitated state. So, detection by the nuclear magnetic resonance instrument was performed before and after the reaction. We measured the T_2 distribution via the low-field NMR system using the following test parameters: time of recycle delay (RD) = 7500 ms; resonant frequency (RF) = 12.71 Hz; number of echoes (NOE) = 46,296; and the time between the 90° pulse and the first acquired echo (Tau) = 0.0127 ms.

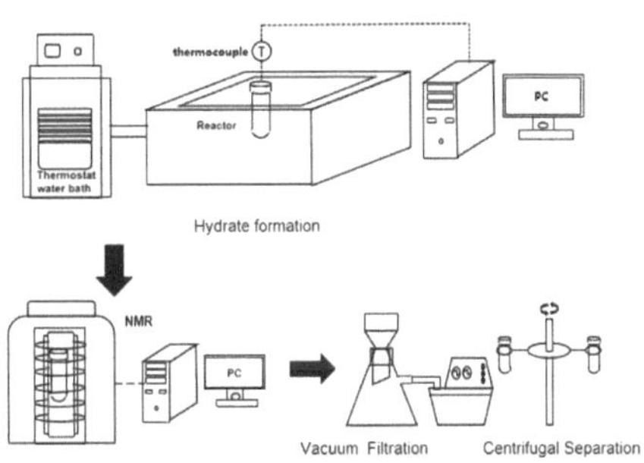

Figure 1. Experimental system.

The second part of the experiment was to detect the hydrate structure. The hydrate generated in the first part of the experiment was vacuumed and then centrifuged for 5 min at −5 °C and 4000 rpm to completely remove the residual liquid in the hydrate. The hydrate was ground into powder under the protection of liquid nitrogen, and then tested by XRD and Raman spectrometer. The measurement temperature for XRD and Raman was set to 250 K to prevent hydrate decomposition during the measurement process. X-ray diffraction analysis was conducted using filtered Cu Kα radiation (λ = 0.154 nm, operated at 40 kV and 40 mA). The 2θ range was set from 5° to 60°, with a step of 1.5° min^{-1}. The Raman analysis was conducted using a 532 nm laser. The experiment was repeated with CP (20 mL) and water (80 mL) antisense as control group.

The purpose of the last part of the experiment was to verify the separation effect. Different concentrations of CP (70–95%) were used to generate hydrate under the above experimental conditions. The separated hydrates decomposed at 20 °C to produce pure CP and water. Because CP is insoluble in water, it was separated via liquid separation and its purity was measured using gas chromatography.

The equation for calculating the purification efficiency based on the proposed hydrate method is expressed as follows [32]:

$$\text{Impurity removal ratio} = \frac{C_0 - C_f}{C_0} \times 100\% \tag{2}$$

where C_0 is the initial concentration of impurity, and C_f is the concentration of impurity in the CP obtained from the decomposition of CP hydrate.

The hydrate conversion rate was calculated using equation [33]:

$$\text{Conversion ratio} = \frac{m_h}{m_0} \times 100\% \tag{3}$$

where m_0 is the mass of hydrate formation theoretically, and m_h is the mass of hydrate formation actually.

3. Results and Discussion

3.1. Nuclear Magnetic Resonance

Magnetic resonance imaging technology is a new optical measurement system often used in medical and biological science research. In addition, many researchers have conducted extensive research and observed the hydrate formation process by using nuclear magnetic resonance (NMR) imaging technology [34].

NMR imaging systems are mainly used to capture images of hydrogen protons in liquid water, and the nuclear-magnetic-signal intensity is proportional to the hydrogen proton content. It is known that the NMR of fluids close to solids can be quite different from that for pure fluids; these effects are strongly linked to the characterization of pore fluids and fluid distributions in porous media through NMR measurements [35]. During hydrate formation, hydrogen protons are mainly distributed in liquid water, and the transverse relaxation time of the hydrogen protons in the hydrate is significantly shorter than that in free water. Therefore, hydrate formation can be determined based on the nuclear-magnetic-signal value of water measured using the NMR system.

NMR data from before and after the reaction were measured, as shown in Figure 2. The 0.01–1 ms section represents the hydrate, and the peak area significantly increases after the reaction. The 10–10,000 ms section represents the water, and the peak area significantly decreases after the reaction [36]. The NMR results showed that the mixture of CP and 22MB reacted with water to form hydrates. Because the agitator used in the experiment was magnetic, measurements could not be made during hydrate formation. According to Kuang [36], in situ hydrate formation experiments can obtain more accurate data. However, electromagnetic stirrers cannot be used in in situ experiments. We attempted to conduct in situ hydrate formation experiments under static conditions, but no hydrate formation occurred within 12 h. So, we had to use experiments that were not in situ. The consumption of water can be calculated according to the change in the nuclear magnetic signal.

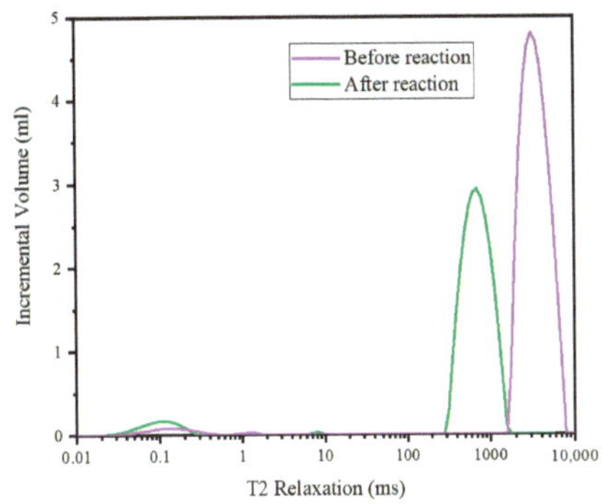

Figure 2. T_2 distribution of hydrate generated from the mixture of CP and 22MB and a real picture of the hydrate.

3.2. X-ray Diffraction

The periodic arrangement of atoms in the crystal creates conditions for light diffraction [37]. The wavelength of X-rays is similar to the distance between atoms, and this produces constructive interference at specific angles. Different diffraction patterns are produced by X-ray scattering from different atomic structures. X-ray diffraction (XRD) techniques use this principle to elucidate the crystalline nature of materials [38].

The hydrate crystals had three main structures: sI, sII and sH [39]. Cubic structure I contains small (0.4–0.55 nm) guests; cubic structure II generally occurs with larger (0.6–0.7 nm) guests; and hexagonal structure H may occur only with mixtures of both small and large (0.8–0.9 nm) molecules [40]. CP could react with water to form sII hydrates, and 22MB could not react with water to form sII hydrates. CP and 22MB can form sH hydrate with other small molecules.

The structural type of the hydrate was determined by analyzing its X-ray diffraction spectrum. Figure 3 shows a comparison between the X-ray diffraction (XRD) patterns of the hydrate generated by the mixture, the CP hydrate, ice and the sII hydrate. The XRD spectra of the ice and sII hydrate were determined by Yousuf [41].

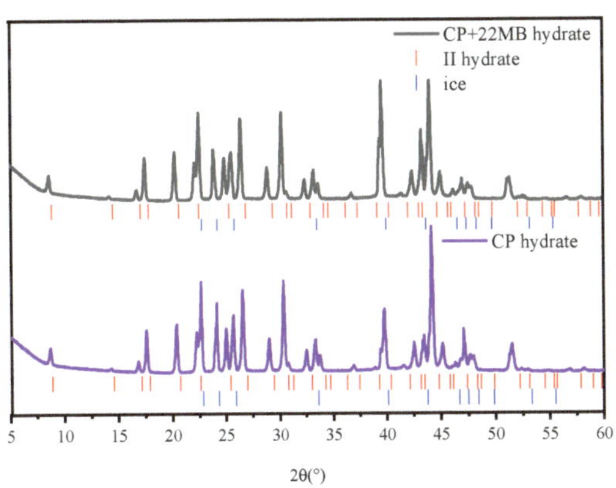

Figure 3. XRD spectra of hydrates of two samples showing characteristic peak positions of ice and sII hydrates in lower half.

The mixture of CP and 22MB reacted with water to form sII hydrate (Figure 3). The XRD wave patterns of the experimental group and the control group were consistent. Compared with the standard peak position, the measured peak position had a certain left deviation, which was caused by slightly different temperature and measurement error. In addition to the hydrate phase, significant ice diffraction peaks appeared in the XRD spectra of the two hydrate samples, owing to water condensation and mixing into the samples when the samples were ground under the protection of liquid nitrogen. Some peaks were not obvious, which was caused by insufficient grinding. No characteristic wave peak of sH hydrate was observed in the XRD spectrum of the mixture hydrate, preliminarily indicating that no hydrate was formed from 22MB.

3.3. Raman Spectroscopy

Raman spectroscopy is considered to be one of the most reliable and versatile tools for analyzing several materials under laboratory and field conditions [42]. Electromagnetic radiation interacts with matter through absorption, transmittance or scattering phenomena. This will cause the photon energy to change. The energy difference between the incoming photon and the outgoing one is called "Raman shift" [43].

CP and 22MB in the hydrated state exhibited different Raman shifts. By analyzing the Raman shift of the sample, the guest molecules in the hydrate could be determined.

The Raman peak position of the sample was the same as the CP hydrate (Figure 4). The Raman peak near 896 cm^{-1} indicated the respiratory ring vibration of CP [44]. The Raman peaks near 2876 and 2983 cm^{-1} corresponded to the C-H symmetric and C-H stretching vibrations of CP, respectively [45]. Based on these three Raman peaks, the formation of CP hydrate in the experiment was determined. According to Lv [46], the Raman peaks near 3160 and 3340 cm^{-1} corresponded to O-H stretching vibration. The Raman peaks near 3160 cm^{-1} corresponded to the coupled symmetric O-H stretching mode of tetrahedrally coordinated hydrogen-bonded water. The Raman peaks near 3340 cm^{-1} corresponded to the O-H stretching mode of incomplete-tetrahedral-coordination hydrogen-bonded water, indicating a water molecular-arrangement disorder. The peak positions of the Raman spectra showed no difference for the formed CP hydrates with and without 22MB, which indicated that the addition of 22MB had no impact on the structure of the produced CP hydrates.

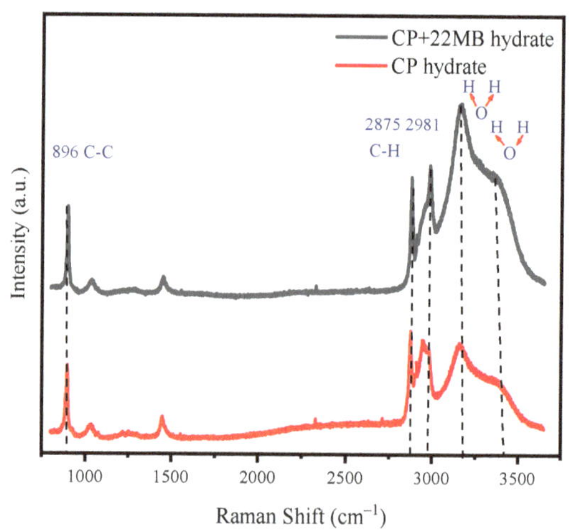

Figure 4. Raman characterization of samples.

3.4. Purification Results

The recovery ratio and purification effect are important parameters to indicate the performance of purification methods. We calculated the reduction rate of impurity (22MB) content in CP to express the purification effect. Because the purified CP was obtained by decomposition of hydrate, we used the hydrate formation ratio to replace the CP recovery ratio. If the purity of CP is too low, it will affect the formation of hydrate, so we tried to purify CP in the concentration range of 70–95%.

The impurity removal ratio determines the quality of the produced CP, which can be monitored by gas chromatography–mass spectrometry and then calculated by Equation (2). For the initially highest mass fraction of CP (95%), the purified CP fraction reached to 98.56%, corresponding to an impurity removal ratio of 71.2%. As shown in Figure 5, with the increase in the initial mass fraction of CP from 70% to 95%, the impurity removal ratio only ranged between 37.6% and 25.3% when using suction filtration alone. With further centrifugal separation, the impurity removal ratio reached from 73.4% to 71.2%. These results demonstrated that the initial CP concentration had a weak effect on the impurity removal ratio, and the solid–liquid separation approach (i.e., suction filtration and centrifugal separation in this study) had a significant influence on impurity removal ration. The mass fraction of CP obtained by ordinary distillation is 80~85% [47]. Hydrate-based separation technology can obtain CP with higher purity than ordinary distillation. According to the detection using XRD and Raman, it was found that the hydrate structure did not contain 22MB, but that a certain amount of 22MB may be attached to the surface of the hydrate. Therefore, efficient solid–liquid separation methods may become the focus of follow-up research.

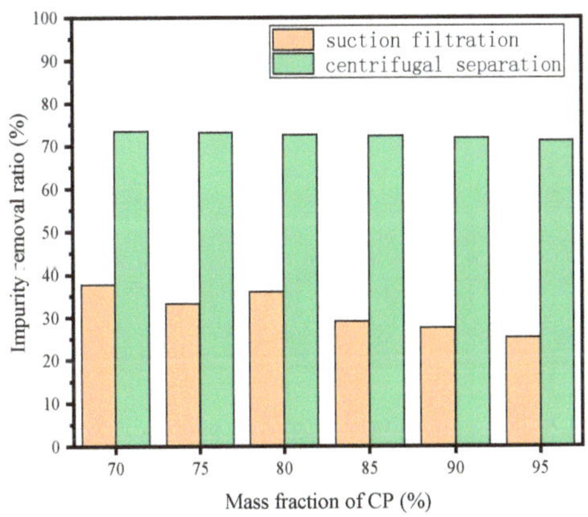

Figure 5. Effect of mass fraction of CP on impurity removal ratio.

As can be seen in Figure 6, the hydrate formation rate was between 40.5% and 80.6%. The hydrate formation ratio increased with the increase in CP purity. This is because the existence of 22MB will inhibit the formation of CP hydrate to a certain extent. The hydrate formation rate of 90% CP was almost the same as that of 95% CP, which means that the lower concentration of 22MB had little effect on hydrate formation. The hydrate conversion ratio calculated by measuring water consumption through T_2 was different from the hydrate conversion ratio measured by weighing. This may have been caused by the formation of condensed water on the surface of the reactor during T_2 measurement. The higher the hydrate formation rate, the more CP will be recovered. When the purity of CP is too low, it is difficult to form hydrates. In Figures 5 and 6, it can be observed that the conversion ratio of hydrates has almost no effect on impurity removal. Conventional technology to obtain high-purity CP requires increasing the number of plates in the distillation column, which will greatly increase energy consumption. In contrast, our technology had excellent results in purifying high-purity CP.

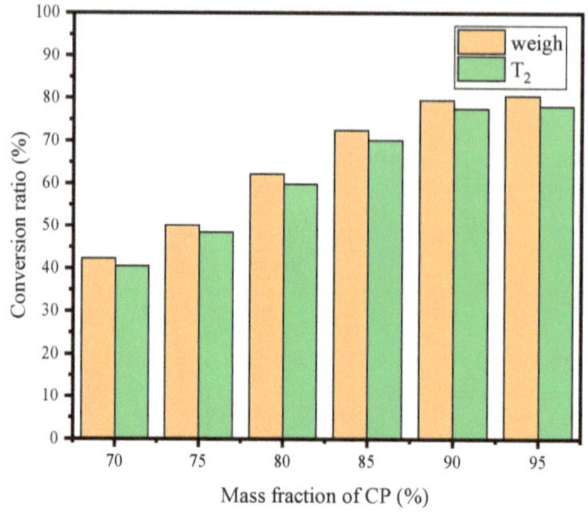

Figure 6. Effect of mass fraction of CP on hydrate conversion ratio.

Currently, as shown in Table 1, ED is commonly used in industry to purify CP. The process consists of two parts, namely, an ED column and a recovery distillation column. If you want to effectively separate CP and 22MB, the number of stages needs to be above 60. This means complex processes and high energy consumption. More importantly, the conventional method requires gasification of CP, which is toxic and explosive. Our technology is relatively safe.

Table 1. Comparison of extractive distillation and hydrate separation technology.

	Extractive Distillation		**Hydrate Separation Technology**	
Process stages	addition of extractant, vaporization (22MB), flow, vaporization (CP), condensation		formation, separation, decomposition	
Device	ED column (42~65 plates)	recovery distillation column (10~36 plates)	formation reactor	decomposition reactor
Temperature (°C)	52.23 (overhead)~114.79 (bottom)	51.78 (overhead)~162.64 (bottom)	2	20
Pressure (kPa)	110 160	110 135	101	101
Additive	N,N-dimethyl formamide (DMF)		water	
Feed: Additives	1:10		1:4	
References	[31,47,48]			

4. Conclusions

This study shows that the hydrate-based process has great potential for treating CP and 22MB mixtures. When the experimental temperature is 2 °C and the electromagnetic stirrer is 700 rpm, this method can effectively deal with CP with a mass fraction of 70–95%. The recovery ratio of CP is 40.5–80.6%, and the impurity removal ratio is more than 71.2%. When 70% CP is to be purified, the recovery ratio of CP is 40.5%. When 95% CP is to be purified, the recovery ratio of CP is 80.6%. This method performs better in the treatment of high-purity CP, which is impossible for ordinary distillation. The volume of ED equipment is too large to handle small amounts of CP. The proposed technique may be valuable for separating high-purity CP from petroleum. This hydrate-based separation technology also has the potential to be expanded to other azeotropes.

Supplementary Materials: The following supporting information can be downloaded at: https://www.mdpi.com/article/10.3390/en16124681/s1, Figure S1: Nuclear Magnetic Resonance System; Figure S2: X-ray diffractometer; Figure S3: Raman spectrometer.

Author Contributions: Conceptualization, X.H. and H.D.; methodology, L.S.; software, X.H., C.Y. and M.L.; validation, L.Z.; investigation, X.H.; data curation, X.H., C.Y. and M.L.; writing—original draft, X.H.; writing—review editing, L.Z.; supervision, L.Y. and J.Z.; funding acquisition, J.Z. and Y.S. All authors have read and agreed to the published version of the manuscript.

Funding: This work was supported by the National Natural Science Foundation of China (grant numbers 52020105007, 52025066, 52006024, and U21B2065) and the Fundamental Research Funds for the Central Universities (grant number DUT22LAB130).

Data Availability Statement: The data presented in this study are available on request from the corresponding author.

Conflicts of Interest: The authors declare no conflict of interest.

References

1. Pereiro, A.B.; Araújo, J.; Esperan, A.J. Ionic liquids in separations of azeotropic systems—A review. *J. Chem. Thermodyn.* **2012**, *46*, 2–28. [CrossRef]
2. Fair, J.R. The workhorse today & tomorrow—Distillation. *Chem. Eng.* **2002**, *109*, 108.
3. Lei, Z.; Li, C.; Chen, B. Extractive Distillation: A Review. *Sep. Purif. Rev.* **2003**, *32*, 121–213. [CrossRef]
4. Mahdi, T.; Ahmad, A.; Nasef, M.M. State-of-the-Art Technologies for Separation of Azeotropic Mixtures. *Sep. Purif. Rev.* **2015**, *44*, 308–330. [CrossRef]
5. Marschner, R.F.; Burney, D.E. Cyclopentane and Neohexane in Petroleum. *Ind. Eng. Chem.* **2002**, *44*, 1406–1408. [CrossRef]
6. Zhou, F.; Xu, A.; Jiang, F. Study on the Process of Cyclopentane Oxidation to Cyclopentanol and Cyclopentanone. *Fine Spec. Chem.* **2020**, *28*, 5–9.
7. Tian, Y.; Sun, J.; Wang, Z. 50 kt/a Fine separation of mixed pentane isomers. *Chem. Prog.* **2005**, *24*, 27–31.
8. Qi, L.; Cui, J.; Xu, R. Selection of Cyclopentane/Neohexane Separation Extracts by COSMO-SAC Method and Process Simulation. *J. Chang. Univ.* **2021**, *33*, 43–49.
9. Fu, M.L. Extractive Distillation of Alkane/Cycloalkane Feed Employing Mixed Solvent. U.S. Patent No. 4,948,470A, 14 August 1990.
10. Sun, Y.B.; Zhao, Q.M.; Gao, Z.R. Development of extraction rectification of cyclopentane/2.2-methyl butane. *Chem. Eng.* **2012**, *26*, 17–19.
11. Fandino, O.; Ruffine, L. Methane hydrate nucleation and growth from the bulk phase: Further insights into their mechanisms. *Fuel* **2014**, *117*, 442–449. [CrossRef]
12. Moridis, G.J.; Collett, T.S.; Boswell, R. *Gas Hydrates as a Potential Energy Source: State of Knowledge and Challenges*; Springer: New York, NY, USA, 2013; Volume 37, pp. 977–1033.
13. Wang, J.; Wang, Q.; Meng, Y. Flow characteristic and blockage mechanism with hydrate formation in multiphase transmission pipelines: In-situ observation and machine learning predictions. *Fuel* **2022**, *330*, 125669. [CrossRef]
14. Park, K.N.; Hong, S.Y.; Lee, J.W.; Kang, K.C.; Lee, Y.C.; Ha, M.G.; Lee, J.D. A new apparatus for seawater desalination by gas hydrate process and removal characteristics of dissolved minerals (Na^+, Mg^{2+}, Ca^{2+}, K^+, B^{3+}). *Desalination* **2011**, *274*, 91–96. [CrossRef]
15. Steinar, G.J. Method for Production of Gas Hydrates for Transportation and Storage. U.S. Patent No. 5,536,893, 16 July 1996.
16. Xue, Q.; Wang, X.; Li, Z. Application progress of hydrate utilization technology. *Chem. Prog.* **2021**, *40*, 722–735.
17. Dong, H.; Wang, J.; Xie, Z. Potential applications based on the formation and dissociation of gas hydrates. *Renew. Sustain. Energy Rev.* **2021**, *143*, 110928. [CrossRef]
18. Sun, L.; Sun, H.; Yuan, C. Enhanced clathrate hydrate formation at ambient temperatures (287.2 K) and near atmospheric pressure (0.1 MPa): Application to solidified natural gas technology. *Chem. Eng. J.* **2023**, *454*, 140325. [CrossRef]
19. Xu, C.G.; Chen, Z.Y.; Cai, J. Study on Pilot-Scale CO_2 Separation from Flue Gas by the Hydrate Method. *Energy Fuels* **2014**, *28*, 1242–1248. [CrossRef]
20. Huang, L.; Liang, M.; Bei, L. Experimental Studies of the Separation of C2 Compounds from CH_4 + C_2H_4 + C_2H_6 + N_2 Gas Mixtures by an Absorption−Hydration Hybrid Method. *Ind. Eng. Chem. Res.* **2013**, *52*, 2707–2713.
21. Zhang, L.W.; Chen, G.J.; Guo, X.Q. The partition coefficients of ethane between vapor and hydrate phase for methane + ethane + water and methane + ethane + THF + water systems. *Fluid Phase Equilibria* **2004**, *225*, 141–144. [CrossRef]
22. Li, S.; Fan, S.; Wang, J. Clathrate Hydrate Capture of CO_2 from Simulated Flue Gas with Cyclopentane/Water Emulsion. *Chin. J. Chem. Eng.* **2010**, *18*, 202–206. [CrossRef]
23. Hassanpouryouzband, A.; Yang, J.; Tohidi, B. Insights into the CO_2 Capture by Flue Gas Hydrate Formation: Gas Composition Evolution in Systems Containing Gas Hydrates and Gas Mixtures at Stable Pressures. *ACS Sustain. Chem. Eng.* **2018**, *6*, 5732–5736. [CrossRef]
24. Czarnota, R.; Knapik, E.; Wojnarowski, P. Carbon Dioxide Separation Technologies. *Comm. Min. PAS* **2019**, *64*, 487–498.
25. Lashof, D.A.; Ahuja, D.R. Relative contributions of greenhouse gas emissions to global warming. *Nature* **1990**, *344*, 529–531. [CrossRef]
26. Zhang, B.; Wu, Q. Thermodynamic Promotion of Tetrahydrofuran on Methane Separation from Low-Concentration Coal Mine Methane Based on Hydrate. *Energy Fuels* **2010**, *24*, 2530–2535. [CrossRef]

27. Ko, G.; Lee, J.; Seo, Y. Separation efficiency and equilibrium recovery ratio of SF_6 in hydrate-based greenhouse gas separation. *Chem. Eng. J.* **2020**, *405*, 126956. [CrossRef]
28. Kim, E.; Ko, G.; Seo, Y. Greenhouse Gas (CHF_3) Separation by Gas Hydrate Formation. *ACS Sustain. Chem. Eng.* **2017**, *5*, 5485–5492. [CrossRef]
29. Wang, X.L.; Chen, G.J.; Yang, L.Y. Study on the recovery of hydrogen from refinery (hydrogen plus methane) gas mixtures using hydrate technology. *Sci. China Ser. B-Chem.* **2008**, *51*, 171–178. [CrossRef]
30. Han, S.; Rhee, Y.W.; Kang, S.P. Investigation of salt removal using cyclopentane hydrate formation and washing treatment for seawater desalination. *Desalination* **2017**, *404*, 132–137. [CrossRef]
31. Wu, L. *Study on Separation Technology of Cyclopentane/Neohexane Mixture*; China University of Petroleum: Beijing, China, 2017.
32. Sun, H.; Sun, L.; Zhao, Y. A combined hydrate-based method for removing heavy metals from simulated wastewater with high concentrations. *J. Environ. Chem. Eng.* **2021**, *9*, 106633. [CrossRef]
33. Zhang, L.; Sun, M.; Wang, T. An in-situ MRI method for quantifying temperature changes during crystal hydrate growths in porous medium. *J. Therm. Sci.* **2022**, *31*, 1542–1550. [CrossRef]
34. Zhang, L.; Dong, H.; Dai, S. Effects of depressurization on gas production and water performance from excess-gas and excess-water methane hydrate accumulations. *Chem. Eng. J.* **2022**, *431*, 133223. [CrossRef]
35. Watson, A.; Chang, C.T. Characterizing porous media with NMR methods. *Prog. Nucl. Magn. Reson. Spectrosc.* **1997**, *31*, 343–386. [CrossRef]
36. Kuang, Y.; Zhang, L.; Song, Y. Quantitative determination of pore-structure change and permeability estimation under hydrate phase transition by NMR. *AIChE J.* **2020**, *66*, 16859. [CrossRef]
37. Chauhan, C.; Bioanal, J.A. Powder XRD Technique and Its Application. *J. Anal. Bioanal. Tech.* **2014**, *5*. [CrossRef]
38. Dann, S.E. *Reactions and Characterization of Solids*; Royal Society of Chemistry: London, UK, 2000; pp. 64–65.
39. Liu, C.; Ye, Y. *Natural Gas Hydrates Experimental Techniques and Their Applications*; Springer: Berlin/Heidelberg, Germany, 2013.
40. Sloan, E. Fundamental principles and applications of natural gas hydrates. *Nature* **2003**, *426*, 353–359. [CrossRef]
41. Yousuf, M.; Qadri, S.B.; Knies, D.L. Novel results on structural investigations of natural minerals of clathrate hydrates. *Appl. Phys. A* **2004**, *78*, 925–939. [CrossRef]
42. Kudelski, A. Analytical applications of Raman spectroscopy. *Talanta* **2008**, *76*, 1–8. [CrossRef]
43. Orlando, A.; Franceschini, F.; Muscas, C. A Comprehensive Review on Raman Spectroscopy Applications. *Chemosensors.* **2021**, *9*, 262. [CrossRef]
44. Lv, Q. Formation Kinetics of Cyclopentane plus Methane Hydrates in Brine Water Systems and Raman Spectroscopic Analysis. *Energy Fuels* **2015**, *29*, 6104–6110. [CrossRef]
45. Lo, C.; Zhang, J.; Somasundaran, P. Raman Spectroscopic Studies of Surfactant Effect on the Water Structure around Hydrate Guest Molecules. *J. Phys. Chem. Lett.* **2010**, *1*, 2676–2679. [CrossRef]
46. Lv, Q.; Li, X. Raman Spectroscopic Studies on Microscopic Mechanism of CP—CH 4 Mixture Hydrate. *Energy Procedia* **2017**, *142*, 3264–3269. [CrossRef]
47. Nuchteera, I.; Thirasak, P.; Chavagorn, M. Cyclopentane Purification from Multicomponent Azeotropic Mixtures. *Comput. Aided Chem. Eng.* **2020**, *48*, 277–282.
48. Peng, K. *Design and Optimization of Cyclopentane Separation Wall Distillation Tower*; Dalian University of Technology: Dalian, China, 2021.

Journal of
Marine Science and Engineering

MDPI

Article

Numerical Simulation of Improved Gas Production from Oceanic Gas Hydrate Accumulation by Permeability Enhancement Associated with Geomechanical Response

Rui Wang [1], Jiecheng Zhang [2,3,*], Tianju Wang [1] and Hailong Lu [4,5,*]

1 CPOE Research Institute of Engineering Technology, Tianjin 300451, China
2 College of Safety and Ocean Engineering, China University of Petroleum, Beijing 102249, China
3 Key Laboratory of Oil and Gas Safety and Emergency Technology, Ministry of Emergency Management, Beijing 102249, China
4 School of Earth and Space Sciences, Peking University, Beijing 100871, China
5 Beijing International Center for Gas Hydrate, School of Earth and Space Sciences, Peking University, Beijing 100871, China
* Correspondence: zhangjc@cup.edu.cn (J.Z.); hlu@pku.edu.cn (H.L.)

Abstract: In the Shenhu Area of the South China Sea, although some numerical studies are conducted on the gas production at well SHSC-4, the geomechanical responses have not been taken into account, and the associated impact of permeability enhancement on gas production has not been thoroughly investigated. In this study, pTOUGH+HYDRATE V1.5 coupled with the RGMS is applied to account for geomechanical responses. Based on actual geological conditions, the reservoir model has five layers: the hydrate-bearing layer (HBL), the three-phase layer (TPL), the free gas layer (FGL), the overburden, and the underburden. The numerical results match the trial production data, validating the numerical model. The analysis shows that gas production from the FGL contributed the most (72.17%) to the cumulative gas production (V_g), followed by the TPL (23.54%) and the HBL (4.29%). The cumulative water-to-gas ratio (R_{wgT}) gradually decreased during gas production, with the HBL exhibiting the highest value. Permeability enhancement can improve gas production, with the FGL being the most responsive to such enhancement. It increased V_g by 87% and reduced R_{wgT} to 85%. To achieve more realistic production schemes and better enhance energy recovery, it is advisable to conduct numerical investigations that incorporate geomechanical considerations due to the intricate nature of hydrate-bearing sediments.

Keywords: permeability enhancement; geomechanical response; coupled geomechanics and flows of fluid and heat; gas hydrate production; Shenhu Area

Citation: Wang, R.; Zhang, J.; Wang, T.; Lu, H. Numerical Simulation of Improved Gas Production from Oceanic Gas Hydrate Accumulation by Permeability Enhancement Associated with Geomechanical Response. *J. Mar. Sci. Eng.* **2023**, *11*, 1468. https://doi.org/10.3390/jmse11071468

Academic Editor: George Kontakiotis

Received: 21 June 2023
Revised: 19 July 2023
Accepted: 21 July 2023
Published: 23 July 2023

1. Introduction

1.1. Background

Natural gas hydrates (NGHs) are commonly referred to as combustible ice, which is a non-stoichiometric cage crystalline compound [1], because of their ice-like appearance and ability to be burned. NGH, being an unconventional source of clean energy, produces the least amount of CO_2 per unit of energy [2]. The potential reserves of hydrated gas are over 1.5×10^{16} m^3 and are widely distributed throughout the earth [3], with over 230 hydrate deposits discovered globally in ocean floors and permafrost zones. To extract methane gas from hydrate reservoirs, the in-situ equilibrium condition (high pressure and low temperature) of NGH must be broken, allowing it to decompose and be produced as fluid. Four methods can be used for gas recovery from hydrate-bearing sediments, including depressurization [4,5], thermal stimulation [6], inhibitor injection [7], and CO_2–CH_4 replacement [8–11]. National programs exist in many countries to research and produce natural gas from gas hydrate deposits in order to discover the commercialization possibility of methane hydrate resources, leading to various studies on the Alaska North Slope [12–15], the Mallik site in Canada [4,16,17], the Black Sea [18,19], the Krishna–Godavari basin in India [20,21], the Ulleung basin in Korea [22,23], the Nankai Trough in Japan [24–26], and the South China Sea [27–29] and Qilian Mountain [30] in China.

In general, both marine and terrestrial gas hydrate accumulations are targeted for exploitation. Extracting methane from hydrates can provide a significant global energy supply as long as the operation is controlled to prevent any leakage. Concerns about potential environmental risks are increasing, particularly regarding methane escape to the seafloor during hydrate exploitation in marine region [31]. In addition to production leakage, the warming atmosphere can disrupt the stability of hydrate fields, leading to the release of sequestered methane into the sediments and soils above. The methane, along with methane-derived carbon, can further contribute to greenhouse warming when it reaches the atmosphere [32]. If this happens in the terrestrial region, it has the potential to cause a catastrophic disaster, even with only around 1% of the global gas-in-place estimated to be in the permafrost-associated NGH [33]. This topic has sparked interest in studies that aim to quantify natural gas hydrates from ice [34] and understand how gas hydrate-bearing permafrost sediments respond to changes in environmental temperature [35]. While both NGH exploitation and global warming can have environmental effects, the focus of this study is to evaluate the production potential of offshore hydrate accumulations in the South China Sea.

The main focus of this study is on the Shenhu area located in the northern part of the South China Sea. Between the years 2007 and 2016, the Guangzhou Marine Geological Survey (GMGS) conducted three gas hydrate drilling expeditions, namely, GMGS1, GMGS3, and GMGS4 [36]. During GMGS3, the team drilled a total of 23 sites for logging-while-drilling (LWD) and 4 sites for core drilling. The results indicated that the average hydrate saturation ranged from 13.7% to 45.2% [36]. Among these drilling sites, the first gas hydrate production site, W17, was selected [29]. This was a significant accomplishment since hydrates in the Shenhu area exist in clayey silt sediments, which pose production challenges due to their low permeability and high levels of clays. The W17 well test demonstrated the possibility of production in such a location, which accounts for 90% of total hydrate reservoirs [37]. In 2017, the China Geological Survey set a new world record for gas hydrate production by performing a successful offshore methane hydrate production test in the Shenhu Area. The team used a single vertical well and depressurization for a production period of 60 days, resulting in a total gas production of 3.09×10^5 ST m^3, averaging 5.15×10^3 ST m^3/d [29]. This achievement marked the most significant amount of gas production and the longest production period in history. Recently, the China Geological Survey performed another production test using a horizontal well for the first time in the Shenhu Area, achieving a 30-day continuous gas production process. The total gas production reached 8.614×10^5 ST m^3, averaging 2.87×10^4 ST m^3/d [27]. The gas production rates in both production tests remain significantly below 5.00×10^5 ST m^3/d, which is necessary for the commercial exploitation of NGHs [1].

The commercialization of methane hydrate resources is a crucial challenge, necessitating the need for enhanced gas recovery from methane hydrate deposits and leading to the exploration of various methods to increase gas production. The second production test in the Shenhu area indicates that horizontal wells have been found to enhance gas production by increasing the contact area between the well and the reservoir. However, the cost of horizontal wells is 1.5 to 2.5 times more than vertical wells [38]. In addition to horizontal wells, permeability enhancement (e.g., hydraulic fracturing), allowing gas to flow more freely towards the wellbore, might be a cost-effective option for methane hydrate reservoirs to stimulate gas production, resulting in a significant increase in gas recovery. This enhanced gas recovery can maximize the productivity of wells and boost overall production rates, ultimately paving the way for the commercialization of methane hydrate resources. Previous studies in the eastern Nankai Trough have demonstrated that permeability anisotropy and permeability improvement can enhance gas recovery [24,39]. Therefore, it appears feasible to improve gas recovery in the South China Sea via permeability enhancement as well.

Conducting field tests to determine gas recovery via different production schemes is impossible, thus necessitating the use of a numerical simulator to simulate production performance. The most commonly used simulator for studying production behavior in gas-hydrate-bearing sediments is TOUGH+HYDRATE (T+H). T+H is capable of describing mass and heat flows, hydrate formation and dissociation, and the inhibitor effect [40]. When compared to other simulators available for hydrate simulations, T+H demonstrates superior predictive results when assessing production feasibility for permafrost and marine hydrates [41]. T+H has been utilized to identify gas recovery from low-permeability hydrate reservoirs via depressurization [42], as well as natural gas hydrate reservoirs in the eastern Nankai Trough [26]. Some studies have also incorporated a geomechanical simulator to analyze geomechanical responses [43–45]. By utilizing these simulators, the effectiveness of permeability enhancement can be determined.

In order to evaluate the potential for gas production resulting from permeability enhancement in the South China Sea, various numerical simulations have been carried out. In the Liwan 3 Area, Zhang et al. [46] focused on a methane hydrate reservoir with the conditions of low permeability located at site LW3-H4-2. Their approach involved the use of an enlarged, highly permeable well wall, which was proposed as a means of promoting gas production. At site SH2 in the Shenhu Area, Li et al. [47] and Sun et al. [48] suggested the stratification split grouting foam mortar method and hydraulic fracturing, respectively, as methods to improve permeability for production enhancement. The latter study demonstrated the influences of horizontal and vertical fractures. Despite the different methods used for permeability enhancement, Yu et al. [49] conducted a study showing that gas production can be significantly enhanced by the complex mechanisms associated with permeability enhancement at well SHSC-4. These models were based on various geological conditions in the South China Sea and explored the impact of different parameters, including permeability, on gas recovery.

Despite the progress made in enhancing gas production from methane hydrate deposits, the effect of permeability associated with geomechanical responses on gas production enhancement has not been thoroughly investigated. Previous studies have predominantly relied on simplistic adjustments of formation porosity and permeability based on pressure and temperature, with a primary focus on understanding the interplay between system flow and thermal aspects. Unfortunately, these studies have overlooked the crucial factors of media deformations and changes in stress fields. Geomechanics is a critical factor in understanding the behavior of hydrate behavior because of the state and phase changes, leading to changes in its porosity, permeability, and flow characteristics [4,50,51]. Specifically, changes in stress and pressure can cause the rock to compact or expand, affecting the pore structure and connectivity. These changes in permeability can significantly impact the flow of gas within the reservoir, influencing production rates and overall productivity.

Before commercial gas production from hydrate deposits can be developed, it is essential to address and understand the geomechanical response of hydrate-bearing sediments. Particularly, potential wellbore instability and casing deformation are significant concerns that must be tackled [52]. The application of mechanical loads to sediments containing hydrates can cause hydrate dissociation and result in a substantial increase in pressure. This pressure increase can negatively impact the wellbore assembly, the hydrate-bearing sediments, and the surrounding formations, posing risks to their integrity and stability [50]. To prevent failures like tensile or buckling failure in the well assembly, it becomes crucial to implement appropriate well design based on numerical simulations accounting for the geomechanical response [53].

To evaluate the production potential in the South China Sea, it is crucial to have a comprehensive understanding of the intricate interactions between permeability, geomechanical responses, and gas production enhancement. To gain a more accurate and holistic understanding of these phenomena, it is imperative to consider the influence of media deformations and stress field changes. Only a geomechanical model can accurately capture the mechanical behavior of subsurface rocks, including (a) the deformation and potential failure of the reservoir media and well assembly; (b) the effects of changes in stresses, fluid pressures, and temperatures on flow properties like porosity and permeability that control production; and (c) the interdependence of system flow, thermal, and geomechanical properties. Hence, the accurate determination of (a) the impact of permeability enhancement associated with geomechanical responses on hydrate development and (b) the effectiveness of permeability enhancement can be achieved.

1.2. Targeted Accumulation

This study focuses on the oceanic hydrate accumulation at well SHSC-4 in the Shenhu area (Figure 1), which is located in the northern South China Sea between the Xisha Trough and Dongsha Islands and is structurally part of the Baiyun Sag of the Zhuer Depression [29]. The seafloor exhibits a complex topography, including various features such as eroded channels, sea valleys, sea mounts, steep slopes, reverse slopes, plateaus, alluvial fans, and slide fans [54]. The continental slope in the northern South China Sea displays both passive and active continental features due to interactions between the Eurasian, Pacific, and Indochina plates. The reservoir lithology is composed of clayey silt, with montmorillonite and illite constituting over a quarter of the total minerals [29]. Based on the logging interpretation and core analysis of SHSC-4, the overlying formation extends from the sea bottom to a depth of 1495 m (201 m below seafloor (mbsf), with a water depth of 1266 m). The NGH system at the test site comprises three intervals, with the first interval ranging from 1495 to 1530 m (201–236 mbsf), and its pore space is filled with solid NGH and liquid water. The second interval spans from 1530 to 1545 m (236–251 mbsf), and its pore space is filled with three phases: solid NGH, free hydrocarbon gas, and liquid water. The third interval spans from 1545 to 1572 m (251–278 mbsf), and its pore space is filled with two phases: free hydrocarbon gas and liquid water.

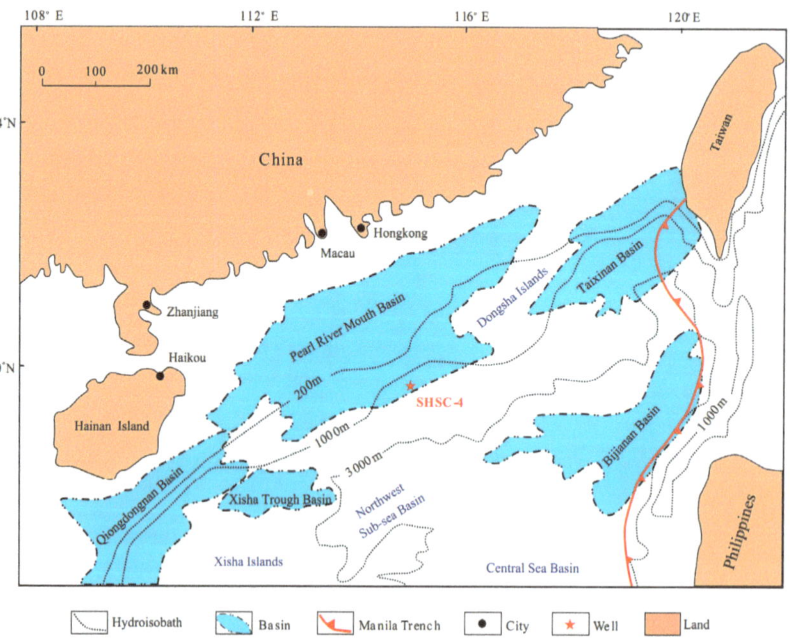

Figure 1. Structural units in the northern South China Sea and the location of SHSC-4 (modified from Li et al. [29]).

1.3. Objectives

The objective of this study is to assess the technical feasibility of gas production and production enhancement by the numerical simulation of oceanic hydrate accumulation in the Shenhu Area of the South China Sea. Specifically, the study evaluates the impact of increased permeabilities in the near-wellbore region on gas production associated with the geomechanical responses. The analysis considers the properties and conditions obtained at well SHSC-4; studies the gas production behavior at well SHSC-4 using a single vertical well by depressurization; assesses the contribution of different layers to gas production; and investigates the feasibility of permeability enhancement, the potential of increased gas production, and the geomechanical response of the geological system during production. To validate the accuracy of the numerical model used in the study, a 60-day simulation result is compared against the actual trial production test conducted at well SHSC-4. The impact of permeability enhancement is evaluated by quantifying the ratios of the cumulative gas production and cumulative water-to-gas ratio, comparing them to the original conditions.

2. Methodology

2.1. Coupled Numerical Simulators

Two parallel simulators, pTOUGH+HYDRATE (pT+H) V1.5 and the Reservoir GeoMechanics Simulator (RGMS) [55,56], based on an MPI (message passing interface) were utilized in this study to simulate the coupled processes of flow, thermal, chemical, and geomechanical processes associated with gas production induced by depressurization. pT+H V1.5 is a parallel version of TOUGH+HYDRATE V1.5 used for conducting numerical investigations of flow and thermal behavior in hydrate-bearing geologic media. The code uses the integral finite difference method (IFDM) [57,58] for space discretization and the Newton–Raphson (NR) iteration for a fully implicit solution. The computational domain is decomposed into subdomains for efficient workload distribution to multiple processes. The RGMS is a parallel geomechanical simulator that accurately characterizes deformations and stresses in subsurface systems by employing parallel strategies in conjunction with the finite element method (FEM). It has the capability to handle grids with Cartesian (2D and 3D) and cylindrical (2D) coordinates. Additionally, the simulator can be used either as a standalone or coupled with T+H, pT+H, and other simulators describing flow and thermal behavior. To explore the geomechanical responses to gas production from hydrate reservoirs, pT+H V1.5 and the RGMS are coupled via the fixed-stress split iterative scheme. The coupling scheme was validated in previous studies [55,56].

2.2. Governing Equations

2.2.1. Flows of Fluid and Heat

The mass and energy balance equation includes the accumulation, flux, and source/sink terms through a control volume in a porous medium [40], which is expressed as

$$\frac{d}{dt}\int_{V_n} M^{\kappa}\mathrm{d}V = \int_{\Gamma_n} \mathbf{F}^{\kappa}\cdot\mathbf{n}\mathrm{d}A + \int_{V_n} q^{\kappa}\mathrm{d}V, \tag{1}$$

where t is the time, V_n is the volume of subdomain n, $\mathrm{d}V$ is the differential volume, M^{κ} is the accumulation of component κ in terms of mass or heat, Γ_n is the surface area of subdomain n, $\mathrm{d}A$ is the differential surface area, $\mathbf{F}^{\kappa}$ is the flux vector of component κ in terms of mass or heat, $\mathbf{n}$ denotes the inward unit normal vector, and q^{κ} is the source/sink term of component κ.

For component κ, the mass accumulation M^{κ} is calculated by

$$M^{\kappa} = \sum_{\beta = A,G,I,H} \phi S_{\beta}\rho_{\beta}X^{\kappa}_{\beta},\ \kappa = w,m,i, \tag{2}$$

where β indicates phase (A refers to aqueous, G refers to gaseous, I refers to solid ice, and H refers to solid hydrate), κ indicates component (w refers to H_2O, m refers to CH_4, and

i refers to water-soluble inhibitor), ϕ represents the porosity, S_β represents the saturation of phase β, ρ_β represents the density of phase β, and X_β^κ represents the mass fraction of component κ in phase β.

The heat accumulation M^θ comprises contributions from both the rock matrix and all the phases and is expressed by

$$M^\theta = \int_{T_0}^{T} (1-\phi)\rho_R C_R dT + \sum\nolimits_{\beta = 1,\cdots,N_\beta} \phi S_\beta \rho_\beta U_\beta + Q_{diss}, \tag{3}$$

where ρ_R is the density of the rock, C_R is the heat capacity of the dry rock, T is the temperature, and U_β is the specific internal energy of phase β. The energy change of hydrate dissociation Q_{diss} is

$$Q_{diss} = \begin{cases} \Delta\left(\phi\rho_H S_H \Delta H^0\right) & \text{for equilibrium dissociation} \\ Q_H \Delta H^0 & \text{for kinetic dissociation} \end{cases}, \tag{4}$$

where $\Delta()$ denotes the quantity change during a given timestep, and ΔH^0 is the specific enthalpy of hydrate dissociation, which can be calculated by Equation (5) from [59]:

$$\Delta H^0 = C_f(C_1 + C_2/T), \tag{5}$$

where the conversion factor C_f is 33.72995 $\text{J}\cdot\text{gmol}\cdot\text{kg}^{-1}\cdot\text{cal}^{-1}$,

$$C_1 = \begin{cases} 13521 \\ 6534 \end{cases}, \text{ and } C_2 = \begin{cases} -4.02 & \text{for } \ 0\,^\circ\text{C} < T_c \le 25\,^\circ\text{C} \\ -11.97 & \text{for } -25\,^\circ\text{C} < T_c \le 0\,^\circ\text{C} \end{cases}. \tag{6}$$

It is important to note that hydrate dissociation can be treated in two ways: (a) a chemical equilibrium reaction (the hydrate is considered a thermodynamic state of the CH_4 and H_2O system) or (b) a kinetic reaction (the hydrate is regarded as a distinct component) [60]. The former was employed in this study. Interested readers can refer to Moridis [40,61] for in-depth information on the specifics of the two models and the associated thermodynamics.

The mass flux of component κ (e.g., H_2O, CH_4, and inhibitor) contributed by the aqueous and gaseous phases is defined as

$$\mathbf{F}^\kappa = \sum_{\beta = A,G} \mathbf{F}_\beta^\kappa = \sum_{\beta = A,G} X_\beta^\kappa \mathbf{F}_\beta,\ \kappa = w, m, i. \tag{7}$$

Following Darcy's law, the mass flux of phase β is defined as

$$\mathbf{F}_\beta = -\mathbf{k}\frac{k_{r\beta}\rho_\beta}{\mu_\beta}(\nabla P_\beta - \rho_\beta \mathbf{g}), \tag{8}$$

in which, for phase β, $k_{r\beta}$ is the relative permeability, μ_β is the viscosity, and P_β is the fluid pressure. Additionally, $\mathbf{k}$ is the absolute permeability tensor, and $\mathbf{g}$ is the gravity vector.

The heat flux is defined as

$$\mathbf{F}^\theta = -\overline{k_\theta}\nabla T + \sum_{\beta = A,G} h_\beta \mathbf{F}_\beta, \tag{9}$$

where h_β is the specific enthalpy of phase β, and $\overline{k_\theta}$ is the composite thermal conductivity considering the properties of medium and fluid.

2.2.2. Geomechanics

The quasi-static momentum conservation equation is based on the underlying assumption that there is always equilibrium between the fluid and rock, which is calculated by

$$\nabla \cdot \boldsymbol{\sigma} + \rho_b \mathbf{g} = \mathbf{0}, \tag{10}$$

where $\boldsymbol{\sigma}$ is the total stress tensor. The bulk density ρ_b is found using

$$\rho_b = (1-\phi)\rho_R + \phi\rho_f, \tag{11}$$

in which ρ_f is the saturation-weighted fluid density, which is calculated using

$$\rho_f = \sum_{\beta} S_\beta \rho_\beta. \tag{12}$$

In accordance with convention, the equations below always consider tensile stress as positive. The relationship that defines the stresses associated with the rock skeleton is expressed as

$$\boldsymbol{\sigma} = \boldsymbol{\sigma}' - \alpha \mathbf{I} P_t = \mathbf{C} : \boldsymbol{\varepsilon} - \alpha \mathbf{I} P_t, \tag{13}$$

where $\mathbf{C}$ is the elasticity tensor, $\boldsymbol{\sigma}'$ is the effective stress tensor, $\mathbf{I}$ is the identity matrix, and $\boldsymbol{\varepsilon}$ is the strain tensor. Biot's coefficient α [62] is defined as

$$\alpha = 1 - \frac{K_{dr}}{K_s}, \tag{14}$$

where K_s is the skeletal grain modulus, and K_{dr} is the drained bulk modulus. The average mobile fluid pressure P_t is calculated using

$$P_t = \frac{\sum_{\beta} S_\beta P_\beta}{\sum_{\beta} S_\beta}. \tag{15}$$

By utilizing the assumption of infinitesimal deformation, the strain tensor is determined using

$$\boldsymbol{\varepsilon} = \frac{1}{2}\left[\nabla \mathbf{u} + (\nabla \mathbf{u})^T\right], \tag{16}$$

in which $\mathbf{u}$ is the displacement vector.

2.2.3. The Coupling Method between Geomechanics and Flows of Fluid and Heat

To investigate the geomechanical response during production, pT+H and the RGMS were coupled with the fixed-stress split iterative scheme so that the changes in fluid pressure, temperature, phase saturations, and associated stresses can be found. The accuracy and reliability of the scheme were verified by comparing its numerical results and the analytical solutions of Terzaghi's problem [63] and McNamee–Gibson's problem [64,65] in prior research [55].

The scheme operates in a sequential manner as shown in Figure 2, where pT+H solves the problem when the stress field is frozen. The strongly nonlinear equations are solved via the NR iteration, and the porosity is calculated using

$$\phi^k = \phi^{k-1} + \frac{\alpha(1+\varepsilon_v) - \phi^{k-1}}{K_{dr}} \delta {P_t}^{k-1}, \tag{17}$$

in which k is the number of the NR iterations, K_{dr} is the drained modulus, ε_v is the volumetric strain, and $\delta {P_t}^{k-1}$ is the difference at the NR iteration k and $k-1$.

Subsequently, the porosity in the RGMS is calculated using

$$\phi^n = \phi_0 + \alpha\varepsilon_v + \frac{(\alpha-\phi_0)(1-\alpha)}{K_{dr}}(P_t - P_{t0}), \tag{18}$$

where subscript n denotes the initial states.

pT+H and the RGMS solve the problem sequentially; do not proceed to the next time step until $\left(\phi^n - \phi^k\right)/\phi^n$ is lower than a predetermined tolerance level.

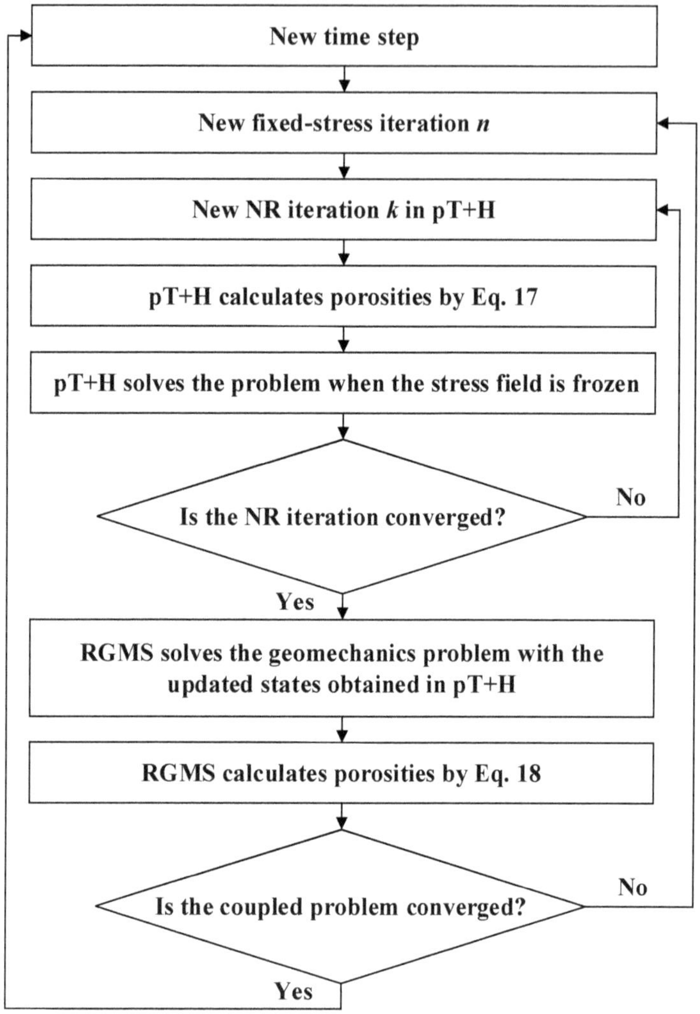

Figure 2. Flow chart of the coupling method between geomechanics and flows of fluid and heat.

The mechanical properties of the hydrate are influenced by the existence of hydrate, which is more pronounced when S_H has a higher value. However, there is a scarcity of research on how the properties are related to the hydrate phase. Rutqvist and Moridis [50] proposed a standard approach that utilizes linear interpolation equations as follows:

$$K_{dr} = K_{dr0}(1 - S_H) + K_{dr1} S_H, \tag{19}$$

and

$$G = G_0(1 - S_H) + G_1 S_H, \tag{20}$$

where G is the shear modulus, subscript 0 denotes $S_H = 0$, and subscript 1 denotes $S_H = 1$.

3. Numerical Model

3.1. The Geologic Model

The geological model was developed from a combination of geophysical surveys and an analysis of core samples retrieved from the well. The model consists of five units:

the overburden (OB), the hydrate-bearing layer (HBL), the three-phase layer (TPL), the free gas layer (FGL), and the underburden (UB). The computational domain used in both the production and geomechanical studies accurately represents the geological model, encompassing the entire OB to the ocean floor and a portion of the UB that was thick enough to fully account for the necessary heat and water exchange with the reservoir, and provides a true zero-displacement boundary for geomechanical computations.

Combining geophysical surveys and core sample analysis from the well, the geological model was developed (Figure 3), comprising five units: the overburden (OB), the hydrate-bearing layer (HBL), the three-phase layer (TPL), the free gas layer (FGL), and the underburden (UB). In pT+H V1.5 and the RGMS, the same computational domain was utilized to represent the geological model, covering the entire OB to the ocean floor and a thick enough portion of the UB to account for water and heat exchange within the reservoir, and to make sure there was no displacement occurring at the bottom boundary for geomechanical computations.

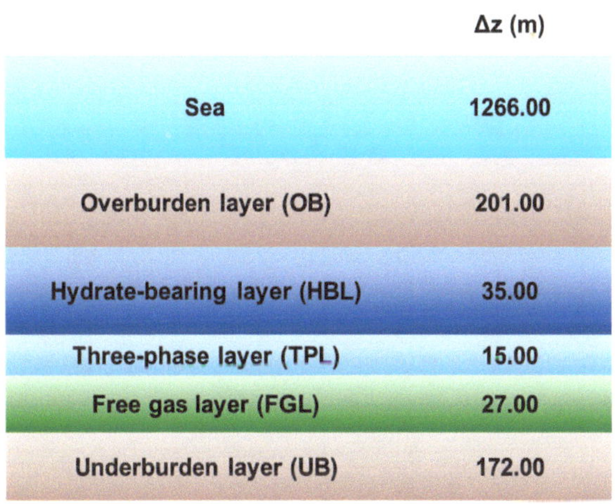

Figure 3. The geological model with layered geometries.

3.2. Domain Discretization

A 2D cylindrical domain with an outer radius of 300 m, as presented in Figure 4, was utilized to investigate gas production with a single vertical well. The domain was divided into 412 segments radially and 232 segments vertically, leading to 95,584 gridblocks. Gridblocks with different lengths were produced due to the high-resolution radial discretization in the vicinity of the well; specifically, the segment length in the radial direction (Δr) of 0.10 m was used for $r_w < r \leq 1$ m, while that of 0.20 m was used for 1 m $\leq$ r $\leq$ 21 m. For distances greater than 21 m but less than 300 m ($r_{\max}$), Δr increased logarithmically for r > 0.20 m. The segment length in the vertical direction (Δz) was 0.5 m in the hydrate accumulation area and was larger in OB and UB. A mesh representation of the domain used in this study is presented in Figure 5a, while a more detailed representation of the grid near the wellbore is given in Figure 5b.

For each element, four equations were designed to account for the mass balance of three components (H_2O, CH_4, and NaCl) as well as the heat balance of the system. Previous research has demonstrated that a chemical equilibrium reaction provides an accurate result during hydrate formation/dissociation in gas production [60]. To account for the scale of the problem, a total of 560,000 equations were formulated, consisting of approximately 380,000 equations in pT+H V1.5 and 180,000 equations in the RGMS. Consequently, the problem sizes necessitated the use of pT+H V1.5 and the RGMS to provide practical solutions.

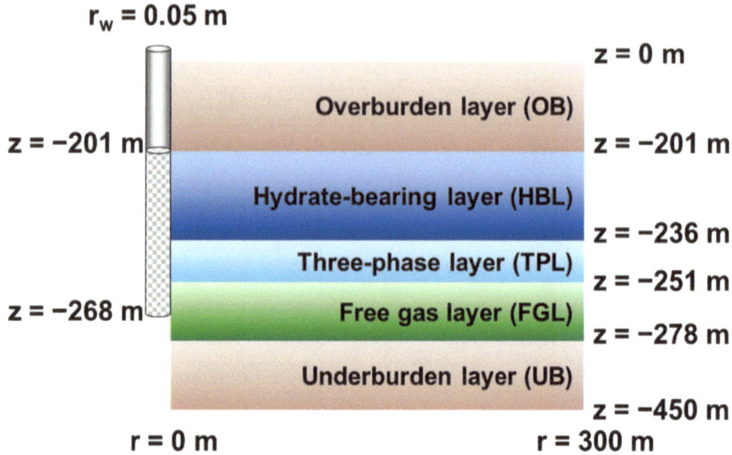

Figure 4. System geometry and configuration of the single vertical well with a radius of 0.05 m, which is perforated from 201 mbsf to 268 mbsf as shown in the checkboard pattern, produced from a cylindrical section at well SHSC-4.

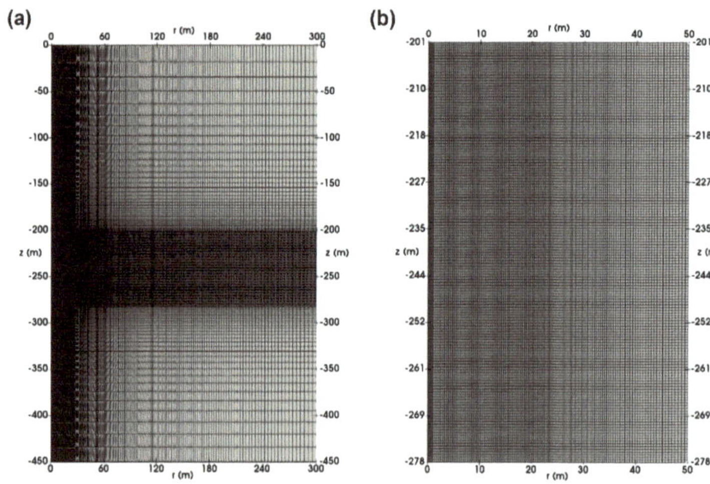

Figure 5. Discretization of (**a**) the entire domain and (**b**) the region near the well.

3.3. Well Description

At the center of the cylindrical domain, the vertical production well was perforated from 201 mbsf to 268 mbsf, covering the HBL, the TPL, and a portion of the FGL (Figure 4). The well was treated as a pseudo-porous medium to use Darcy's flow concepts within the wellbore, of which the properties are as follows: the vertical permeability was 5×10^{-9} m^2, the porosity was 1, the capillary pressure was 0, the irreducible gas saturation was 0.005, and the relative permeabilities had linear relationships with phase saturations. This approach was able to simulate the pressure drop in a steel wellbore, which was validated in a previous study [66]. The bottomhole pressure (P_{bh}) was 3 MPa [67] at a gridblock above the topmost well gridblock.

3.4. System Properties

Table 1 provides the properties used in pT+H V1.5 that are based on the previous studies [29,49,67,68]. The properties used in the RGMS are based on the previous studies [53,69] as shown in Table 2.

Table 1. Properties, conditions, and models used in pT+H V1.5.

Properties, Conditions, Models	Values
Initial pressure at the bottom of TPL	14.93 MPa
Initial temperature at the bottom of TPL	14.82 °C
Gas composition	100% CH_4
Initial saturation of HBL	$S_H = 0.34$
Intrinsic permeabilities of HBL	$k_h = 2.86 \times 10^{-15}$ m^2 = 2.9 mD; $k_z = k_h$
Porosity ϕ of HBL	0.35
Initial saturation of TPL	$S_H = 0.31$, $S_G = 0.078$
Intrinsic permeabilities of TPL	$k_h = 1.48 \times 10^{-15}$ m^2 = 1.5 mD; $k_z = k_h$
Porosity ϕ of TPL	0.33
Initial saturation of FGL	$S_G = 0.078$
Intrinsic permeabilities of FGL	$k_h = 7.30 \times 10^{-15}$ m^2 = 7.4 mD; $k_z = k_h$
Porosity ϕ of FGL	0.32
Intrinsic permeabilities of OB	$k_h = 9.87 \times 10^{-18}$ m^2 = 0.01 mD; $k_z = k_h$
Porosity ϕ of OB	0.10
Intrinsic permeabilities of UB	$k_h = 9.87 \times 10^{-18}$ m^2 = 0.01 mD; $k_z = k_h$
Porosity ϕ of UB	0.10
Dry thermal conductivity	$k_{\theta d} = 1$ W·m^{-1}·K^{-1}
Specific heat C_R	1000 J kg^{-1}·K^{-1}
Grain density ρ_R	2650 kg·m^{-3}
Composite thermal conductivity model [40]	$\overline{k_\theta} = k_{\theta d} + (\sqrt{S_A} + \sqrt{S_H})(k_{\theta w} - k_{\theta d}) + \phi S_I k_{\theta I}$
Relative permeability model EPM#2 [40]	$k_{rA} = \max\left\{0, \min\left\{\left[\frac{S_A - S_{irA}}{1 - S_{irA}}\right]^n, 1\right\}\right\}$; $k_{rG} = \max\left\{0, \min\left\{\left[\frac{S_G - S_{irG}}{1 - S_{irA}}\right]^{n_G}, 1\right\}\right\}$; $k_{rH} = 0$
S_{irA}, S_{irG}, n, n_G [69]	0.65; 0.03; 3.50; 2.50
Capillary pressure model [70]	$P_{cap} = -P_0\left[(S^*)^{-\frac{1}{\lambda}} - 1\right]^{1-\lambda}$; $S^* = \frac{S_A - S_{irA}}{S_{mxA} - S_{irA}}$
λ, P_0, S_{irA}, S_{mxA} of HBLs	0.45; 10^4 Pa; 0.65; 1.0
Porosity–permeability relationship [71]	$\frac{k}{k_0} = \exp\left[\gamma\left(\frac{\phi}{\phi_0} - 1\right)\right]$
Empirical permeability reduction factor γ [71]	29.0

Table 2. Properties used in RGMS.

Properties	Values
Young's modulus of HBL	E = 200 MPa at $S_H = 0$; E = 1.4 GPa at $S_H = 1$
Young's modulus of TPL	E = 200 MPa at $S_H = 0$; E = 1.4 GPa at $S_H = 1$
Young's modulus of FGL	E = 200 MPa
Young's modulus of OB	E = 70 MPa
Young's modulus of UB	E = 200 MPa
Poisson's ratio of HBL	$\nu = 0.15$
Poisson's ratio of TPL	$\nu = 0.15$
Poisson's ratio of FGL	$\nu = 0.45$
Poisson's ratio of OB	$\nu = 0.45$
Poisson's ratio of UB	$\nu = 0.45$
Biot's coefficient	$\alpha = 0.99$

3.5. Initial Conditions

Table 1 presents the temperature and pressure values at the bottom boundary of the TPL, which conform to the geothermal and hydrostatic gradients, respectively, as is typical of most hydrate deposits. Following the initialization process outlined by Moridis and Reagan [66], the simulation with the initial conditions continued until the entire domain reached a steady state without any change in various distributions. For geomechanical simulation, the initial total stress field was obtained by assuming that the initial effective stress field was zero, without requiring the application of the overburden pressure.

3.6. Model Validation

In order to guarantee the prediction accuracy of gas production at well SHSC-4, it is imperative to first validate the numerical model built earlier. As stated in the Introduction section, a trial production test was conducted at well SHSC-4, lasting 60 days, with a total gas production of 3.09×10^5 ST m^3 [29]. Using the constructed model, a simulation was performed to replicate the 60-day production, taking into account the geomechanical responses. The simulation resulted in a total gas production of 3.08×10^5 ST m^3, as depicted in Figure 6. This successful replication serves as validation for the constructed model.

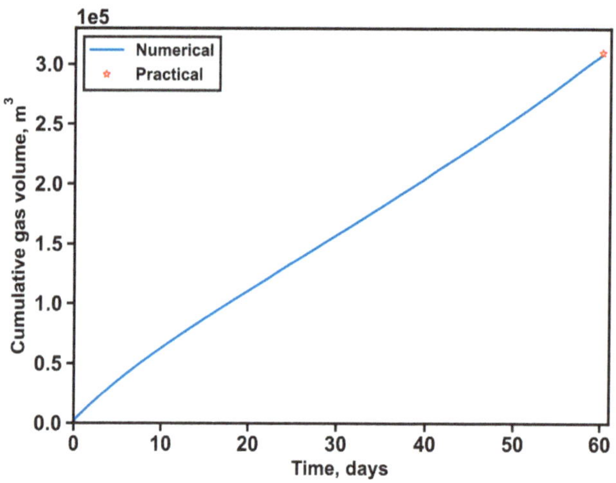

Figure 6. Comparisons of numerical simulation result and field test data.

3.7. Simulations Cases

The base case was based on the original formation information obtained from the first test production at well SHSC-4 (Figure 4). In addition to the base case, this study investigates whether the cumulative gas production can be improved by increasing the permeabilities of the HBL, TPL, and FGL, individually. As illustrated in Figure 7, the permeability was increased to k_s within the region from 0 to the stimulated radius (r_s) within a specific layer. To assess the effectiveness of permeability enhancement, a permeability enhancement ratio ($f_k = k_s/k_0$) was proposed, in which k_0 is the original permeability of the layer. There are four different values of r_s (0.3 m, 0.5 m, 1 m, and 2 m) and three different values of f_k (2, 4, and 8), combined with three layers, resulting in a total of thirty-six cases.

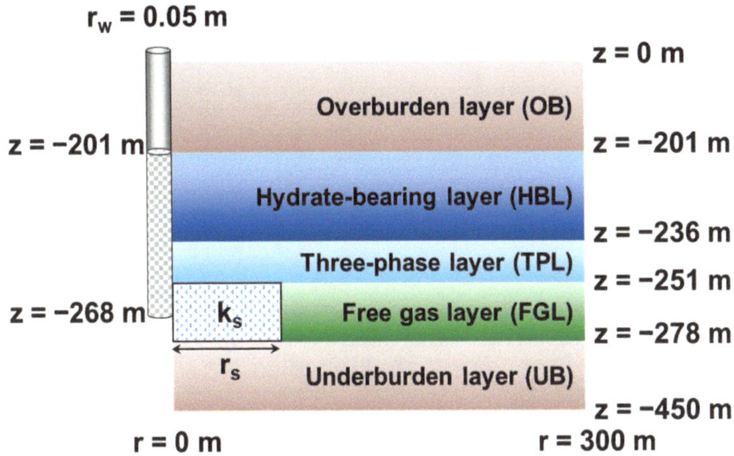

Figure 7. The illustration of the permeability enhancement region within a layer.

4. Results and Discussion

In the analysis of the production potential associated with the geomechanical response at well SHSC-4, the monitoring included various parameters related to flow and geomechanics both with and without permeability enhancement, which were obtained from pT+H and the RGMS. The monitored flow-related parameters were pressure (P), temperature (T), gas saturation (S_G), hydrate saturation (S_H), the production rates of CH_4 and H_2O (Q_g and Q_w, respectively), and the cumulative production of CH_4 and H_2O (V_g and M_w, respectively), The water-to-gas ratio was also monitored, both instantaneously ($R_{wg} = Q_w/Q_g$) and cumulatively ($R_{wgT} = M_w/V_g$). Geomechanics-related parameters monitored were radial and vertical displacements (u_r and u_z, respectively) at key locations. To evaluate the influence of permeability enhancement, the key parameters are V_g and R_{wgT}. Specifically, more gas and less water are desired after permeability enhancement, so larger V_g and smaller R_{wgT} values are better.

4.1. Base Case

4.1.1. Fluid Production

Figure 8a shows Q_g produced at the well from the HBL, TPL, and FGL and all layers in the base case. The value of Q_g has an initial peak after production begins, followed by a decline and minor fluctuations within a certain range in the subsequent production period. The initial peak is caused by the rapid dissociation of hydrates near the wellbore region and a subsequent surge in gas production rate after the bottomhole pressure drops. In addition, the free gas in the TPL and FGL contributes to the initial peak of Q_g. The average gas production over the entire production period is 0.074 ST m^3/s, which is far below the gas production rate of 0.579 ST m^3/s (=5.00 × 10^5 ST m^3/d) required for the commercial exploitation of NGHs [1]. Compared with hydrate deposits in Mount Elbert, Alaska North Slope, where there exists a lag time before substantial gas production [14], the hydrate deposit at well SHSC-4 does not exhibit such a lag phenomenon but instead has the highest gas production rate in the early stage of production, indicating that this class of hydrate deposit is conducive to exploitation. The contribution of each layer to the total gas production rate was ranked from highest to lowest as FGL, TPL, and HBL, indicating that the FGL is the primary source of gas production. This suggests that the FGL is the most important layer for gas production in the studied area.

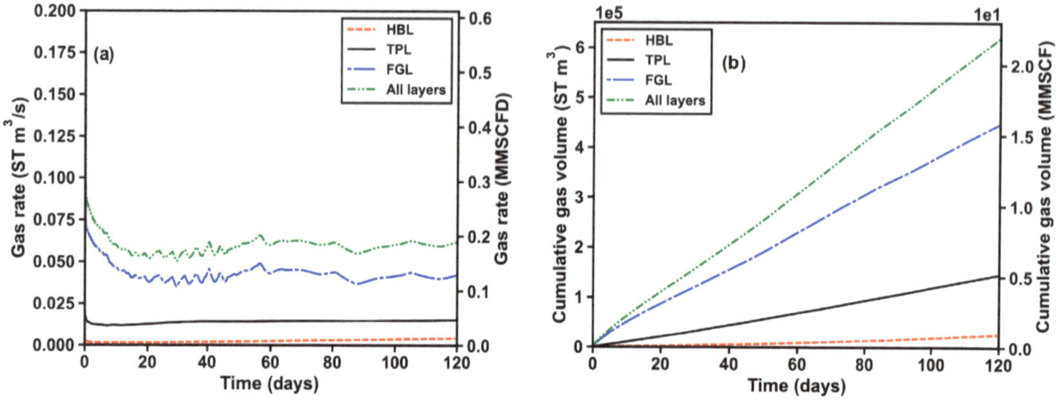

Figure 8. Evolution of (**a**) the volumetric rate of the total CH4 production (Q_g) and (**b**) the cumulative volumes of CH_4 produced at the well (V_g) in the base case.

Figure 8b shows V_g produced at the well from the HBL, TPL, and FGL and all layers in the base case. As shown in the figure, in the later production period, since Q_g fluctuates within a certain range, V_g, which is the integral of gas production over time, shows a nearly linear relationship with time. After 120 days of production, the FGL, TPL, and HBL accounted for 72.17%, 23.54%, and 4.29% of the total cumulative gas production, respectively. This also indicates that the FGL is the most important layer for gas production because it has the highest contribution to the total cumulative gas production. It is anticipated that the FGL will exhibit the most pronounced response to permeability enhancement.

Figure 9 shows R_{wg} and R_{wgT} produced at the well from the HBL, TPL, and FGL and all layers in the base case. Apart from directly evaluating production via Q_g and V_g, R_{wg} and R_{wgT} can also be used to indirectly characterize production performance. In real practice, more gas and less water are desired, so smaller R_{wg} and R_{wgT} values are better. R_{wg} produced from the HBL, TPL, and FGL and all layers decreased gradually during production. Among the three layers, the HBL has the highest R_{wg} with the smallest contribution to gas production observed in Figure 8. Moreover, R_{wg} produced from the HBL is tens of times higher than those from TPL and FGL. If permeability enhancement is carried out within HBL, gas production may increase, while water production may also increase. R_{wgT} reaches a short-term peak in the first two days of production and then shows a decreasing trend throughout the entire production period. R_{wgT} produced from the FGL and all layers tends to stabilize in the later period of production. Due to the large R_{wgT} produced from HBL, the total R_{wgT} was far higher than those from TPL and FGL.

Figure 10 shows u_r and u_z at key locations in the base case. As P_{bh} is lower than the pressure of the formation, the reservoir is "squeezed" and moves toward the vertical well in the radial direction, but the compaction is not significant. Although the location with the largest radial displacement occurs at $(r, z) = (1\ \text{m}, -201\ \text{m})$, the absolute value does not exceed 0.01 m when the simulation ends. In the vertical direction, the subsidence at the top of the HBL and the uplift at the bottom are observed. As the gridblock that was set to the bottomhole pressure is closer to the top of the HBL, the subsidence at the top of HBL is more obvious, with a maximum level of no more than 0.08 m. Overall, the displacement within the formation is not significant.

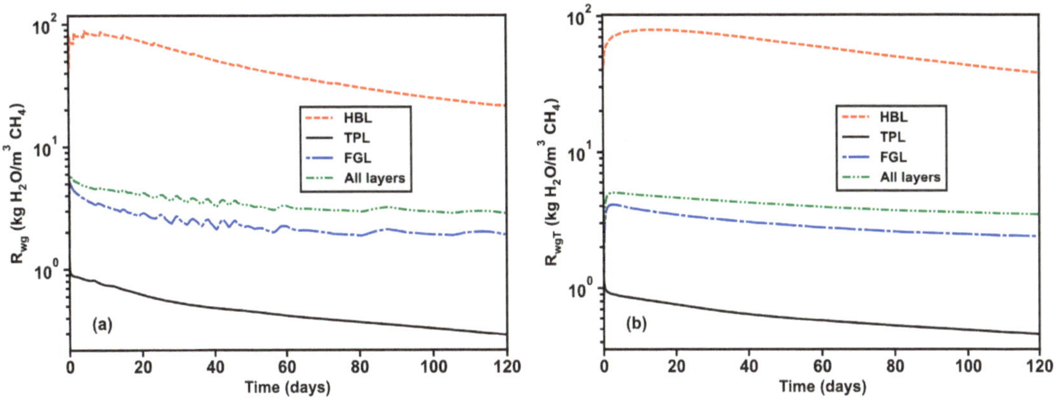

Figure 9. Evolution of (**a**) the instantaneous water-to-gas ratio at the well (R_{wg}) and (**b**) the cumulative water-to-gas ratio at the well (R_{wgT}) in the base case.

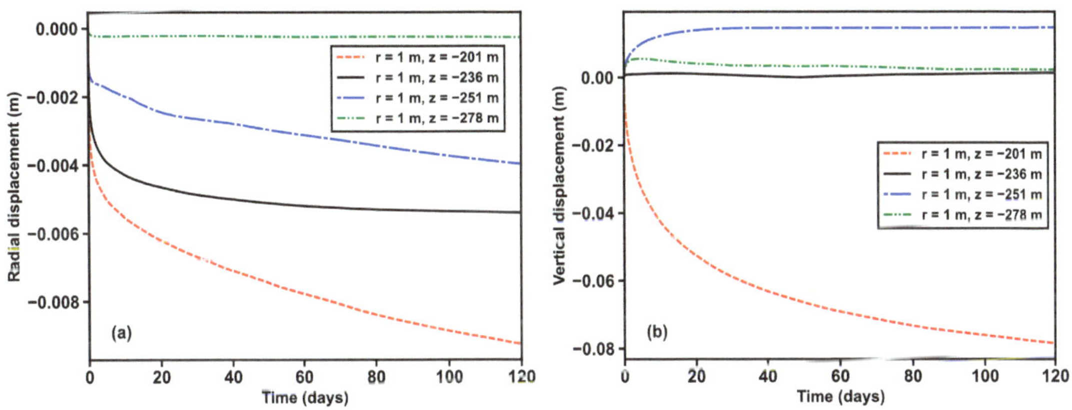

Figure 10. Evolution of (**a**) u_r and (**b**) u_z at key locations in the base case.

4.1.2. Spatial Distributions

Figure 11 illustrates the spatial evolution of P in the base case. The cylindrical region of pressure drop gradually expands over production time, with those in the HBL and FGL being larger than that in the TPL. This is mainly due to the higher permeability of the HBL and FGL compared to that of the TPL. Additionally, as the hydrate saturation is highest in the hydrate layer and the effective permeability is lower, fluid replenishment becomes difficult, resulting in the most significant pressure drop. Although the cylindrical region of the pressure drops in the TPL and FGL is relatively small during production, the rates of pressure drop are faster than that in the HBL. Particularly, the pressure drops in the FGL and the lower part of the TPL are very significant. In previous studies [49,67], the pressure drop area obtained using T+H was only not significant in the FGL, which differs from the results in this study. It is evident that considering mechanical response leads to different hydrate production mechanisms.

Figure 12 illustrates the spatial evolution of T in the base case. As hydrate dissociation is endothermic, the low-temperature region generally indicates the location of hydrate dissociation. In this figure, the low-temperature region is found in the FGL and the lower part of the TPL, coinciding with the significant pressure drop area and indicating a large amount of hydrate dissociates. Unlike production lasting over 1000 days in previous studies [49,67], the numerical simulation in this study was halted due to the temperature in the low-temperature region reaching 0.01 °C after 120 days of production.

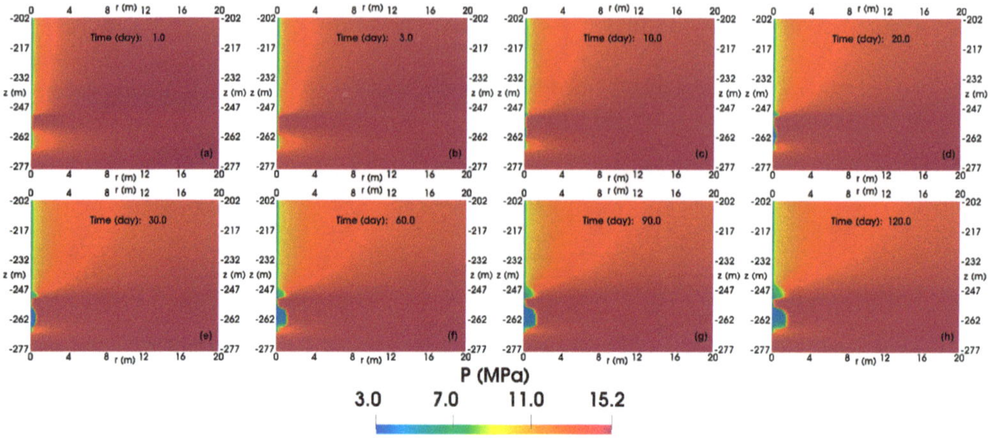

Figure 11. Evolution of the spatial distributions of pressure (MPa) in the reservoir of the base case. (**a**) 1-day production, (**b**) 3-day production, (**c**) 10-day production, (**d**) 20-day production, (**e**) 30-day production, (**f**) 60-day production, (**g**) 90-day production, and (**h**) 120-day production.

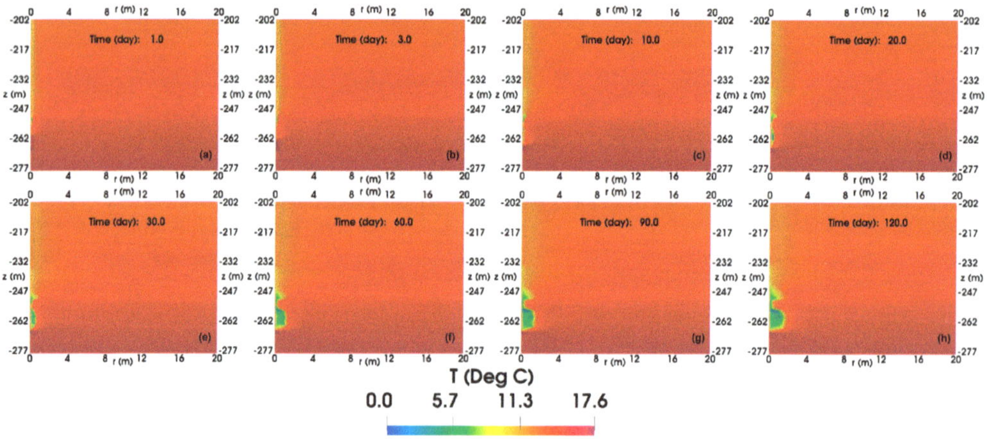

Figure 12. Evolution of the spatial distributions of temperature (°C) in the reservoir of the base case. (**a**) 1-day production, (**b**) 3-day production, (**c**) 10-day production, (**d**) 20-day production, (**e**) 30-day production, (**f**) 60-day production, (**g**) 90-day production, and (**h**) 120-day production.

Figure 13 illustrates the spatial evolution of S_H in the base case. Hydrate dissociation occurs in the area where the significant pressure drop is shown in Figure 11 and the low temperature is presented in Figure 12. Hydrates gradually dissociate during production, but the unevenness of hydrate dissociation progress in each layer becomes apparent. The dissociation rate of the HBL is relatively uniform, while the lower part of the TPL undergoes hydrate dissociation, followed by hydrate formation. Moreover, in the FGL, the hydrate forms and dissociates alternatively. This phenomenon may be caused by the Joule–Thomson cooling effect, the capillary effect, the "upstream weighting" approach applied in the simulator [72], and the equilibrium model used in this study. The cyclic process of hydrate formation and dissociation in the FGL ultimately led to the temperature reaching 0.01 °C, resulting in the simulation stopping.

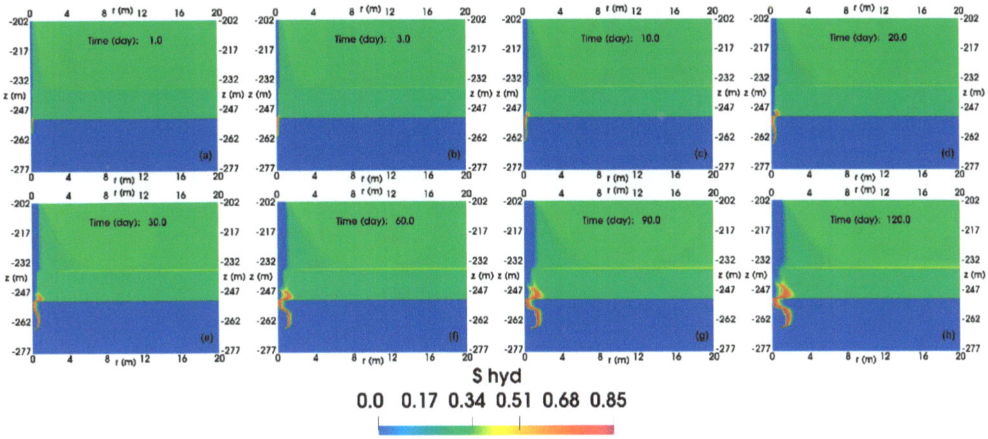

Figure 13. Evolution of the spatial distributions of hydrate saturation in the reservoir of the base case. (**a**) 1-day production, (**b**) 3-day production, (**c**) 10-day production, (**d**) 20-day production, (**e**) 30-day production, (**f**) 60-day production, (**g**) 90-day production, and (**h**) 120-day production.

Figure 14 illustrates the spatial evolution of S_G in the base case. The evolution of gas saturation in the base case is also presented. The gas saturation in the HBL and TPL gradually expands, and some gas migrates from the TPL to the HBL. Gas dissociated from hydrate in the lower part of the TPL migrates toward the wellbore radially and toward the HBL vertically at a very slow rate due to the low permeability of the TPL (1.5 mD), resulting in gas accumulation in the lower part of the TPL. In the FGL, a large amount of gas flows into the wellbore because of the significantly lower wellbore pressure compared to the formation pressure and higher formation permeability. The Joule–Thomson effect caused by the rapidly migrating gas may have caused the low temperature in the FGL, meeting the conditions for hydrate generation. Thus, hydrates are formed, hindering the radial gas migration in the FGL and causing the gas to accumulate on the side away from the wellbore.

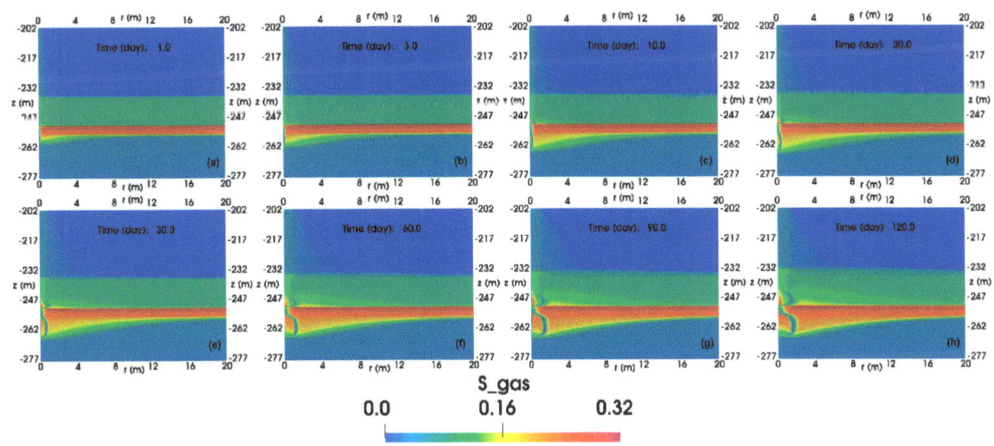

Figure 14. Evolution of the spatial distributions of gas saturation in the reservoir of the base case. (**a**) 1-day production, (**b**) 3-day production, (**c**) 10-day production, (**d**) 20-day production, (**e**) 30-day production, (**f**) 60-day production, (**g**) 90-day production, and (**h**) 120-day production.

4.2. Effect of Permeability Enhancement

4.2.1. Fluid Production

In the following tables, the values marked in green are the minimum, while the values marked in red are the maximum. Tables 3 and 4 present V_g with permeability enhancement and its ratio relative to that in the base case after 120-day production, respectively. For reference, the V_g in the base case ($V_{g,0}$) is 620,668 ST m^3 following 120 days of production. The permeability enhancement of three different layers (i.e., the HBL, TPL, and FGL) shows that the larger the values of r_s and k_f, the more significant the increase in gas production. When (k_f, r_s) = (8, 2 m), the V_g values, predicted by improving the permeabilities of the HBL, TPL, and FGL, are 711,590, 706,541, and 1,160,649 ST m^3, respectively. Compared to the base case, the production was increased by 15%, 15%, and 87% corresponding to the modification in the HBL, TPL, and FGL, respectively. In order to reach higher production, the permeability of the FGL should be enhanced.

Table 3. The cumulative gas production with permeability enhancement (V_g) after 120-day production.

Cumulative Gas Production (ST m^3)						
			r_s (m)			
			0.3	0.5	1	2
k_f	HBL	2	631,851	637,193	642,373	645,883
		4	639,688	648,514	661,047	679,954
		8	646,531	657,912	677,975	711,590
	TPL	2	641,762	648,207	657,112	667,160
		4	654,888	667,240	685,330	702,448
		8	660,201	676,255	698,987	706,541
	FGL	2	687,226	696,737	714,312	731,513
		4	712,884	740,743	811,315	855,335
		8	758,690	788,555	924,427	1,160,649

Table 4. The ratios of cumulative gas production with permeability enhancement (V_g) to that in the base case ($V_{g,0}$) after 120-day production.

$V_g/V_{g,0}$						
			r_s (m)			
			0.3	0.5	1	2
k_f	HBL	2	1.019	1.028	1.036	1.042
		4	1.032	1.046	1.066	1.097
		8	1.043	1.061	1.094	1.148
	TPL	2	1.035	1.046	1.060	1.076
		4	1.056	1.076	1.106	1.133
		8	1.065	1.091	1.128	1.140
	FGL	2	1.109	1.124	1.152	1.180
		4	1.150	1.195	1.309	1.380
		8	1.224	1.272	1.491	1.872

Tables 5 and 6 present R_{wgT} with permeability enhancement and its ratio relative to that in the base case after 120-day production, respectively. For reference, the cumulative water–gas ratio ($R_{wgT,0}$) is 2.84 kg H_2O/m^3 CH_4 following 120 days of production. The results of permeability enhancement in the HBL show that the larger the values of r_s and k_f, the larger the R_{wgT}. The increase in permeability near the wellbore area in the HBL results in a greater increase in water production than gas production, leading to a larger R_{wgT}. The results of permeability enhancement in the TPL and FGL show that the larger the values of r_s and k_f, the smaller the R_{wgT}. When (k_f, r_s) = (8, 2 m), the R_{wgT} values, determined by permeability enhancement in the HBL, TPL, and FGL are 4.03, 3.17, and 2.41 kg H_2O/m^3 CH_4, respectively. The resulting ratios of R_{wgT} to $R_{wgT,0}$ are 1.42, 1.12, and 0.85 when the permeabilities of the HBL, TPL, and FGL are increased, respectively. To reduce the amount of separated water required for unit gas production, the FGL should be treated to enhance its permeability.

Table 5. The cumulative water-to-gas ratio with permeability enhancement (R_{wg}) after 120-day production.

Cumulative Water-to-Gas Ratio (kg H_2O/m^3 CH_4)						
			r_s (m)			
			0.3	0.5	1	2
k_f	HBL	2	3.599	3.624	3.676	3.760
		4	3.683	3.739	3.834	3.939
		8	3.725	3.800	3.920	4.029
	TPL	2	3.373	3.349	3.318	3.282
		4	3.325	3.283	3.226	3.172
		8	3.308	3.256	3.192	3.173
	FGL	2	3.308	3.307	3.289	3.221
		4	3.294	3.268	3.159	2.969
		8	3.234	3.236	3.066	2.406

Table 6. The ratios of cumulative water-to-gas ratio with permeability enhancement (R_{wgT}) to that in the base case ($R_{wgT,0}$) after 120-day production.

$R_{wgT}/R_{wgT,0}$						
			r_s (m)			
			0.3	0.5	1	2
k_f	HBL	2	1.268	1.277	1.296	1.325
		4	1.298	1.318	1.351	1.388
		8	1.313	1.339	1.382	1.420
	TPL	2	1.189	1.180	1.169	1.157
		4	1.172	1.157	1.137	1.118
		8	1.166	1.148	1.125	1.118
	FGL	2	1.166	1.165	1.159	1.135
		4	1.161	1.152	1.113	1.046
		8	1.140	1.140	1.080	0.848

The radial and vertical displacements with permeability enhancement are not significant, and therefore, further discussion is not included in this section.

4.2.2. Spatial Distributions

In this section, the figures depicting the spatial distributions of P, T, S_H, and S_G with permeability enhancement after 120 days of production are arranged in a manner where the r_s values increase from left to right and the k_f values increase from top to bottom. The highest V_g and the lowest R_{wgT} are achieved after increasing the permeabilities of the FGL, and only the spatial distributions with permeability enhancement in the FGL are discussed.

Figure 15 shows the spatial distributions of P with permeability enhancement in the FGL after 120 days of production.

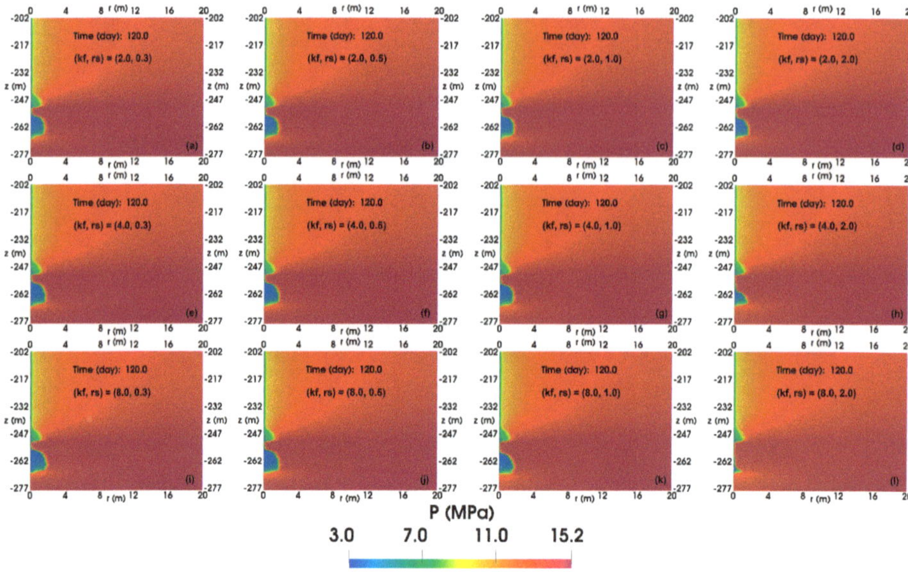

Figure 15. The spatial distributions of pressure (MPa) in the reservoir with permeability enhancement in FGL after 120-day production. **(a)** (k_f, r_s) = (2, 0.3 m), **(b)** (k_f, r_s) = (2, 0.5 m), **(c)** (k_f, r_s) = (2, 1.0 m), **(d)** (k_f, r_s) = (2, 2.0 m), **(e)** (k_f, r_s) = (4, 0.3 m), **(f)** (k_f, r_s) = (4, 0.5 m), **(g)** (k_f, r_s) = (4, 1.0 m), **(h)** (k_f, r_s) = (4, 2.0 m), **(i)** (k_f, r_s) = (8, 0.3 m), **(j)** (k_f, r_s) = (8, 0.5 m), **(k)** (k_f, r_s) = (8, 1.0 m), and **(l)** (k_f, r_s) = (8, 2.0 m).

In cases where the value of k_f is small, augmenting r_s has an insignificant effect on the area of pressure drop, and the spatial distributions of P remain relatively unchanged compared to the base case. Conversely, when the value of k_f is large, the increasing r_s results in a narrower area of pressure drop in the upper section of the FGL and the lower section of the TPL, which is closer to the wellbore.

Figure 16 shows the spatial distributions of T with permeability enhancement in the FGL after 120 days of production. When the value of k_f is small, the varying r_s has a negligible effect on the spatial distributions of T compared to the base case, which is comparable to the area of pressure drop presented in Figure 15. However, when k_f equals 4, an increase in r_s results in a shrinkage of the low-temperature area within the FGL, with a gradual shift of the lowest temperature from the upper to the middle section of the FGL. Furthermore, when (k_f, r_s) = (8, 2 m), the low-temperature region within the FGL becomes exceedingly small.

Figures 17 and 18 show the spatial distributions of S_H and S_G with permeability enhancement in the FGL after 120 days of production, respectively. In Figure 17, as the k_f and r_s values increase, the formation and dissociation of gas hydrate transpire in closer proximity to the wellbore, thereby facilitating the production of gas dissociated from gas hydrate. Furthermore, Figure 18 demonstrates that the formation of gas hydrate is less likely to obstruct the flow of gas, resulting in less gas accumulating on the side of the gas hydrate that is farther from the wellbore. These two figures collectively suggest that augmenting k_f and r_s values is more conducive to gas production. This assertion is supported by Tables 3 and 4, which indicate that larger k_f and r_s values yield higher cumulative gas production.

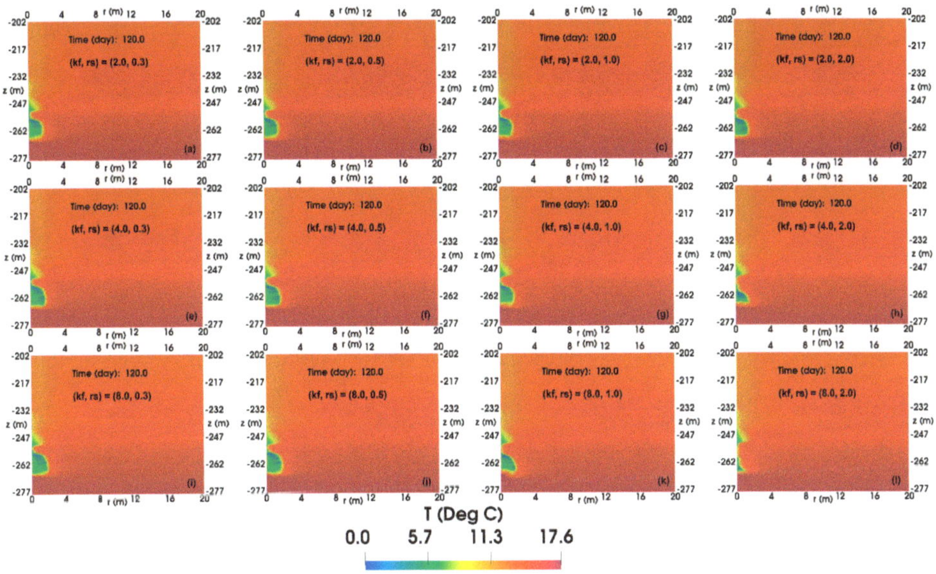

Figure 16. The spatial distributions of temperature (°C) in the reservoir with permeability enhancement in FGL after 120-day production. **(a)** (k_f, r_s) = (2, 0.3 m), **(b)** (k_f, r_s) = (2, 0.5 m), **(c)** (k_f, r_s) = (2, 1.0 m), **(d)** (k_f, r_s) = (2, 2.0 m), **(e)** (k_f, r_s) = (4, 0.3 m), **(f)** (k_f, r_s) = (4, 0.5 m), **(g)** (k_f, r_s) = (4, 1.0 m), **(h)** (k_f, r_s) = (4, 2.0 m), **(i)** (k_f, r_s) = (8, 0.3 m), **(j)** (k_f, r_s) = (8, 0.5 m), **(k)** (k_f, r_s) = (8, 1.0 m), and **(l)** (k_f, r_s) = (8, 2.0 m).

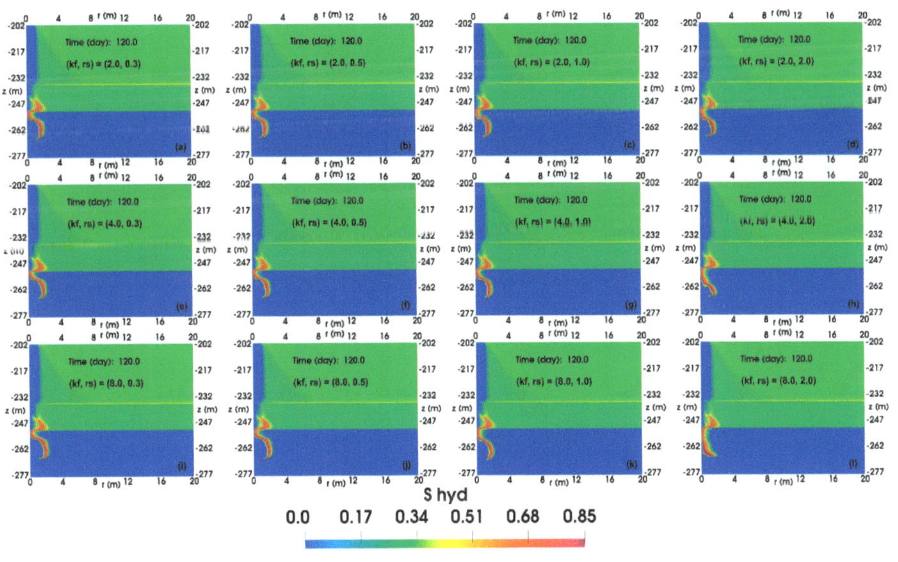

Figure 17. The spatial distributions of hydrate saturation in the reservoir with permeability enhancement in FGL after 120-day production. **(a)** (k_f, r_s) = (2, 0.3 m), **(b)** (k_f, r_s) = (2, 0.5 m), **(c)** (k_f, r_s) = (2, 1.0 m), **(d)** (k_f, r_s) = (2, 2.0 m), **(e)** (k_f, r_s) = (4, 0.3 m), **(f)** (k_f, r_s) = (4, 0.5 m), **(g)** (k_f, r_s) = (4, 1.0 m), **(h)** (k_f, r_s) = (4, 2.0 m), **(i)** (k_f, r_s) = (8, 0.3 m), **(j)** (k_f, r_s) = (8, 0.5 m), **(k)** (k_f, r_s) = (8, 1.0 m), and **(l)** (k_f, r_s) = (8, 2.0 m).

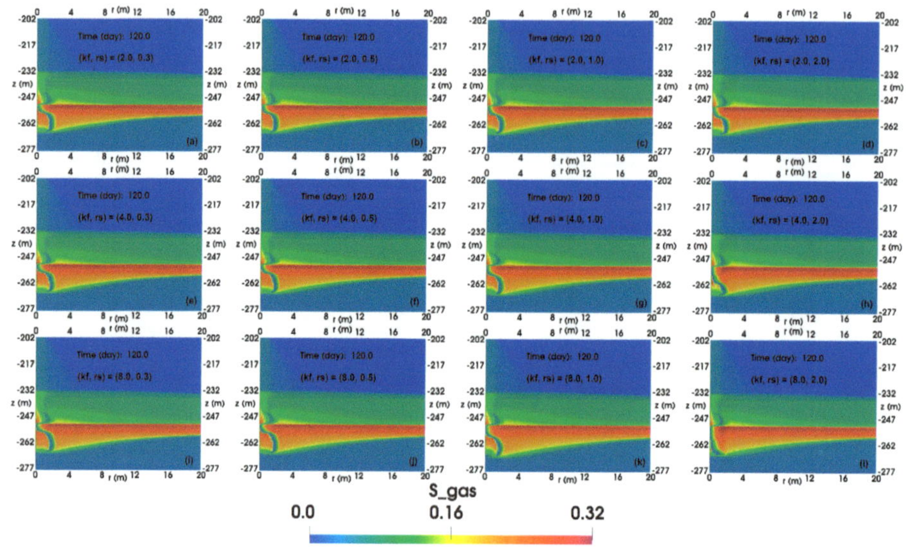

Figure 18. The spatial distributions of gas saturation in the reservoir with permeability enhancement in FGL after 120-day production. (**a**) (k_f, r_s) = (2, 0.3 m), (**b**) (k_f, r_s) = (2, 0.5 m), (**c**) (k_f, r_s) = (2, 1.0 m), (**d**) (k_f, r_s) = (2, 2.0 m), (**e**) (k_f, r_s) = (4, 0.3 m), (**f**) (k_f, r_s) = (4, 0.5 m), (**g**) (k_f, r_s) = (4, 1.0 m), (**h**) (k_f, r_s) = (4, 2.0 m), (**i**) (k_f, r_s) = (8, 0.3 m), (**j**) (k_f, r_s) = (8, 0.5 m), (**k**) (k_f, r_s) = (8, 1.0 m), and (**l**) (k_f, r_s) = (8, 2.0 m).

5. Conclusions

The following conclusions can be drawn from this study:

1. To evaluate the effectiveness of permeability enhancement considering the geomechanical responses in the Shenhu area, a coupled simulation using pTOUGH+HYDRATE V1.5 and the RGMS (Reservoir Geomechanics Simulator) is implemented.
2. Based on the geophysical surveys and analysis of core samples at well SHSC-4 located in the Shenhu area of the northern South China Sea, the established numerical simulation model is accurate, and the simulation results are highly consistent with the trial production data, ensuring the reliability of the outcomes obtained in this study.
3. In the base case, the formation and dissociation of gas hydrates in the free gas layer (FGL) alternate, ultimately resulting in a low-temperature region near 0 °C and leading to the cessation of the simulation after 120 days of production. The cumulative gas production reached 6.2×10^5 ST m^3.
4. In the base case, the FGL contributes the most to gas production, accounting for 72.17% of the cumulative gas production (V_g), followed by the three-phase layer (TPL), accounting for 23.54% of the cumulative gas production, and the hydrate-bearing layer (HBL) contributes the least, accounting for only 4.29% of the cumulative gas production.
5. In the base case, the cumulative water-to-gas ratio (R_{wg}) from the HBL, TPL, and FGL gradually decreases during the production of gas hydrates. R_{wgT} from the HBL, which contributes the least to gas production, is the highest, with a value several times those from TPL and FGL.
6. In the base case, the gas production obtained without permeability enhancement is insufficient for commercial production. Permeability enhancement can be an option used to increase gas production.

7. After increasing the permeabilities of the HBL, TPL, and FGL with the same permeability enhancement ratio (f_k) and the same simulated radius (r_s), the improvement effect of modifying the FGL is the best, with a maximum increase of 87%. The required mass of water separated from a unit of gas is the lowest when applying permeability enhancement in the FGL, with a minimum value of 85% of the original separation mass.
8. The results of modifying the FGL show that the higher the degree of permeability enhancement, the deeper the impact of permeability enhancement and the closer the formation and dissociation of gas hydrates are to the wellbore, making it more difficult for gas to be obstructed by the formation of gas hydrates, which is more conducive to production.
9. Although permeability enhancement is attempted in this study, it did not extend the production period as the simulation still ends due to low temperature in the FGL. Future research should focus on exploring methods to prevent such low temperatures from occurring in the FGL.
10. The results obtained by considering geomechanical responses differ from previous numerical studies that only considered flow and thermal behaviors. This indicates that neglecting geomechanical responses may result in an incorrect natural gas hydrate production scheme. Therefore, future numerical studies should take geomechanical responses into consideration to obtain more realistic results.
11. In future work, it is imperative to discover production schemes that effectively mitigate the occurrence of a low-temperature region after 120 days of production, which currently causes disruptions in numerical simulations, thus enabling the extension of the observation period. Moreover, new production schemes combined with permeability enhancement should be explored to facilitate the achievement of production rates that meet the necessary threshold for the commercial exploitation of natural gas hydrates.

Author Contributions: Conceptualization, J.Z.; methodology, J.Z.; software, J.Z.; formal analysis, J.Z.; writing—original draft preparation, J.Z.; writing—review and editing, R.W., T.W., and H.L.; supervision, H.L.; funding acquisition, H.L.; project administration, R.W. and T.W. All authors have read and agreed to the published version of the manuscript.

Funding: This research was funded by CNPC Engineering Technology R&D Company Limited, grant number 2021DJ4902.

Institutional Review Board Statement: Not applicable.

Informed Consent Statement: Not applicable.

Data Availability Statement: Not applicable.

Conflicts of Interest: The authors declare no conflict of interest.

Nomenclature

$\Delta()$	Change in the quantity in parentheses
ΔH^0	Specific enthalpy of hydrate dissociation/formation ($J \cdot kg^{-1}$)
∇	Del operator
C_R	Heat capacity of the dry rock ($J \cdot kg^{-1} \cdot K^{-1}$)
dA	Differential surface (m^2)
dV	Differential volume (m^3)
E	Young's modulus (Pa)
G	Shear modulus (Pa)
G_0	Shear modulus when the hydrate saturation is zero (Pa)
G_1	Shear modulus when the hydrate saturation is one (Pa)
h_β	Specific enthalpy of phase β ($J \cdot kg^{-1}$)
K_{dr}	Drained bulk modulus (Pa)
K_{dr0}	Drained modulus when the hydrate saturation is zero (Pa)
K_{dr1}	Drained modulus when the hydrate saturation is one (Pa)
K_s	Skeletal grain modulus (Pa)

k_r	Radial permeability (m^2)
$k_{r\beta}$	Relative permeability of phase β
k_v	Vertical permeability (m^2)
$\overline{k_\theta}$	Composite thermal conductivity of the medium/fluid ensemble ($W{\cdot}m^{-1}{\cdot}K^{-1}$)
$k_{\theta d}$	Formation thermal conductivity under desaturated conditions ($W{\cdot}m^{-1}{\cdot}K^{-1}$)
$k_{\theta w}$	Formation thermal conductivity under fully liquid-saturated conditions ($W{\cdot}m^{-1}{\cdot}K^{-1}$)
$k_{\theta I}$	Thermal conductivity of ice phase ($W{\cdot}m^{-1}{\cdot}K^{-1}$)
M_A	Cumulative mass of aqueous phase
M_G	Cumulative mass of gaseous phase
M^θ	Heat accumulation term
M^κ	Mass accumulation of component κ ($kg{\cdot}m^{-3}$)
P	Pressure (Pa)
P_t	Average mobile fluid pressure (Pa)
$P_{t,0}$	Initial equivalent pore pressure (Pa)
P_β	Pressure of phase β (Pa)
Q_g	Volumetric rate of CH_4 well production
Q_w	Water mass production rate
q^κ	Source/sink term of component κ ($kg{\cdot}m^{-3}{\cdot}s^{-1}$)
r	Radial direction
R_{wg}	Instantaneous water-to-gas ratio
R_{wgT}	Cumulative water-to-gas ratio
S_β	Saturation of phase β
T	Temperature (K or °C)
t	Time (s)
u_r	Radial displacement (m)
u_z	Vertical displacement (m)
U_β	Specific internal energy of phase β ($J{\cdot}kg^{-1}$)
V_g	Cumulative volume of CH_4 produced at the well
V_n	Volume of the subdomain (m^3)
X_β^κ	Mass fraction of component κ in phase β
z	Direction along the z-axis
α	Biot's coefficient
Γ_n	Surface of subdomain n (m^2)
γ	Empirical permeability reduction factor
ε_v	Current volumetric strain
$\varepsilon_{v,0}$	Initial volumetric strain
μ_β	Viscosity of phase β (Pa·s)
ν	Poisson's ratio
ρ_b	Bulk density ($kg{\cdot}m^{-3}$)
ρ_f	Fluid density ($kg{\cdot}m^{-3}$)
ρ_R	Rock density ($kg{\cdot}m^{-3}$)
ρ_β	Density of phase β ($kg{\cdot}m^{-3}$)
ϕ	Reservoir porosity
ϕ_0	Initial porosity
$\mathbf{F}^\kappa$	Flux vector of component κ ($kg{\cdot}m^{-2}{\cdot}s^{-1}$)
$\mathbf{F}_\beta$	Flux vector of phase β ($kg{\cdot}m^{-2}{\cdot}s^{-1}$)
$\mathbf{F}_\beta^\kappa$	Flux vector of component κ in phase β ($kg{\cdot}m^{-2}{\cdot}s^{-1}$)
$\mathbf{g}$	Gravitational acceleration vector ($m{\cdot}s^{-2}$)
$\mathbf{k}$	Absolute permeability tensor (m^2)
$\mathbf{u}$	Displacement vector (m)
$\boldsymbol{\varepsilon}$	Strain tensor
$\boldsymbol{\sigma}$	Total stress tensor (Pa)
$\boldsymbol{\sigma}'$	Effective stress tensor (Pa)

References

1. Sloan, E.D. Fundamental Principles and Applications of Natural Gas Hydrates. *Nature* **2003**, *426*, 353–359. [CrossRef]
2. Chong, Z.R.; Yang, S.H.B.; Babu, P.; Linga, P.; Li, X.-S. Review of Natural Gas Hydrates as an Energy Resource: Prospects and Challenges. *Appl. Energy* **2016**, *162*, 1633–1652. [CrossRef]

3. Makogon, Y.F.; Holditch, S.A.; Makogon, T.Y. Natural Gas-Hydrates—A Potential Energy Source for the 21st Century. *J. Pet. Sci. Eng.* **2007**, *56*, 14–31. [CrossRef]
4. Rutqvist, J.; Moridis, G.J.; Grover, T.; Collett, T. Geomechanical Response of Permafrost-Associated Hydrate Deposits to Depressurization-Induced Gas Production. *J. Pet. Sci. Eng.* **2009**, *67*, 1–12. [CrossRef]
5. Sun, J.; Ning, F.; Li, S.; Zhang, K.; Liu, T.; Zhang, L.; Jiang, G.; Wu, N. Numerical Simulation of Gas Production from Hydrate-Bearing Sediments in the Shenhu Area by Depressurising: The Effect of Burden Permeability. *J. Unconv. Oil Gas Resour.* **2015**, *12*, 23–33. [CrossRef]
6. Yin, Z.; Moridis, G.; Chong, Z.R.; Tan, H.K.; Linga, P. Numerical Analysis of Experiments on Thermally Induced Dissociation of Methane Hydrates in Porous Media. *Ind. Eng. Chem. Res.* **2018**, *57*, 5776–5791. [CrossRef]
7. Li, G.; Li, X.-S.; Tang, L.-G.; Zhang, Y. Experimental Investigation of Production Behavior of Methane Hydrate under Ethylene Glycol Injection in Unconsolidated Sediment. *Energy Fuels* **2007**, *21*, 3388–3393. [CrossRef]
8. Ohgaki, K.; Takano, K.; Sangawa, H.; Matsubara, T.; Nakano, S. Methane Exploitation by Carbon Dioxide from Gas Hydrates—Phase Equilibria for CO_2-CH_4 Mixed Hydrate System. *J. Chem. Eng. Jpn.* **1996**, *29*, 478–483. [CrossRef]
9. Kvamme, B.; Zhao, J.; Wei, N.; Sun, W.; Saeidi, N.; Pei, J.; Kuznetsova, T. Hydrate Production Philosophy and Thermodynamic Calculations. *Energies* **2020**, *13*, 672. [CrossRef]
10. Kvamme, B. Mechanisms for CH_4/CO_2 Swapping in Natural Sediments. *Fluids* **2022**, *7*, 260. [CrossRef]
11. Kvamme, B. Thermodynamics and Kinetic Mechanisms for CH_4/CO_2 Swapping in Natural Sediments. *Energy Fuels* **2022**, *36*, 6374–6396. [CrossRef]
12. Moridis, G.J.; Silpngarmlert, S.; Reagan, M.T.; Collett, T.; Zhang, K. Gas Production from a Cold, Stratigraphically-Bounded Gas Hydrate Deposit at the Mount Elbert Gas Hydrate Stratigraphic Test Well, Alaska North Slope: Implications of Uncertainties. *Mar. Pet. Geol.* **2011**, *28*, 517–534. [CrossRef]
13. Collett, T.S.; Lee, M.W.; Agena, W.F.; Miller, J.J.; Lewis, K.A.; Zyrianova, M.V.; Boswell, R.; Inks, T.L. Permafrost-Associated Natural Gas Hydrate Occurrences on the Alaska North Slope. *Mar. Pet. Geol.* **2011**, *28*, 279–294. [CrossRef]
14. Myshakin, E.M.; Ajayi, T.; Anderson, B.J.; Seol, Y.; Boswell, R. Numerical Simulations of Depressurization-Induced Gas Production from Gas Hydrates Using 3-D Heterogeneous Models of L-Pad, Prudhoe Bay Unit, North Slope Alaska. *J. Nat. Gas Sci. Eng.* **2016**, *35*, 1336–1352. [CrossRef]
15. Anderson, B.J.; Kurihara, M.; White, M.D.; Moridis, G.J.; Wilson, S.J.; Pooladi-Darvish, M.; Gaddipati, M.; Masuda, Y.; Collett, T.S.; Hunter, R.B.; et al. Regional Long-Term Production Modeling from a Single Well Test, Mount Elbert Gas Hydrate Stratigraphic Test Well, Alaska North Slope. *Mar. Pet. Geol.* **2011**, *28*, 493–501. [CrossRef]
16. Moridis, G.J.; Collett, T.S.; Dallimore, S.R.; Satoh, T.; Hancock, S.; Weatherill, B. Numerical Studies of Gas Production from Several CH4 Hydrate Zones at the Mallik Site, Mackenzie Delta, Canada. *J. Pet. Sci. Eng.* **2004**, *43*, 219–238. [CrossRef]
17. Uddin, M.; Wright, F.; Dallimore, S.; Coombe, D. Gas Hydrate Dissociations in Mallik Hydrate Bearing Zones A, B, and C by Depressurization: Effect of Salinity and Hydration Number in Hydrate Dissociation. *J. Nat. Gas Sci. Eng.* **2014**, *21*, 40–63. [CrossRef]
18. Bazaluk, O.; Sai, K.; Lozynskyi, V.; Petlovanyi, M.; Saik, P. Research into Dissociation Zones of Gas Hydrate Deposits with a Heterogeneous Structure in the Black Sea. *Energies* **2021**, *14*, 1345. [CrossRef]
19. Klymenko, V.; Ovetskyi, S.; Martynenko, V.; Vytyaz, O.; Uhrynovskyi, A. An Alternative Method of Methane Production from Deposits of Subaquatic Gas Hydrates. *Min. Miner. Depos.* **2022**, *16*, 11–17. [CrossRef]
20. Lin, J.-S.; Uchida, S.; Myshakin, E.M.; Seol, Y.; Rutqvist, J.; Boswell, R. Assessing the Geomechanical Stability of Interbedded Hydrate-Bearing Sediments under Gas Production by Depressurization at NGHP-02 Site 16. *Mar. Pet. Geol.* **2019**, *108*, 648–659. [CrossRef]
21. Moridis, G.J.; Reagan, M.T.; Queiruga, A.F.; Boswell, R. Evaluation of the Performance of the Oceanic Hydrate Accumulation at Site NGHP-02-09 in the Krishna-Godavari Basin during a Production Test and during Single and Multi-Well Production Scenarios. *Mar. Pet. Geol.* **2019**, *108*, 660–696. [CrossRef]
22. Zhang, J.; Moridis, G.J.; Blasingame, T.A. Effect of Geomechanics and of Grid Discretization on the Predictions of Production from Natural Hydrate Deposits and of the Associated Geomechanical System Response. *Gas Sci. Eng.* **2023**, *112*, 204942. [CrossRef]
23. Moridis, G.J.; Kim, J.; Reagan, M.T.; Kim, S.-J. Analysis of Short- and Long-Term System Response During Gas Production from a Gas Hydrate Deposit at the UBGH2-6 Site of the Ulleung Basin in the Korean East Sea. *Can. J. Chem. Eng.* **2023**, *101*, 735–763. [CrossRef]
24. Yu, T.; Guan, G.; Abudula, A.; Yoshida, A.; Wang, D.; Song, Y. Enhanced Gas Recovery from Methane Hydrate Reservoir in the Nankai Trough, Japan. *Energy Procedia* **2019**, *158*, 5213–5218. [CrossRef]
25. Zhu, H.; Xu, T.; Yuan, Y.; Xia, Y.; Xin, X. Numerical Investigation of the Natural Gas Hydrate Production Tests in the Nankai Trough by Incorporating Sand Migration. *Appl. Energy* **2020**, *275*, 115384. [CrossRef]
26. Gu, Y.; Sun, J.; Qin, F.; Ning, F.; Cao, X.; Liu, T.; Qin, S.; Zhang, L.; Jiang, G. Enhancing Gas Recovery from Natural Gas Hydrate Reservoirs in the Eastern Nankai Trough: Deep Depressurization and Underburden Sealing. *Energy* **2023**, *262*, 125510. [CrossRef]
27. Ye, J.; Qin, X.; Xie, W.; Lu, H.; Ma, B.; Qiu, H.; Liang, J.; Lu, J.; Kuang, Z.; Lu, C.; et al. The Second Natural Gas Hydrate Production Test in the South China Sea. *China Geol.* **2020**, *3*, 197–209. [CrossRef]
28. Wu, N.; Li, Y.; Wan, Y.; Sun, J.; Huang, L.; Mao, P. Prospect of Marine Natural Gas Hydrate Stimulation Theory and Technology System. *Nat. Gas Ind. B* **2021**, *8*, 173–187. [CrossRef]

29. Li, J.; Ye, J.; Qin, X.; Qiu, H.; Wu, N.; Lu, H.; Xie, W.; Lu, J.; Peng, F.; Xu, Z.; et al. The First Offshore Natural Gas Hydrate Production Test in South China Sea. *China Geol.* **2018**, *1*, 5–16. [CrossRef]
30. Sun, Y.; Li, B.; Guo, W.; Lü, X.; Zhang, Y.; Li, K.; Wang, P.; Jin, G.; Jia, R.; Qu, L. Comparative Analysis of the Production Trial and Numerical Simulations of Gas Production from Multilayer Hydrate Deposits in the Qilian Mountain Permafrost. *J. Nat. Gas Sci. Eng.* **2014**, *21*, 456–466. [CrossRef]
31. Liu, H.; Zhan, L.; Zhang, J.; Shang, S.; Lu, H. Numerical Investigation on Environmental Effect Associated with Gas-Hydrate Exploitation. *Geoenergy Sci. Eng.* **2023**, *227*, 211857. [CrossRef]
32. Ruppel, C.D.; Kessler, J.D. The Interaction of Climate Change and Methane Hydrates. *Rev. Geophys.* **2017**, *55*, 126–168. [CrossRef]
33. Ruppel, C. Permafrost-Associated Gas Hydrate: Is It Really Approximately 1 % of the Global System? *J. Chem. Eng. Data* **2015**, *60*, 429–436. [CrossRef]
34. Farahani, M.V.; Hassanpouryouzband, A.; Yang, J.; Tohidi, B. Development of a Coupled Geophysical–Geothermal Scheme for Quantification of Hydrates in Gas Hydrate-Bearing Permafrost Sediments. *Phys. Chem. Chem. Phys.* **2021**, *23*, 24249–24264. [CrossRef] [PubMed]
35. Farahani, M.V.; Hassanpouryouzband, A.; Yang, J.; Tohidi, B. Insights into the Climate-Driven Evolution of Gas Hydrate-Bearing Permafrost Sediments: Implications for Prediction of Environmental Impacts and Security of Energy in Cold Regions. *RSC Adv.* **2021**, *11*, 14334–14346. [CrossRef] [PubMed]
36. Wu, S.; Wang, J. On the China's Successful Gas Production Test from Marine Gas Hydrate Reservoirs. *Chin. Sci. Bull.* **2018**, *63*, 2–8. [CrossRef]
37. Boswell, R.; Collett, T.S. Current Perspectives on Gas Hydrate Resources. *Energy Environ. Sci.* **2011**, *4*, 1206–1215. [CrossRef]
38. Joshi, S.D. Cost/Benefits of Horizontal Wells. In Proceedings of the SPE Western Regional/AAPG Pacific Section Joint Meeting, Long Beach, CA, USA, 19–24 May 2003; p. 2. [CrossRef]
39. Feng, Y.; Chen, L.; Suzuki, A.; Kogawa, T.; Okajima, J.; Komiya, A.; Maruyama, S. Enhancement of Gas Production from Methane Hydrate Reservoirs by the Combination of Hydraulic Fracturing and Depressurization Method. *Energy Convers. Manag.* **2019**, *184*, 194–204. [CrossRef]
40. Moridis, G. *User's Manual of the TOUGH+ Core Code v1.5: A General-Purpose Simulator of Non-Isothermal Flow and Transport through Porous and Fractured Media*; LBNL-6871E.; Lawrence Berkeley National Lab. (LBNL): Berkeley, CA, USA, 2014. [CrossRef]
41. Chibura, P.E.; Zhang, W.; Luo, A.; Wang, J. A Review on Gas Hydrate Production Feasibility for Permafrost and Marine Hydrates. *J. Nat. Gas Sci. Eng.* **2022**, *100*, 104441. [CrossRef]
42. Sun, X.; Luo, T.; Wang, L.; Wang, H.; Song, Y.; Li, Y. Numerical Simulation of Gas Recovery from a Low-Permeability Hydrate Reservoir by Depressurization. *Appl. Energy* **2019**, *250*, 7–18. [CrossRef]
43. Yuan, Y.; Gong, Y.; Xu, T.; Zhu, H. Multiphase Flow and Geomechanical Responses of Interbedded Hydrate Reservoirs during Depressurization Gas Production for Deepwater Environment. *Energy* **2023**, *262*, 125603. [CrossRef]
44. Zhao, X.; Wang, Z.; Zhao, Y.; Zuo, J.; Li, P.; Liang, W.; Wang, B.; Chen, X.; Lei, H.; Jin, G. Coupled Thermal–Hydrodynamic–Mechanical Numerical Simulation of Natural Gas Hydrate Horizontal Well Depressurization Production: Method and Application in the South China Sea. *Nat. Gas Ind. B* **2022**, *9*, 548–560. [CrossRef]
45. Xia, Y.; Xu, T.; Yuan, Y.; Xin, X.; Zhu, H. Geomechanical Response Induced by Multiphase (Gas/Water) Flow in the Mallik Hydrate Reservoir of Canada. *SPE J.* **2022**, *27*, 434–451. [CrossRef]
46. Zhang, J.; Li, X.; Chen, Z.; Li, Q.; Li, G.; Lv, T. Numerical Simulation of the Improved Gas Production from Low Permeability Hydrate Reservoirs by Using an Enlarged Highly Permeable Well Wall. *J. Pet. Sci. Eng.* **2019**, *183*, 106404. [CrossRef]
47. Li, B.; Ma, X.; Zhang, G.; Guo, W.; Xu, T.; Yuan, Y.; Sun, Y. Enhancement of Gas Production from Natural Gas Hydrate Reservoir by Reservoir Stimulation with the Stratification Split Grouting Foam Mortar Method. *J. Nat. Gas Sci. Eng.* **2020**, *81*, 103473. [CrossRef]
48. Sun, J.; Ning, F.; Liu, T.; Liu, C.; Chen, Q.; Li, Y.; Cao, X.; Mao, P.; Zhang, L.; Jiang, G. Gas Production from a Silty Hydrate Reservoir in the South China Sea Using Hydraulic Fracturing: A Numerical Simulation. *Energy Sci. Eng.* **2019**, *7*, 1106–1122. [CrossRef]
49. Yu, T.; Guan, G.; Wang, D.; Song, Y.; Abudula, A. Numerical Investigation on the Long-Term Gas Production Behavior at the 2017 Shenhu Methane Hydrate Production Site. *Appl. Energy* **2021**, *285*, 116466. [CrossRef]
50. Rutqvist, J.; Moridis, G.J. Numerical Studies on the Geomechanical Stability of Hydrate-Bearing Sediments. *SPE J.* **2009**, *14*, 267–282. [CrossRef]
51. Moridis, G.J.; Kim, J.; Reagan, M.T.; Kim, S.-J. Feasibility of Gas Production from a Gas Hydrate Accumulation at the UBGH2-6 Site of the Ulleung Basin in the Korean East Sea. *J. Pet. Sci. Eng.* **2013**, *108*, 180–210. [CrossRef]
52. Moridis, G.J.J.; Collett, T.S.S.; Pooladi-Darvish, M.; Hancock, S.; Santamarina, C.; Boswell, R.; Kneafsey, T.; Rutqvist, J.; Kowalsky, M.B.B.; Reagan, M.T.T.; et al. Challenges, Uncertainties, and Issues Facing Gas Production from Gas-Hydrate Deposits. *SPE Reserv. Eval. Eng.* **2011**, *14*, 76–112. [CrossRef]
53. Rutqvist, J.; Moridis, G.J.; Grover, T.; Silpngarmlert, S.; Collett, T.S.; Holdich, S.A. Coupled Multiphase Fluid Flow and Wellbore Stability Analysis Associated with Gas Production from Oceanic Hydrate-Bearing Sediments. *J. Pet. Sci. Eng.* **2012**, *92–93*, 65–81. [CrossRef]
54. Wang, J.; Liang, J.; Zong, X.; Gong, Y.; Wan, T. Differentiated distribution of methane hydrate in the Shenhu area of the northern South China Sea and controlling factors. *Mar. Geol. Front.* **2015**, *31*, 24.

55. Zhang, J.; Moridis, G.J.; Blasingame, T.A. Message Passing Interface (MPI) Parallelization of Iteratively Coupled Fluid Flow and Geomechanics Codes for the Simulation of System Behavior in Hydrate-Bearing Geologic Media. Part 1: Methodology and Validation. *SPE Reserv. Eval. Eng.* **2022**, *25*, 600–620. [CrossRef]
56. Zhang, J.; Moridis, G.J.; Blasingame, T.A. Message Passing Interface (MPI) Parallelization of Iteratively Coupled Fluid Flow and Geomechanics Codes for the Simulation of System Behavior in Hydrate-Bearing Geologic Media. Part 2: Parallel Performance and Application. *SPE Reserv. Eval. Eng.* **2022**, *25*, 621–640. [CrossRef]
57. Narasimhan, T.N.; Witherspoon, P.A. An Integrated Finite Difference Method for Analyzing Fluid Flow in Porous Media. *Water Resour. Res.* **1976**, *12*, 57–64. [CrossRef]
58. Narasimhan, T.N.; Witherspoon, P.A.; Edwards, A.L. Numerical Model for Saturated-Unsaturated Flow in Deformable Porous Media: 2. The Algorithm. *Water Resour. Res.* **1978**, *14*, 255–261. [CrossRef]
59. Kamath, V.A. *Study of Heat Transfer Characteristics during Dissociation of Gas Hydrates in Porous Media*; University of Pittsburgh: Pittsburgh, PA, USA, 1984.
60. Kowalsky, M.B.; Moridis, G.J. Comparison of Kinetic and Equilibrium Reaction Models in Simulating Gas Hydrate Behavior in Porous Media. *Energy Convers. Manag.* **2007**, *48*, 1850–1863. [CrossRef]
61. Moridis, G.J. *User's Manual for the Hydrate v1.5 Option of TOUGH+ v1.5: A Code for the Simulation of System Behavior in Hydrate-Bearing Geologic Media*; LBNL-6869E; Lawrence Berkeley National Lab. (LBNL): Berkeley, CA, USA, 2014. [CrossRef]
62. Biot, M.A. General Theory of Three-Dimensional Consolidation. *J. Appl. Phys.* **1941**, *12*, 155–164. [CrossRef]
63. Terzaghi, K.; Peck, R.B.; Mesri, G. *Soil Mechanics in Engineering Practice*, 3rd ed.; John Wiley & Sons: New York, NY, USA, 1996.
64. Mcnamee, J.; Gibson, R.E. Displacement functions and linear transforms applied to diffusion through porous elastic media. *Q. J. Mech. Appl. Math.* **1960**, *13*, 98–111. [CrossRef]
65. Mcnamee, J.; Gibson, R.E. Plane strain and axially symmetric problems of the consolidation of a semi-infinite clay stratum. *Q. J. Mech. Appl. Math.* **1960**, *13*, 210–227. [CrossRef]
66. Moridis, G.J.; Reagan, M.T. Strategies for Gas Production from Oceanic Class 3 Hydrate Accumulations. In Proceedings of the Offshore Technology Conference, Houston, Texas, USA, 30 April–3 May 2007. [CrossRef]
67. Qin, X.; Liang, Q.; Ye, J.; Yang, L.; Qiu, H.; Xie, W.; Liang, J.; Lu, J.; Lu, C.; Lu, H.; et al. The Response of Temperature and Pressure of Hydrate Reservoirs in the First Gas Hydrate Production Test in South China Sea. *Appl. Energy* **2020**, *278*, 115649. [CrossRef]
68. Shang, S.; Gu, L.; Zhan, L.; Qiu, H.; Lu, H. Application of Horizontal Well to Gas Production from a Hydrate Reservoir with Free Gas and High Irreducible Water. *J. Nat. Gas Sci. Eng.* **2021**, *94*, 104102. [CrossRef]
69. Sun, J.; Zhang, L.; Ning, F.; Lei, H.; Liu, T.; Hu, G.; Lu, H.; Lu, J.; Liu, C.; Jiang, G.; et al. Production Potential and Stability of Hydrate-Bearing Sediments at the Site GMGS3-W19 in the South China Sea: A Preliminary Feasibility Study. *Mar. Pet. Geol.* **2017**, *86*, 447–473. [CrossRef]
70. van Genuchten, M.T. A Closed-Form Equation for Predicting the Hydraulic Conductivity of Unsaturated Soils. *Soil Sci. Soc. Am. J.* **1980**, *44*, 892–898. [CrossRef]
71. Rutqvist, J.; Wu, Y.; Tsang, C.; Bodvarsson, G. A Modeling Approach for Analysis of Coupled Multiphase Fluid Flow, Heat Transfer, and Deformation in Fractured Porous Rock. *Int. J. Rock Mech. Min. Sci.* **2002**, *39*, 429–442. [CrossRef]
72. Moridis, G.J.; Kowalsky, M.B.; Pruess, K. Depressurization-Induced Gas Production from Class 1 Hydrate Deposits. *SPE Reserv. Eval. Eng.* **2007**, *10*, 458–481. [CrossRef]

energies

Article

Controlling Factors of Vertical Geochemical Variations in Hydrate-Rich Sediments at the Site GMGS5-W08 in the Qiongdongnan Basin, Northern South China Sea

Huaxin Liu [1,2], Meijun Li [1,2,*], Hongfei Lai [3,*], Ying Fu [1,2], Zenggui Kuang [3] and Yunxin Fang [3]

1 National Key Laboratory of Petroleum Resources and Engineering, China University of Petroleum, Beijing 102249, China; huaxin_liu0805@163.com (H.L.)
2 College of Geosciences, China University of Petroleum, Beijing 102249, China
3 National Engineering Research Center of Gas Hydrate Exploration and Development, Guangzhou Marine Geological Survey, Ministry of Natural Resources, Guangzhou 511458, China
* Correspondence: meijunli@cup.edu.cn (M.L.); laihfei@sina.com (H.L.)

Abstract: Large amounts of natural gas hydrates have been discovered in the Qiongdongnan Basin (QDNB), South China Sea. The chemical and stable carbon isotopic composition shows that the hydrate-bound gas was a mixture of thermogenic and microbial gases. It is estimated that microbial gas accounts for 40.96% to 60.58%, showing a trend of decrease with the increase in burial depth. A significant amount of gas hydrates is thought to be stored in the mass transport deposits (MTDs), exhibiting vertical superposition characteristics. The stable carbon isotopic values of methane ($\delta^{13}C_1$) in the MTD1, located near the seabed, are less than −55‰, while those of the methane below the bottom boundary of MTD3 are all higher than −55‰. The pure structure I (sI) and structure II (sII) gas hydrates were discovered at the depths of 8 mbsf and 145.65 mbsf, respectively, with mixed sI and sII gas hydrates occurring in the depth range 58–144 mbsf. In addition, a series of indigenous organic matters and allochthonous hydrocarbons were extracted from the hydrate-bearing sediments, which were characterized by the origin of immature terrigenous organic matter and low-moderate mature marine algal/bacterial materials, respectively. More allochthonous (migrated) hydrocarbons were also discovered in the sediments below the bottom boundary of MTD3. The gas hydrated is "wet gas" characterized by a low $C_1/(C_2 + C_3)$ ratio, from 2.55 to 43.33, which was mainly derived from a deeply buried source kitchen at a mature stage. There is change in the heterogeneity between the compositions of gas and biomarkers at the site GMGS5-W08 along the depth and there is generally a higher proportion of thermogenic hydrocarbons at the bottom boundary of each MTDs, which indicates a varying contribution of deeply buried thermogenic hydrocarbons. Our results indicate that the MTDs played a blocking role in regulating the vertical transportation of hydrate-related gases and affect the distribution of gas hydrate accumulation in the QDNB.

Keywords: gas hydrate; biomarkers; extractable organic matter; sediment; mass transport deposits; Qiongdongnan Basin

Citation: Liu, H.; Li, M.; Lai, H.; Fu, Y.; Kuang, Z.; Fang, Y. Controlling Factors of Vertical Geochemical Variations in Hydrate-Rich Sediments at the Site GMGS5-W08 in the Qiongdongnan Basin, Northern South China Sea. *Energies* **2024**, *17*, 412. https://doi.org/10.3390/en17020412

Academic Editors: Roland W Lewis, Lunxiang Zhang and Hongsheng Dong

Received: 15 November 2023
Revised: 9 December 2023
Accepted: 10 January 2024
Published: 14 January 2024

1. Introduction

Natural gas hydrate was formed by methane-dominated hydrocarbon gas and water in the gas hydrate stability zone in low-temperature and high-pressure conditions [1–4]. They preferentially occurred in shallowly buried subsea sediments or in permeable rocks in the land tundra [4]. The total carbon fixed in the form of natural gas hydrate is enormous, which is almost twice as much as the conventional fossil fuels [5]. Recently, natural gas hydrates have received increasing attention worldwide because of their significant carbon storage capability and their potential as a new energy resource [6]. A total of six natural gas hydrate drilling expeditions were conducted in the South China Sea (SCS) by the Guangzhou Marine Geological Survey (GMGS) between 2007 and 2019 [7–11]. During the fifth natural gas hydrate drilling expedition (GMGS5) in 2018, a complex gas hydrate system that was characterized by abundant visible vein-like and massive gas hydrates was discovered in the Songnan low-uplift block of the western Qiongdongnan Basin (QDNB) [8,10–12].

In the past five years, a series of geological and geochemical investigations have been conducted to interpret the accumulation process and mechanism of the typical leakage-type gas hydrates at the site GMGS5-W08 in the QDNB [8,12–14]. These hydrates predominantly originate from deeply buried coaly type thermogenic gas sources via effective faults [8,12–14]. Organic geochemical and Raman spectroscopic results suggest that the gas hydrates at the site GMGS5-W08 are a mixture of structure I (sI) and structure II (sII) gas hydrates derived from both the thermogenic and microbial gas sources. Variations in the proportions of these gases along the depth profile result in distinct sI and sII hydrate distributions [11,13,15]. The microbial gas and coaly type thermogenic gas migrated upward to the gas hydrates stability zone through the combination of faults, gas chimneys, and pipe-like gas pathways [10,13,16–19]. Gas chimneys, acting as migration conduits, are formed due to high-pressure conditions from Oligocene thermogenic and Miocene microbial gas accumulations [10,12–14,18,20,21]. In the deep-water sedimentary system, including central channel sedimentary systems, curved channels, and mass transport deposits (MTDs) [22]. Three sets of MTDs have been identified using seismic and logging data and are characterized via vertical superposition and fracture development. The three MTDs are buried at depths of 0–58 mbsf, 58–98 mbsf, and 98–143 mbsf, respectively, and are potential storage spaces for leakage-type gas hydrates. The bottom boundaries of each MTD are shown in the Figure 1b,c.

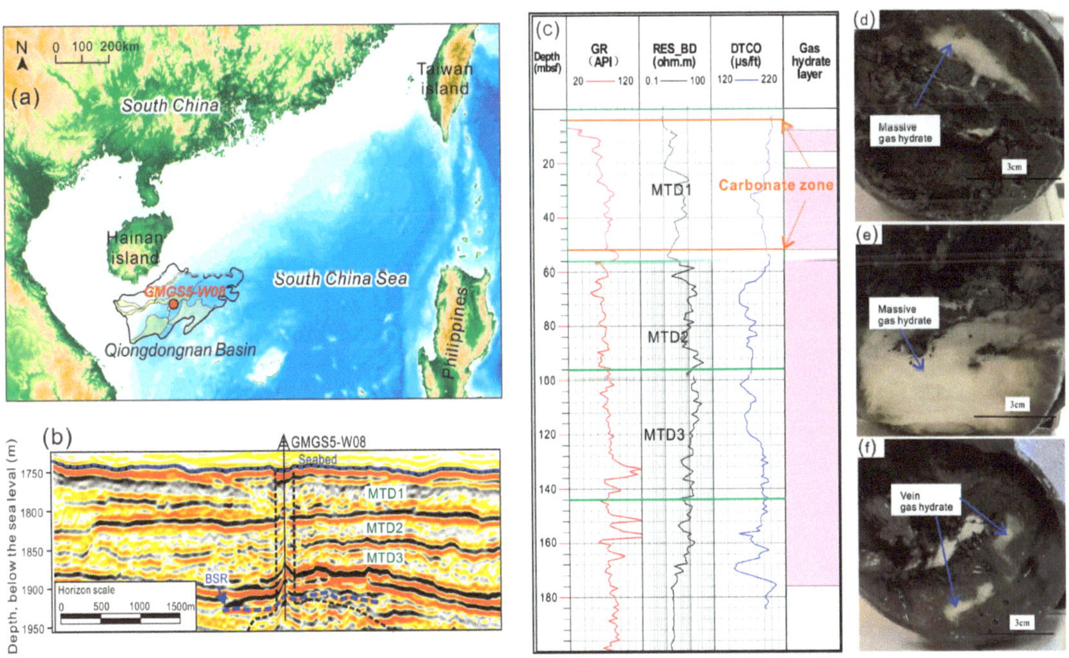

Figure 1. Regional geographic map of the northern South China Sea and Qiongdongnan Basin (QDNB). (**a**) Geographic location of the QDNB in the northwestern South China Sea and the location of gas hydrate drilling site GMGS5-W08 in this study [7,8,12,13]; (**b**) representative seismic profiles at site GMGS5-W08 [7,8,12,13], MTDn, mass transport deposit (n = 1–3); (**c**) well logging profiles of site GMGS5-W08; (**d**–**f**) representative core photos of gas hydrates at the site GMGS5-W08.

Notably, the geochemical characteristics of hydrate-related gas and sediments at the site GMGS5-W08 show a strong vertical heterogeneity. For example, the proportion of methane within the hydrate gas gradually decreased with the increase in buried depth [7]. However, the characteristics of the vertical geochemical heterogeneity and its controlling factors have not yet been fully understood. On the basis of the systematic geochemical analyses of hydrate-related gas and sediment samples from 0 mbsf to 188 mbsf, and with the relevant results of previous studies, this paper aims to study the origin of hydrate-related hydrocarbons and the hydrocarbons' evolution during migration at the site GMGS5-W08. Our results reveal a comprehensive control of the effect on natural gas hydrate geochemistry via multiple factors, including gas sources, gas migration, and overlay cap rocks, which may be highly significant for the exploration of natural gas hydrates in this region.

2. Geological Setting

The QDNB is a Cenozoic extensional basin located in the southeast of the Hainan Island (Figure 1a) [13–24]. The basin underwent the Eocene–Oligocene rifting and the Neogene–Quaternary depression structural evolution stages, in which four significant tectonic events, i.e., the Shenhu, Zhujiang, Nanhai, and Dongsha movements, have been further identified [19,25–27]. The deep-water area in the QDNB has been divided into five third-order tectonic units, including the Songnan-Baodao, Lingshui, and Beijiao sags, and the Songnan and Lingnan low uplifts. The QDNB was filled by eight formations (from bottom to top), including the Eocene, lower Oligocene Yacheng Formation, upper Oligocene Lingshui Formation, lower Miocene Sanya Formation, middle Miocene Meishan Formation, upper Miocene Huangliu Formation, Pliocene Yinggehai Formation, and Quaternary Ledong Formation (Figure 2) [19,21,27]. Two sets of source rocks, i.e., the Oligocene marine and continental transitional source rocks (Yacheng and Lingshui Formations), and the Miocene marine source rock (Sanya and Meishan Formations), have been found in the QDNB. Several natural gas fields with huge reserves have been reported in the Meishan and Huangliu Formation sandstone reservoirs of the Miocene submarine Central Canyon System, such as the LS17-2, L18-1, and L25-1 gas fields [28–33]. The coal-bearing mudstones in the Yacheng and Lingshui Formations were the main source rocks [30,31,34,35], and the thick marine mudstones within the Miocene and Pliocene formations are considered to be the regional cap rocks of the conventional petroleum system [23,31,33,36].

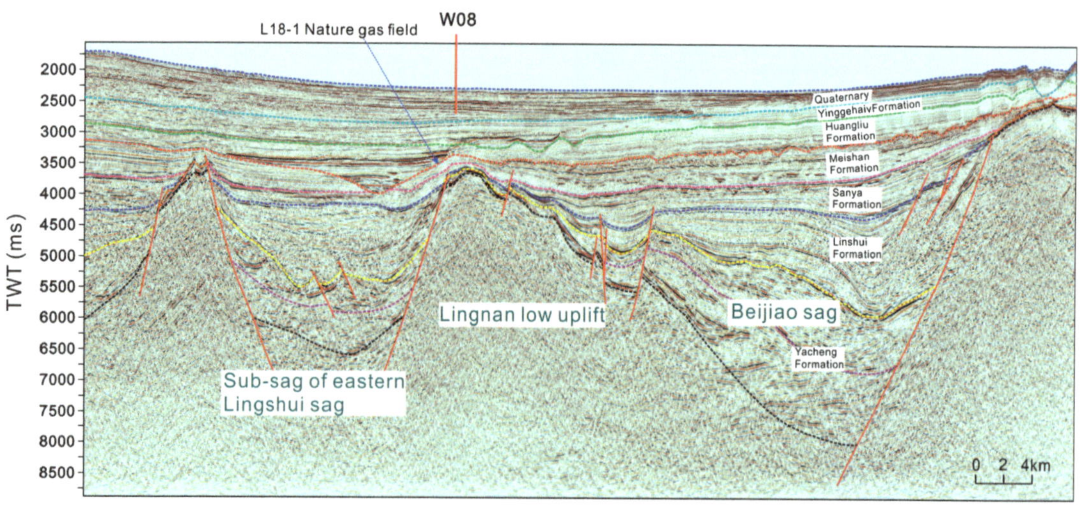

Figure 2. The representative seismic profiles crossing the site GMGS5-W08 showing the faults and stratigraphic sequences in the Qiongdongnan Basin.

During the GMGS5 and GMGS6 expeditions, a large amount of visible vein, nodular, and massive gas hydrates were discovered in the QDNB [7,12–14,19,22]. A large number of gas hydrate samples were drilled and obtained at site GMGS5-W08 in 2018 [7,8,12].

3. Samples and Methods

3.1. Sample Collection

A total of 20 hydrate-related gas samples were obtained from 0 mbsf to 188 mbsf at the site GMGS-W08 in the QDNB, including 4 samples during the depressurization of pressurized cores and 16 hydrate-bound gas samples. All the hydrate-related gas samples were obtained and stored following the experimental method reported by Lai et al. [14].

In addition, a total of 11 core samples of clayey silt sediment were collected at the site GMGS5-W08, with a burial depth ranging from 0 mbsf to 188 mbsf. All sediment samples were wrapped in aluminum foil and stored at a temperature of −20 °C.

3.2. Geochemical Experiments of Gas Samples

The gas compositions were analyzed using an Agilent 7890A (Santa Clara, CA, USA) gas chromatograph equipped with an HP-PLOT Q capillary column and flame ionization detector, following the standards set by Lai et al. [14]. In addition, the stable carbon and hydrogen isotopes were determined using a gas chromatography−isotope ratio mass spectrometry system (GC-IR-IRMS, Thermo Finnigan MAT 253, San Jose, CA, USA) with the same capillary column as used in the Agilent 7890A gas chromatograph.

3.3. Geochemical Experiments of Sediments

3.3.1. Total Organic Carbon Analyses

The 11 rock core samples were initially dried by freezing at a temperature of −80 °C. After freeze drying, they were crushed to 80 mesh using an agate mortar. These powdered samples were stored at −20 °C for subsequent analysis and testing. To remove inorganic carbon, the powdered samples were treated with hydrochloric acid (1.5 mol/L). They were then washed with deionized water for 48 h and subsequently dried at 80 °C for 24 h in an oven. The total organic carbon (TOC) content was determined using an LECO CS-230 carbon/sulfur analyzer.

3.3.2. Rock-Eval Pyrolysis Analyses

Rock-Eval pyrolysis, following the procedure outlined by Espitali'e et al. [37] and conducted using the OGE-VI rock pyrolyzer, was employed to assess the bulk geochemical properties of 11 sedimentary rock samples. This analysis focused on four key parameters: S_1, S_2, S_3, and T_{max}. S_1 quantifies extractable hydrocarbon compounds released during pyrolysis, S_2 measures the hydrocarbons generated and retained post pyrolysis, S_3 signifies CO_2 emissions during pyrolysis, and T_{max} indicates the maximum temperature for hydrocarbon generation during organic matter thermal maturation.

3.3.3. Extractable Organic Matter (EOM) and Gas Chromatography–Mass Spectrometry (GC–MS)

The 11 samples underwent a 72 h extraction in a Soxhlet apparatus using a dichloromethane/methanol (DCM/MeOH) solvent mixture (9:1, *v*/*v*). Initially, asphalt removal was achieved with petroleum ether at room temperature over 12 h. Subsequently, a silica/alumina column separation was employed to isolate saturated and aromatic hydrocarbon fractions from the deasphalting extract, using petroleum ether and a mixture of petroleum ether and dichloromethane (2:1, *v*/*v*), respectively. Analysis of these fractions was performed using Agilent 5975i and Agilent 5977i GC–MS systems, following the procedures and conditions described by Lai et al. [14]. All geochemical experiments on the core samples were conducted at the National Key Laboratory of Petroleum Resources and Engineering, China University of Petroleum (Beijing), Beijing, China.

4. Results

4.1. The Geochemical Characteristics of Natural Gas

Table 1 presents the compositional analysis of three gas hydrate samples and 17 collected gas data, detailing both gaseous hydrocarbon compositions and stable carbon isotope compositions. The predominant component is methane (C_1), ranging from 70.1% to 97.5%, followed by moderate levels of ethane (C_2) (2.2% to 18.5%) and propane (C_3) (0.05% to 11.3%). Higher carbon number hydrocarbons (C_{3+}) are present in limited quantities, ranging from 0% to 4.10%. Notably, the $C_1/(C_2 + C_3)$ ratio varies from 2.55 to 43.33. In addition, the values of stable carbon isotopes of C_1 ($\delta^{13}C_1$), C_2 ($\delta^{13}C_2$), and C_3 ($\delta^{13}C_3$), range from −63.7‰ to −49.3‰, −29.2‰ to −22‰, and −24.6‰ to −20.6‰, respectively. Furthermore, the hydrogen isotope value of C_1 (δD_1) ranges from −181‰ to −171‰.

Table 1. Molecular and isotopic compositions of gas hydrates from GMGS5-W08.

Depth (mbsf)	Gaseous Hydrocarbon Composition (%)				$C_1/(C_2 + C_3)$ Molar Ratio	Isotopic Composition (‰)				Reference
	C_1	C_2	C_3	C_{3+}		$\delta^{13}C_1$	$\delta^{13}C_2$	$\delta^{13}C_3$	δD_1	
45	97.28	2.59	0.05	0.08	36.85	−58.3	−25.1	−20.2	n.d.	This study
81.23	96.37	3.27	0.25	0.11	27.38	−58.8	−24.5	−20.2	n.d.	
146	85.19	10.43	3.5	0.88	6.12	−51.2	−26.6	−22.7	n.d.	
8	97.5	2.2	0.05	0.25	43.33	−59.5	−26	−22.4	−187	Data from [14]
17.07	96.95	2.6	0.1	0.35	33.32	−57.6	−26.6	−20.6	−169	
60	95.6	3.4	0.7	0.3	23.32	−60.9	−28.4	−24.6	−181	
63.18	96.2	3.3	0.39	0.11	26.07	−54.6	−22	n.d.	−178	
79	94.17	5.11	0.72	0	16.15	−58.6	−25.1	−21.6	−170	
112.3	94.22	5.02	0.75	0.01	16.33	−53.8	−23.3	−20.7	−175	
145.65	73.62	18.51	7.87	0	2.79	−49.3	−27.5	−23	n.d.	
158	76.1	15.43	8.48	0	3.18	−50.4	−26.9	−22.2	−183	
187.1	89.22	6.71	4.08	0	8.27	−50.4	−26.7	−23.2	−180	
63.73	96.7	2.7	0.4	0.2	31.19	−63.7	−28.8	−23.7	n.d.	Data from [11]
64.7	97.2	2.5	0.2	0.1	36	−61.1	−29.2	n.d.	n.d.	
170.2	72.5	12.1	11.3	4.1	3.1	−52	−27	−24.4	n.d.	
87.35	92.2	5.3	1.65	0.85	13.27	n.d.	n.d.	n.d.	n.d.	Data from [8]
89.66	90.5	6.3	2.24	0.96	10.6	n.d.	n.d.	n.d.	n.d.	
90.23	89.9	6.5	2.45	1.15	10.04	n.d.	n.d.	n.d.	n.d.	
142.39	87.4	9.3	1.94	1.36	7.78	n.d.	n.d.	n.d.	n.d.	
164.49	70.1	17.6	9.88	2.42	2.55	n.d.	n.d.	n.d.	n.d.	

Note: n.d., no data.

4.2. Bulk Geochemical Properties of Sediments

Table 2 presents the TOC contents and pyrolysis parameters of the 11 sediment samples. The TOC contents exhibit remarkably low values, ranging from 0.29% to 0.71%, with an average of 0.48%. The measured pyrolysis parameters, including S_1, S_2, and T_{max}, also exhibit notably low values, with averages of 0.06 mg/g, 0.23 mg/g, and 367 °C, respectively. In contrast, the values of S_3 are remarkably high, ranging from 1.12 to 1.57 mg/g, with an average of 1.29 mg/g. The calculated pyrolysis parameters hydrocarbon index (HI), oxygen index (OI), and production index (PI), range from 25.76 to 67.78 mg HC/g TOC, 221.75 to 444.14 mg CO_2/g TOC, and 0.05 to 0.36, respectively.

Table 2. Geochemical results of TOC, Rock-Eval pyrolysis and EOM of the hydrate-bearing sediments at the site GMGS5-W08.

Depth (mbsf)	Lithology	TOC (wt.%)	T_{max} (°C)	S_1 (mg/g)	S_2 (mg/g)	S_3 (mg/g)	HI	OI	PI	Organic Extracts (mg/g)	Saturate %	Aromatic %	Resein %	Asphaleten %
2.50–2.85	Silty clay	0.63	370	0.05	0.28	1.50	44.11	236.29	0.15	0.99	5.15	1.46	5.92	87.47
9.10–9.15	Silty clay	0.71	376	0.02	0.37	1.57	52.26	221.75	0.05	0.23	21.67	13.61	21.11	43.61
45.20–45.30	Silty clay	0.56	387	0.08	0.34	1.42	60.67	253.39	0.19	0.21	19.83	17.53	15.80	46.84
87.15–87.25	Silty clay	0.55	366	0.09	0.25	1.23	45.45	223.64	0.26	0.33	25.47	12.17	12.17	50.19
112.20–112.25	Silty clay	0.58	359	0.12	0.23	1.34	39.40	229.57	0.34	0.25	14.78	16.36	17.15	51.72
145.55–145.65	Silty clay	0.36	374	0.09	0.16	1.26	44.26	348.55	0.36	0.36	7.88	4.44	1.21	86.46
148.00–148.14	Silty clay	0.45	374	0.10	0.19	1.24	42.53	277.59	0.34	0.31	3.68	11.95	8.28	76.09
148.60–148.70	Silty clay	0.49	374	0.10	0.19	1.12	38.61	227.60	0.34	0.24	13.20	3.37	16.57	66.85
164.10–164.20	Silty clay	0.45	373	0.10	0.18	1.13	40.00	251.11	0.36	0.25	13.20	3.37	16.57	66.85
176.20–176.30	Silty clay	0.40	375	0.09	0.16	1.24	39.92	309.38	0.36	0.36	18.18	4.71	11.45	65.66
187.20–187.30	Silty clay	0.44	382	0.11	0.20	1.18	45.19	266.61	0.35	0.37	19.44	5.30	13.13	62.12

Note: TOC, total organic carbon content, wt.%. T_{max}, the temperature at which the S_2 yield reaches a peak. S_1, the amount of free hydrocarbons (HC) in a sample before the analysis, mg HC/g rock. S_2, the amount of hydrocarbons that result from cracking of kerogen in a sample, mg HC/g rock. S_3, the amount of carbon dioxide produced by pyrolysis of organic matter in a sample. HI, hydrocarbon index, HI = S_2/TOC × 100, mg HC/g TOC. OI, oxygen index, OI = S_3/TOC × 100, mg CO_2/g TOC. PI, production index, PI = $S_1/(S_1 + S_2)$.

4.3. The Compositions of Biomarkers

4.3.1. Normal Alkanes and Acyclic Isoprenoids

The total ion current (TIC) of the saturated hydrocarbon displays a distribution pattern of *n*-alkanes ranging from nC_{14} to nC_{35} (Figure 3a–d). Significant odd-to-even preference (OEP, 0.65 < OEP < 2.26) and high carbon preference index (CPI, 0.86 < CPI < 2.02) are observed in the nC_{23} to nC_{35} range (Table 3). In addition, the ratios of pristane to phytane (Pr/Ph), Pr/nC_{17}, and Ph/nC_{18} vary between 0.22 to 1.30, 0.65 to 1.00, and 0.61 to 0.77, respectively.

Table 3. Selected biomarker parameters relative to the composition of n-alkanes, terpanes, and steranes of EOM of samples at the site GMGS5-W08.

Depth (mbsf)	*n*-Alkanes and Isoprenoid Ratios								Terpanes				Steranes		
	Max. Peak	Pr/Ph	Pr/nC_{17}	Ph/nC_{18}	$\sum nC_{22+}/\sum nC_{21-}$	CPI	OEP	TAR	C_{23}TT/C_{21}TT	C_{24}TeT/C_{26}TT	Ts/(Ts + Tm)	St.C_{31}22S/(22S + 22R)	St.C_{29}20S/(20S + 20R)	St.$C_{29}\beta\beta/(\beta\beta+\alpha\alpha)$	C_{27} Diasterane/St.C_{27}
2.50–2.85	nC_{16}, nC_{31}	0.22	0.85	4.78	0.76	1.21	1.23	0.47	2.14	0.86	0.44	0.38	0.42	0.41	0.67
9.10–9.15	nC_{16}, nC_{31}	1.11	0.74	0.66	1.08	2.02	2.26	1.70	1.32	1.11	0.43	0.39	0.39	0.36	0.75
45.20–45.30	nC_{16}, nC_{31}	1.01	0.71	0.61	1.09	1.88	2.09	1.56	1.39	0.89	0.43	0.40	0.31	0.30	0.42
87.15–87.25	nC_{16}, nC_{31}	0.88	0.77	0.70	1.03	0.86	0.71	0.23	2.24	0.72	0.60	0.32	0.41	0.39	0.70
112.20–112.25	nC_{16}, nC_{31}	0.88	0.65	0.65	0.66	1.55	1.82	0.58	1.50	0.67	0.70	0.51	0.58	0.47	0.58
145.55–145.65	nC_{17}	1.24	0.83	0.72	0.64	0.83	0.65	0.13	1.63	0.53	0.73	0.50	0.32	0.35	0.59
148.00–148.14	nC_{17}	1.21	0.68	0.63	0.73	0.83	0.68	0.19	1.65	0.49	0.71	0.50	0.43	0.35	0.68
148.60–148.70	nC_{17}	1.09	0.73	0.73	0.75	1.81	2.19	0.75	1.45	0.51	0.70	0.53	0.33	0.34	0.64
164.10–164.20	nC_{17}	1.30	1.00	0.77	0.88	0.89	0.89	0.35	1.39	0.52	0.70	0.56	0.27	0.27	0.29
176.20–176.30	nC_{17}	1.11	0.73	0.66	0.49	1.37	1.56	0.29	1.38	0.54	0.71	0.54	0.30	0.34	0.52
187.20–187.30	nC_{17}	1.19	0.76	0.69	0.50	1.46	1.73	0.37	1.44	0.59	0.69	0.55	0.34	0.33	0.65

Note: Max. peak, the main (max.) peak of n-alkanes. Pr, pristane. Ph, phytane. CPI, carbon preference index, $\{[(nC_{23} + nC_{25} + nC_{27}) + (nC_{25} + nC_{27} + nC_{29})]/(nC_{24} + nC_{26} + nC_{28})\}/2$ [38]. OEP, odd-to-even predominance, $(nC_{25} + 6 \times nC_{27} + nC_{29})/(4 \times nC_{26} + 4 \times nC_{28})$ [38]. TAR = $(nC_{27} + nC_{29} + nC_{31})/(nC_{15} + nC_{17} + nC_{19})$ [39]. C_iTT, C_i tricyclic terpane (i = 21–26). C_{24}TeT, C_{24} tetracyclic terpane. Ts, 18α(H)-22,29,30-trisnorhopane. Tm, 17α(H)-22,29,30-trisnorhopane. St.C_{27-29}, C_{27-29} regular sterane.

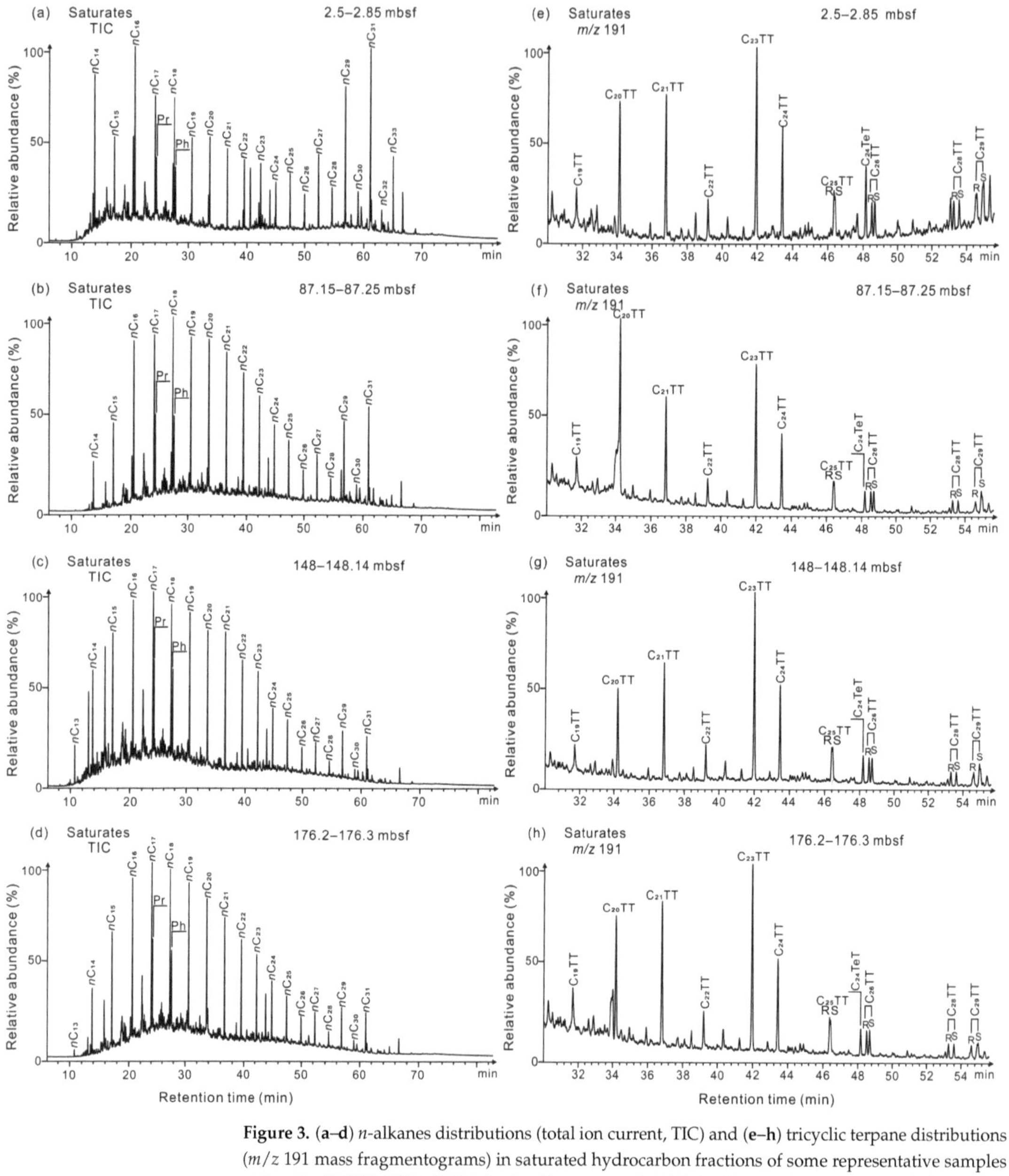

Figure 3. (**a–d**) *n*-alkanes distributions (total ion current, TIC) and (**e–h**) tricyclic terpane distributions (m/z 191 mass fragmentograms) in saturated hydrocarbon fractions of some representative samples at the site GMGS5-W08. C_nTT, $C_n$tricyclic terpane. C_{24}TeT, C_{24} Tetracyclic terpane, similarly to other compounds. R, S, the configuration of compounds.

4.3.2. Tricyclic Terpanes, Hopanes, and Steranes

A series of tricyclic terpanes (TTs), tetracyclic terpanes (TeT) (Figure 3e–h), and hopane series (Figure 4a–d) in EOM are detected in the m/z 191 mass chromatograms. The ratios of C_{23}TT to C_{21}TT and C_{24}TeT to C_{26}TT range from 1.32 to 2.24 and 0.49 to 1.11, respectively (Table 3). Within the distribution patterns of the hopane series, a dominance of C_{30} hopane (C_{30}H) and C_{29}H, and a high abundance of biological configuration isomers

17β(H), 21β(H)-C_{29-31}H (ββ-C_{29-31}H) were observed (Figure 4a–d). In addition, the 18α(H)-22,29,30-trisnorhopane (Ts) is slightly less abundant than the 17α(H)-22,29,30-trisnorhopane (Tm), with Ts/(Ts + Tm) ratios ranging from 0.40 to 0.70 (Table 3). All the samples have an extremely low abundance of gammacerane (Ga), oleanane (OL), and C_{33-35} homohopanes (Figure 4a–d).

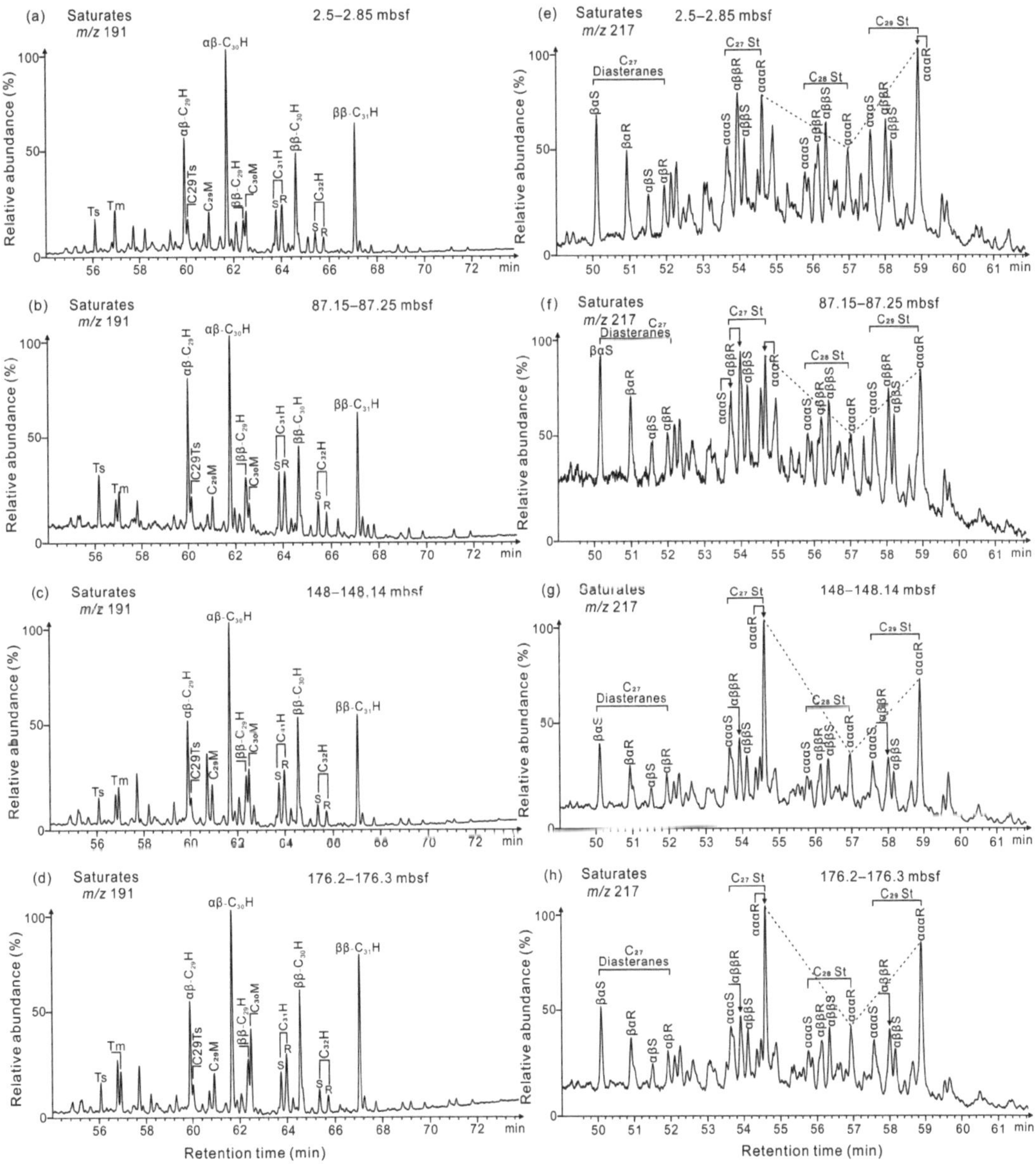

Figure 4. (**a**–**d**) Triterpane distributions (*m*/*z* 191 mass fragmentograms) and (**e**–**h**) sterane distributions (*m*/*z* 217 mass fragmentograms) in saturated hydrocarbon fractions of representative samples at the site GMGS5-W08. C_{30}H, C_{30}Hopane, C_{29}M, C_{29}Moretane, similarly to other compounds. R, S, the configuration of compounds. C_{27}St, Regular C_{27} Sterane, C_{28}S, Regular C_{28} Sterane, C_{29}St, Regular C_{29} Sterane.

In the m/z 217 mass chromatograms (Figure 4e–h), numerous sterane homologues have been identified. The C_{27} diasteranes are present in relatively high abundance, with C_{27} diasteranes/C_{27} regular steranes ratios ranging from 0.29 to 0.75 and an average of 0.59 (Table 3). Furthermore, the distribution pattern of C_{27} to C_{29} regular steranes appears to follow a distinctive V-shaped to L-shaped trend (Figure 4e–h). The $\alpha\alpha\alpha$ C_{29}20S/(20S + 20R) index and $C_{29}\beta\beta/(\beta\beta + \alpha\alpha)$ index vary between 0.27 to 0.58 and 0.27 to 0.47, respectively (Table 3).

4.3.3. Aromatic Distributions

The aromatic hydrocarbons in sediment samples extract are mainly composed of phenanthrene (Phen) (m/z 178), methyl phenanthrenes (m/z 192) (Figure 5a–d), naphthalene (m/z 128), trimethyl naphthaleneon (m/z 170) (Figure 5e–h), chrysene, fluoranthene, perilene, and some other components were detected in all EOM of sediments. The equivalent vitrinite reflectance (%R_c and %R_{cb}), calculated based on the methylphenanthrene index (MPI-1) [40] and the trimethylnaphthalene ratio (TNR-2) [40], ranges from 0.79% to 0.86% and 0.77% to 0.80%, respectively (Table 4).

In addition, the ratios of (3- + 2-MP)/(2-MP + 3-MP + 1-MP + 9-MP) (F_1) and 2-MP/(2-MP + 3-MP + 1-MP + 9-MP) (F_2) [41] are relatively low, each measuring below 0.54 and 0.32, respectively (Table 4). The dibenzothiophenes to phenanthrene ratio (DBT/Phen) varies between 1.22 and 2.12.

Table 4. Selected biomarker parameters and ratios of the samples calculated from phenanthrene, methyl-phenanthrenes, trimethyl-naphthaleneon, methyl-phenanthrene, and dibenzothiophenes distributions at the site GMGS5-W08.

Depth (mbsf)	MPI-1	%R_c	TNR-2	%R_{cb}	F_1	F_2	DBT/Phen
2.50–2.85	0.65	0.79	0.61	0.77	0.50	0.28	1.94
9.10–9.15	0.62	0.77	0.61	0.77	0.50	0.30	1.22
45.20–45.30	0.65	0.79	0.61	0.77	0.50	0.29	1.55
87.15–87.25	0.72	0.83	0.65	0.79	0.54	0.32	1.69
112.20–112.25	0.77	0.86	0.66	0.80	0.54	0.32	2.12
145.55–145.65	0.76	0.85	0.64	0.78	0.53	0.31	2.08
148.00–148.14	0.74	0.85	0.65	0.79	0.53	0.31	2.03
148.60–148.70	0.74	0.85	0.65	0.79	0.54	0.32	1.90
164.10–164.20	0.72	0.83	0.64	0.78	0.54	0.32	1.68
176.20–176.30	0.75	0.85	0.64	0.78	0.53	0.31	1.88
187.20–187.30	0.76	0.86	0.66	0.79	0.53	0.32	1.94

Note: MPI-1, the methylphenanthrene index, [1.5 × (2-MP + 3-MP)]/(P + 1-MP + 9-MP) [40]. MP, methyl-phenanthrenes. %R_c, calculated vitrinite reflectance, %R_c = 0.6 × MPI-1 + 0.4 [40]. TNR-2, the trimethylnaphthalene ratio, (1,3,7-TMN + 2,3,6-TMN)/(1,3,5-TMN + 1,3,6-TMN + 1,4,6-TMN). TMN, trimethyl-naphthaleneon. %R_{cb} = 0.4 + 0.6 × TNR-2. F_1 = (2-MP + 3-MP)/(2-MP + 3-MP + 1-MP + 9-MP) [42]. F_2 = 2-MP/(2-MP + 3-MP + 1-MP + 9-MP) [41]. DBT, dibenzothiophene. Phen, phenanthrene.

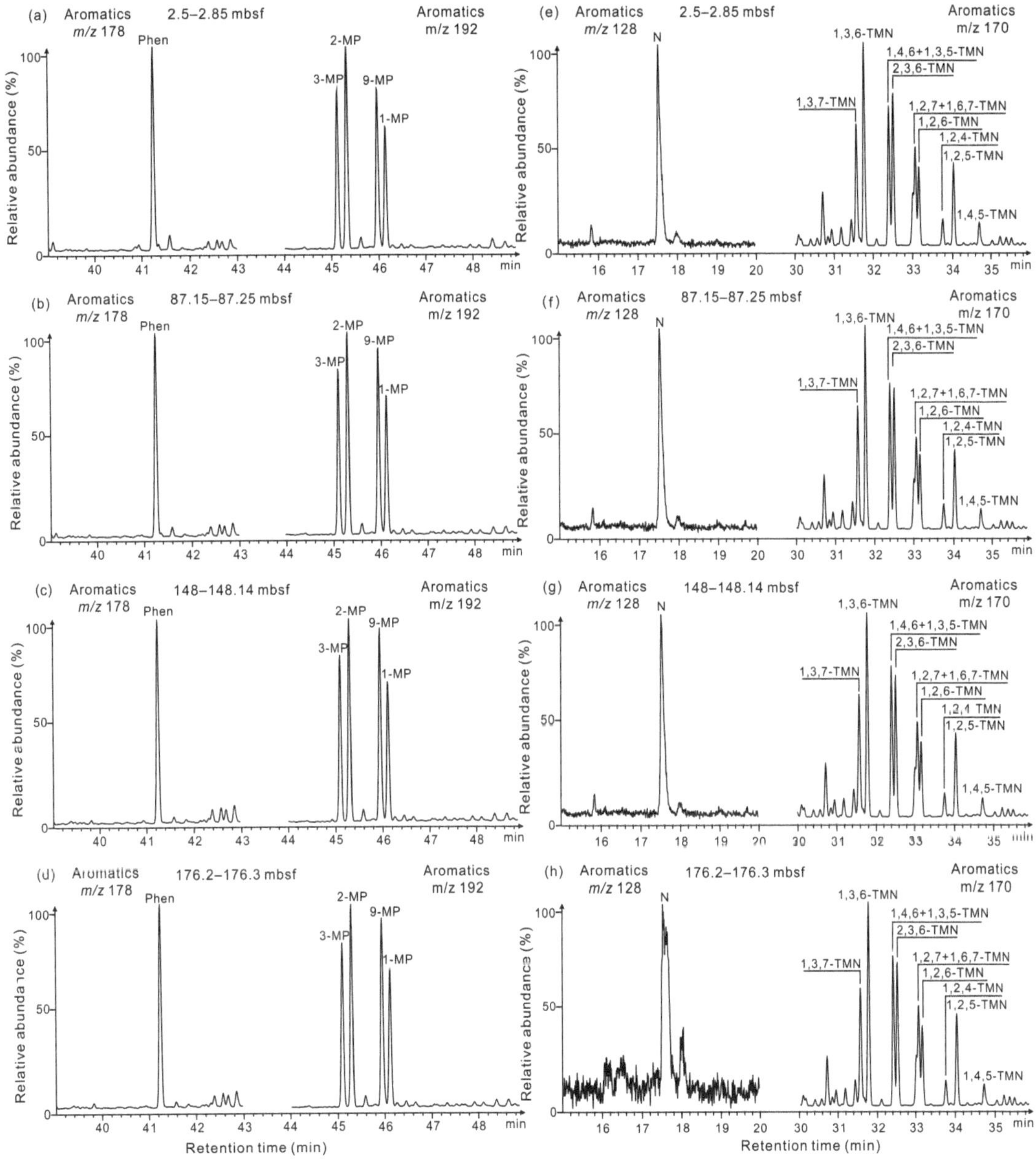

Figure 5. (**a–d**) Phenanthrene distributions (*m*/*z* 178 mass fragmentograms), methyl-phenanthrenes distributions (*m*/*z* 192 mass fragmentograms), (**e–h**) naphthalene distributions (*m*/*z* 128 mass fragmentograms) and trimethyl-naphthaleneon distributions (*m*/*z* 170 mass fragmentograms) in aromatic hydrocarbon fractions of representative samples at the site GMGS5-W08. hen, phenanthrene, and 3-, 2-, 9-, 1-MP are the 3-, 2-, 9-, 1-methyl-phenanthrenes, respectively. N, naphthalene, and 1,3,7-, 1,3,6-, 1,4,6 + 1,3,5-, 2,3,6-, 1,2,4-, 1,2,5-, 1,4,5-TMN are the 1,3,7-, 1,3,6-, 1,4,6 + 1,3,5-, 2,3,6-, 1,2,4-, 1,2,5-, 1,4,5-trimethyl-naphthaleneon, respectively.

5. Discussion

5.1. Origin of Gas Hydrates

The revised genetic diagrams of $C_1/(C_2 + C_3)$ versus $\delta^{13}C_1$ and $\delta^{13}C_1$ versus δD_1 proposed by Milkov and Etiope [43] can be applied to divide types of natural gas hydrates. Figure 6 shows the gas hydrates and pressurized core degassing gas samples are mainly of early mature thermogenic gas.

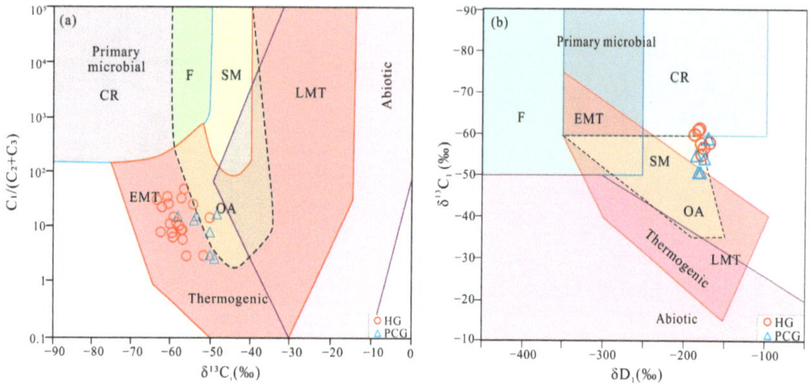

Figure 6. Genetic diagrams showing the genetic types of the hydrate-related gas samples at the site GMGS5-W08. (**a**) Plot of $\delta^{13}C_1$ versus $C_1/(C_2 + C_3)$. The diagram is modified from [43] (**b**) plot of δD_1 versus $\delta^{13}C_1$. The diagram is modified from [43]. CR, CO_2 reduction; F, methyl-type fermentation; SM, secondary microbial gas; OA, oil-associated thermogenic gas; EMT, early mature thermogenic gas; LMT, late-mature thermogenic gas.

All samples are primarily consisted of C_1 (Table 1, Figure 7a), with varying amounts of C_2 and C_3, and trace amounts of C_{3+}. The presence of C_3 and C_{3+} gases in hydrate-related gas suggests a contribution from thermogenic gases, as microbes typically produce C_1 exclusively. The highest C_1 content in the samples was 97.7%, which contrasts with the C_1 content of up to 99% in pure microbial gas from the Shenhu area in the South China Sea [15]. Interestingly, the proportion of C_1 gradually decreases along the vertical profile (Figure 7a,b), while the $\delta^{13}C_1$ values increase with increasing burial depth (Figure 7c). In contrast, $\delta^{13}C_2$ and $\delta^{13}C_3$ values remain stable with depth (Figure 7d,e). These features suggest that the contribution of microbial gas within the hydrate-related gas decreases with increasing burial depth from top to bottom. It is generally accepted that the $\delta^{13}C_1$ values for microbial origin typically fall below −55‰, whereas thermogenic $\delta^{13}C_1$ exhibits values higher than −55‰ [13,14,26,44,45]. Unlike the purely microbial gas in the Shenhu area, which has a maximum $\delta^{13}C_1$ of −66‰, the range of $\delta^{13}C_1$ within the samples is between 63.7‰ and −49.3‰ [15]. Therefore, it can be inferred that the vertical variations in natural gas composition and stable carbon isotopic are mainly caused by the different mixing ratios of microbial gas and thermogenic gas [13,14].

Based on the principle of mass conservation and stable carbon isotope fractionation of methane, the relative proportions of microbial and thermogenic gas in gas hydrates can be estimated [13,14,46]. In this study, we used the measured $\delta^{13}C_1$ values of −28‰ for pure thermogenic C_1 [32,47] and −80‰ for pure biogenic C_1 [13,14] to calculate the contribution of microbial C_1. The calculation results show that the proportion of microbial C_1 in the gas samples from site GMGS5-W08 ranges from 40.96% to 60.58%, progressively decreasing with increasing burial depth (Figure 7f).

The C_{3+} gases, typically originating from thermogenic natural gas sources, play a crucial role in the formation of sII gas hydrates [11,48,49]. Variations in the proportion of microbial gases along the vertical profile influence the composition of gas hydrates, particularly C_1 content. This leads to the formation of distinct zones with different gas hydrate crystal types along the depth profile. Pure sI and sII gas hydrates were found at depths of 8 mbsf and 145.65 mbsf, respectively, while a mixture of sI and sII gas hydrates was identified within the depth range of 58 to 144 mbsf. The gas sample taken at 8 mbsf exhibited high concentrations of C_1 and C_2 hydrocarbon gases but lacked C_3 and C_{3+} hydrocarbons, with a $\delta^{13}C_1$ value below −55‰. Conversely, the gas sample retrieved from a burial depth of 145.65 mbsf displayed a lower C_1 content and higher C_{2+} content, with a $\delta^{13}C_1$ value exceeding −55‰. Ultimately, based on the crystal structure of gas hydrates, the hydrate deposits can be divided into three zones from top to bottom: Zone sI (0–58 mbsf), Zone sI + sII (58–144 mbsf), and Zone sII (below 144 mbsf) (Figure 7). The boundaries between the zones are located at the bottom boundaries of MTD1 (58 mbsf) and MTD3 (144 mbsf).

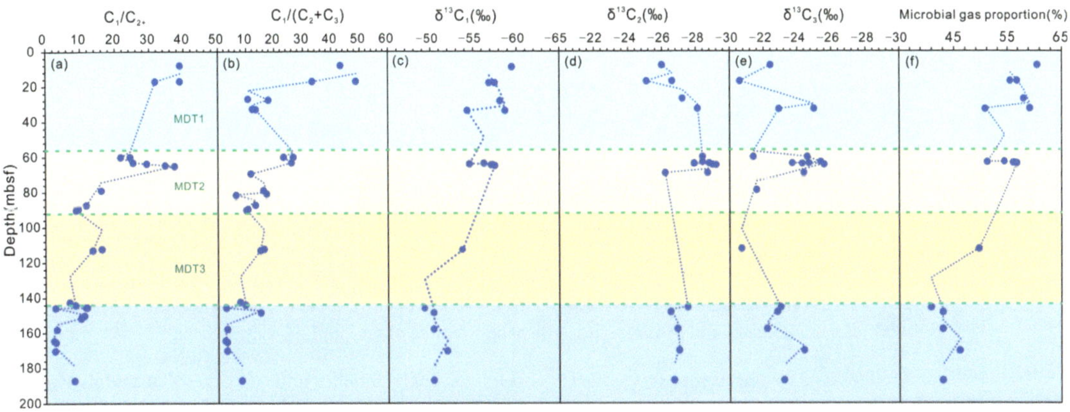

Figure 7. Vertical variation in the geochemical characteristics of the hydrate-related gas samples at the site GMGS5-W08. (**a**) Vertical distribution of sample C_1/C_{2+} values; (**b**) Vertical distribution of sample $C_1/(C_2+ C_3)$ values; (**c**) Vertical distribution of sample δC_1(‰)values; (**d**) Vertical distribution of sample δC_2(‰)values; (**e**) Vertical distribution of sample δC_3(‰)values; (**f**) Microbial gas proportion = ($\delta^{13}C_{\text{mixed gas sample}} - \delta^{13}C_{\text{pure thermogenic gas}}$)/($\delta^{13}C_{\text{pure primary microbial gas}} - \delta^{13}C_{\text{pure thermogenic gas}}$). The diagram is modified from [14].

In summary, the varying contribution rates of thermogenic gases along the vertical profile result in differences in the vertical distribution of sI and sII gas hydrates. The greatest variation in thermogenic gas proportions occurs at the bottom boundaries of MTD1 and MTD3. Below MTD3, thermogenic gases become the predominant component of hydrates. The distribution of gas components is evidently influenced by the presence of MTDs, indicating that the MTDs may play a sealing role in preventing gas migration upward.

5.2. *Profile Variations in Geochemical Characteristics of EOM*

5.2.1. Basic Geochemical Characteristics

Most sediment samples have TOC and HI values lower than 0.5% and 50 mg HC/g TOC, respectively (Table 2, Figure 8a). In addition, the values of the pyrolysis parameters S_1 and S_2 are extremely low (Table 2, Figure 8b), indicating a very low organic carbon in the indigenous sediments and a minor potential for microbial gas generation.

In contrast, the values of S_3 are relatively high (Table 2), which represents the significant production of carbon dioxide during organic matter pyrolysis and reflects a high oxygen content in the organic materials. This is likely due to the low thermal maturation levels. The T_{max} values are significantly low, ranging from 359 to 387 °C (Table 2, Figure 8c). Resins and asphaltics are the major components in the extracts of the eleven samples, ranging from 1.21% to 21.11% and 43.61% to 87.47%, respectively (Table 2). The T_{max} values and percentage distribution of the four fractions indicate a very low thermal maturation level of the sediment samples.

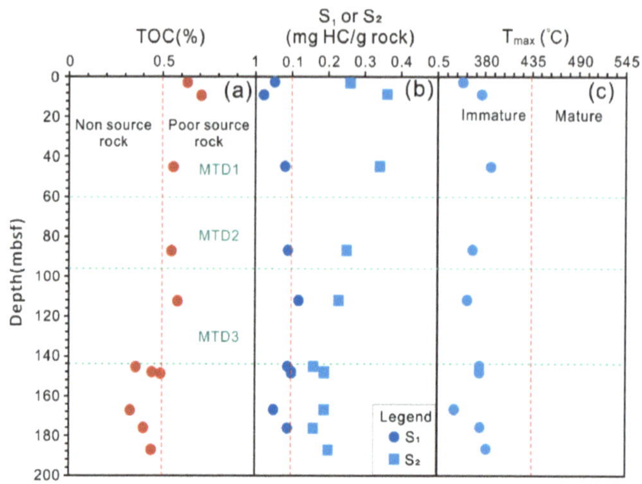

Figure 8. Vertical variation in TOC and Rock-Eval pyrolysis of the hydrate-bearing sediments at the site GMGS5-W08. (**a**) Vertical distribution of sample TOC values; (**b**) Vertical distribution of sample S_1 values and S_2 values; (**c**) Vertical distribution of sample T_{max} values.

In summary, the basic geochemical characteristics indicate that the abundance of EOM in the sediment is low and that the EOM is in an immature stage.

5.2.2. Organic Matter Source

The distribution of *n*-alkanes is a common method for tracing the origin of organic matter. Short-chain *n*-alkanes (<nC_{20}) typically indicate contributions from algae and microorganisms, whereas long-chain n-alkanes (>nC_{25}) are associated with terrigenous higher plants [50–52]. In this study, samples buried less than 58 mbsf predominantly exhibit long-chain *n*-alkanes (Figure 3a), while others show a prevalence of short-chain *n*-alkanes (Figure 3b–d). Near the seabed, we observe a bimodal distribution pattern (with maximum peaks around nC_{16}, and nC_{31}) (Figure 3a). In contrast, deeper-buried sediment samples display a unimodal pattern.

Furthermore, the TIC of saturated hydrocarbons demonstrates an even/odd predominance, particularly in the nC_{27}-nC_{33} range, suggesting an origin from immature terrestrial organic matter. Additional *n*-alkane parameters, including nC_{22+}/nC_{21-} and the terrigenous/aquatic index (TAR), help identify the sources of organic matter (Table 3, Figure 9a,b). The TAR index is the ratio of $(nC_{27} + nC_{29} + nC_{31})/(nC_{15} + nC_{17} + nC_{19})$, reflecting the contribution of leaf waxes and algae to n-alkanes, reveals the input of terrigenous organic matter (Table 3, Figures 3a–d and 9). Higher nC_{22+}/nC_{21-} and TAR values within MTD1 compared to MTD3 (Figure 9) indicate an increasing contribution of marine aquatic organisms to organic matter with increasing burial depth.

These sediment samples consist of unconsolidated clays with relatively shallow burial depths (0–200 mbsf) and low in situ temperatures (3.5–21 °C) [8], indicating an immature diagenesis phase. Therefore, the immature terrigenous organic matter is likely sourced from the terrigenous organic matters.

In addition to immature EOM, our analysis revealed several biomarker compounds in mass chromatograms, originating from mature source rocks. These include isoprenoids (Figure 3a–d), tricyclic terpanes (Figure 3e–h), hopanes (Figure 4a–d), diasteranes, regular steranes (Figure 4e–h), and some other compounds. These biomarkers indicate the presence of allochthonous hydrocarbons in the extracts, helping identify the allochthonous hydrocarbons origin. The distribution of steranes is widely used to infer the input of organic matter sources. For example, high concentrations of C_{29} steranes suggest higher plant organic matter input, while C_{27} steranes indicate eukaryotic algae and animal sources [53,54]. We detected C_{27} to C_{29} regular steranes in all extracts, with high C_{27} and C_{29} abundance, implying contributions from terrigenous and aquatic organisms. The C_{27} to C_{29} sterane distribution shifts from a V-shaped (above 85 mbsf) to an L-shaped pattern (below 145 mbsf), indicating decreasing terrigenous influence with burial depth and a gradual reduction in terrestrial plant contributions.

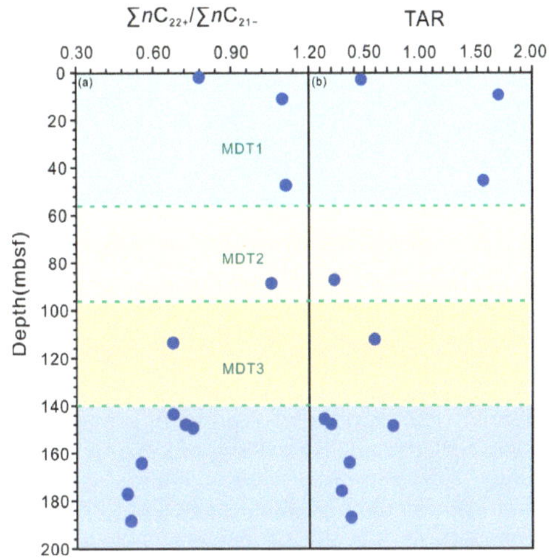

Figure 9. Vertical changes in showing various EOM inputs at the site GMGS5-W08. (**a**) Vertical distribution of sample $\sum nC_{22+}/\sum nC_{21-}$ values; (**b**) Vertical distribution of sample TAR values.

The plot of Ph/nC_{18} versus Pr/nC_{17} (Figure 10) and the ternary diagram of C_{27}–C_{29} regular steranes (Figure 11) depict the source of EOM in our samples. In MTD1, terrestrial higher plants are the primary source, whereas MTD2 and MTD3 samples contain a mix of EOM with a substantial contribution from aquatic organisms.

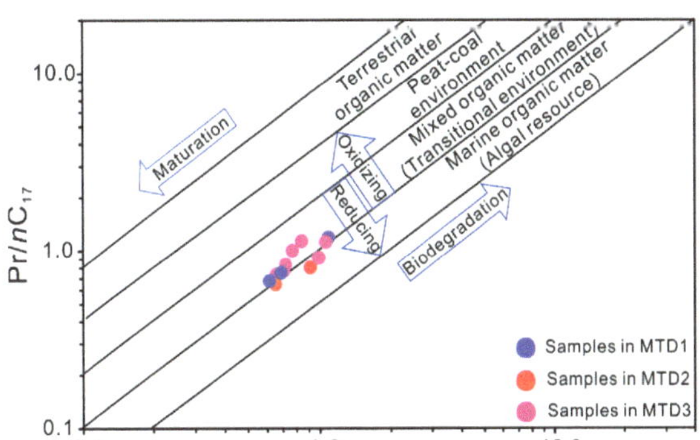

Figure 10. Phytane to n-C_{18} alkane (Ph/nC_{18}) versus pristane to n-C_{17} alkane (Pr/nC_{17}) showing depositional conditions and type of organic matter at the site GMGS5-W08. The diagram is modified from [55].

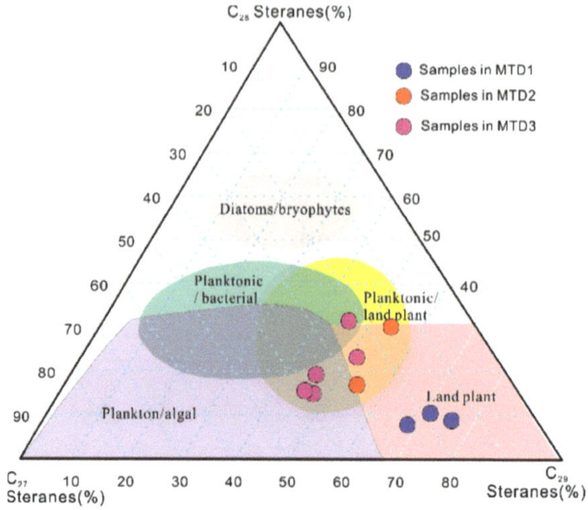

Figure 11. Ternary diagram of regular steranes (C_{27}, C_{28}, and C_{29}) at the site GMGS5-W08 showing the relationship between sterane compositions and organic matter input. The diagram is modified from [56].

In summary, indigenous organic matter primarily comprises immature terrigenous material, while allochthonous hydrocarbons mainly originate from planktonic algae, bacteria, and other microorganisms. Moreover, the contribution of terrigenous organic matter decreases notably at the bottom boundary of MTD1 and MTD3 (Figure 9), coinciding with the presence of gas hydrates zones. Below the bottom boundary of MTD3, organic matter from mature source rocks and deeply buried thermogenic gas accumulate. Changes in the parameters of these source rocks along the depth profile indicate that MTDs obstruct the upward migration of allochthonous hydrocarbons and gas hydrates, especially thermogenic gas.

5.2.3. Depositional Environment of Organic Matter

Parameters derived from the distribution of TTs have historically played a pivotal role in assessing depositional environments and organic matter conditions [57–59]. In this study, all sediment samples exhibit a substantial abundance of C_{23}TT, with C_{23}TT/C_{21}TT ratios consistently exceeding 1.2 (Table 3). This observation implies that EOM in the shallowly buried sediments of GMGS5-W08 predominantly originates from marine and saline environments, as opposed to terrestrial freshwater environments. In addition, C_{24}TeT/C_{26}TT ratios can be used to determine the depositional environment [60], with C_{24}TeT/C_{26}TT ratios consistently higher than 0.5 (Table 3, Figure 3b), suggesting an overfilled and balanced-fill marine depositional environment. Moreover, parameters such as Pr/Ph and DBT/Phen are commonly used to indicate the depositional environment. As shown in Figure 12, our results illustrate that reducing conditions within the depositional environment become more pronounced with increasing burial depth. Notably, significant variations in these ratios become apparent at the bottom of the MTDs boundaries.

The ternary diagram of C_{19+20}TT, C_{21}TT, and C_{23}TT by Xiao et al. [59] (Figure 13) and the plot of Ph/Pr versus DBT/Phen (Figure 14) indicate that the EOM in the samples is sourced from marine sediments.

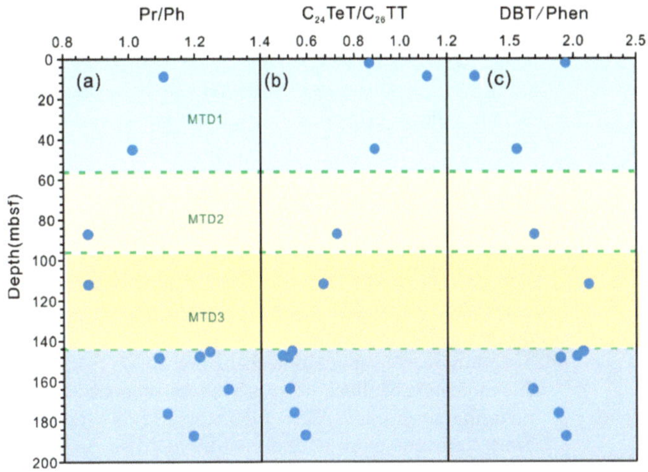

Figure 12. Vertical variation in showing various organic matter depositional environment at the site GMGS5-W08. (**a**) Vertical distribution of sample Pr/Ph values; (**b**) Vertical distribution of sample $C_{24}TeT/C_{26}TT$ values; (**c**) Vertical distribution of sample DBT/Phen values.

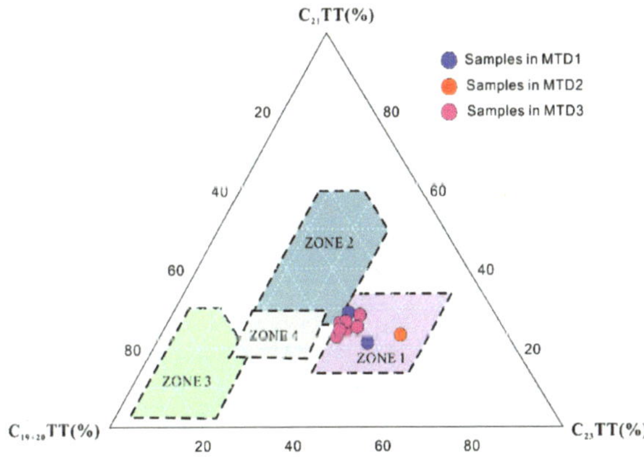

Figure 13. Ternary diagram of $C_{19+20}TT$, $C_{21}TT$, and $C_{23}TT$ to discriminate the various depositional environments at the site GMGS5-W08. The diagram is modified from [59].

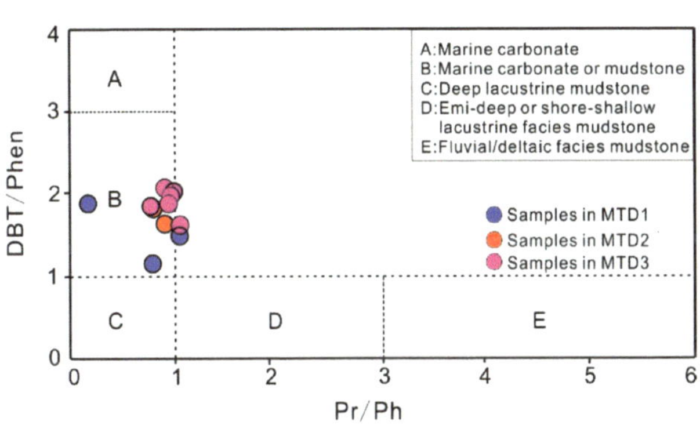

Figure 14. DBT/Phen versus Pr/Ph to discriminate the various depositional environments at the site GMGS5-W08.

In summary, the biomarker parameters suggest that the allochthonous hydrocarbons in the extracts primarily originate from source rocks within marine depositional environments. Meanwhile, the parameters indicate an anoxic condition below the bottom boundary of MTD3 and a sub-anoxic condition above the bottom boundary of MTD1. These observations indicate a significant correlation between shifts in sedimentary environments and the MTDs.

5.2.4. Maturity of Sediment Organic Matter

All samples exhibit immaturity, as evidenced by the T_{max} values ranging from 359 °C to 382 °C (Table 2, Figure 8c). The moderate to high OEP, mostly higher than 1.5 (Table 3), indicates low thermal maturity. This is further supported by the presence of squalene, a high abundance of C_{29}–C_{31} ββ-hopanes (Figure 5a–d), and relatively abundant C_{29}–C_{31} moretanes (Figure 5a–d).

Moreover, a set of biomarker compounds from mature source rocks serves as valuable indicators of allochthonous hydrocarbons maturity. The thermal maturity indicators, including the C_{31} 22S/(22S + 22R) ratio, Ts/(Ts + Tm) ratio, %R_c, %R_{cb}, F_1, and F_2, collectively confirm the presence of mature EOM sources. Furthermore, these six parameters demonstrate an increase in organic matter maturity with increasing burial depth, with significant variations observed at the bottom boundary of MTD1 and MTD3 (Figure 15). The six maturity parameters within MTD1 indicate that the EOM is in the immature-mature stage. The values of the maturity parameters increase with increase in depth of burial, and within MTD3 the EOM is in the mature stage. In addition, the pole ααα C_{29}20S/(20S + 20R) versus C_{29} ββ/(ββ + αα) and F_1 versus F_2 present low-mature to mature level and mature level, respectively (Figure 16).

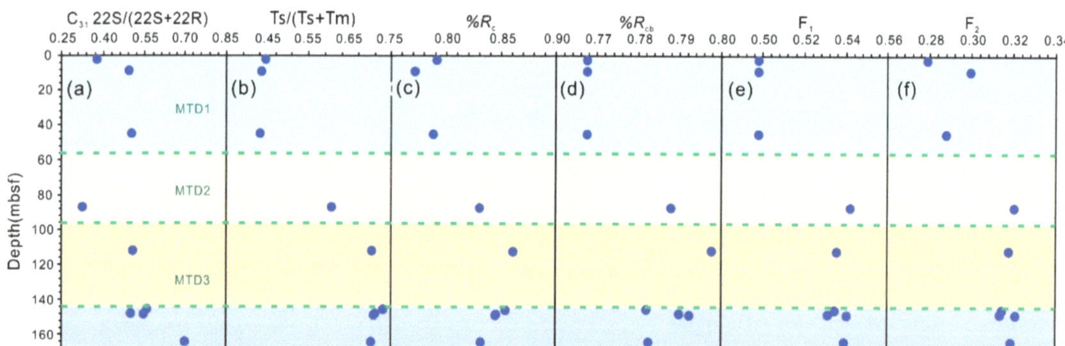

Figure 15. Vertical variation in maturity of the hydrate-bearing sediments at the site GMGS5-W08. (**a**) Vertical distribution of sample C_{31}22S/(22S+22R) values; (**b**) Vertical distribution of sample Ts/(Ts+Tm) values; (**c**) Vertical distribution of sample %R_c values; (**d**) Vertical distribution of sample %R_{cb} values (**e**) Vertical distribution of sample F_1 values; (**f**) Vertical distribution of sample F_2 values.

In general, the EOM within clay sediment extracts at relatively shallow burial depths (0–200 mbsf) is expected to be in an immature stage. However, certain biomarker parameters related to saturated and aromatic hydrocarbons exhibit distinct characteristics indicative of low to moderate organic matter maturity. This suggests the influence of allochthonous hydrocarbons sources originating from deeply buried source rocks. Variations in maturity parameters along the vertical profile suggest that MTDs serve as impediments, obstructing the upward migration of allochthonous hydrocarbons.

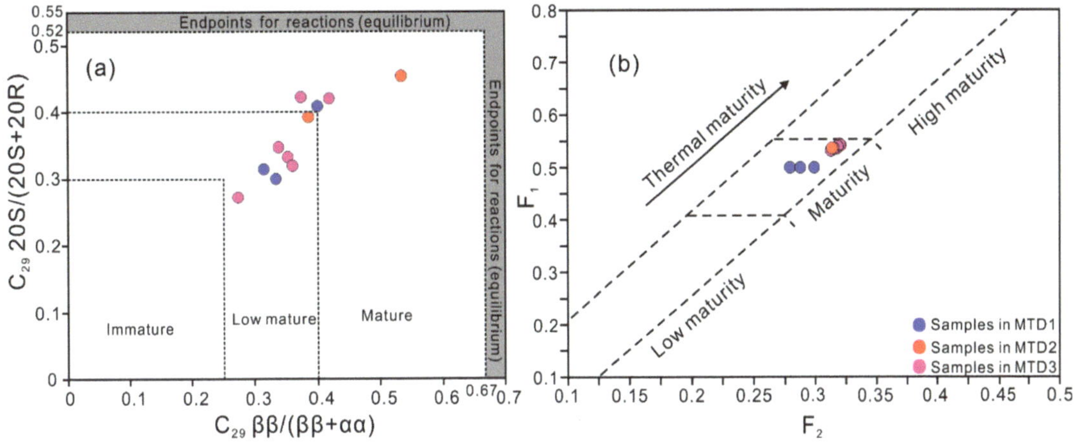

Figure 16. Diagram showing the maturity of the hydrate-bearing sediments at the site GMGS5-W08. (**a**) Plot of C_{29} $\beta\beta/(\beta\beta + \alpha\alpha)$ versus C_{29} 20S/(20S + 20R). (**b**) Plot of F_1 versus F_2.

In summary, the proportion of thermogenic gas and biomarker parameters associated with source input, depositional environment, and organic matter maturity consistently display variations along the vertical profile. These observations strongly indicate the upward migration of gases and hydrocarbons sourced from deeply buried reservoirs, with MTDs playing a significant role as barriers impeding this migration process.

5.3. Implications for Gas Hydrate Exploration

The preliminary findings of this study indicate that MTDs act as barriers, hindering the upward migration of both gas and EOM. Our research indicates significant changes in the contribution of thermogenic gas in the hydrate-related gas and contributions of allochthonous hydrocarbons in hydrate-bearing sediment extracts at the bottom boundaries of MTD1 and MTD3. These variations become more pronounced with burial depth. Moreover, the proportion of thermogenic gas may affect the distribution of sII gas hydrates along the vertical profile. Below the boundary of MTD3, the proportion of thermogenic gas is over 50%, indicating a significant accumulation of sII gas hydrates.

In addition, allochthonous hydrocarbons are present in the sediment extracts. The geochemical characteristics of the allochthonous hydrocarbons in the extracts indicate that they are originated from marine source rocks of low to moderate maturity, and the organic matter is derived from mixed organic matter with a significant contribution from aquatic organisms. The distribution is influenced by the presence of MTDs. The highest concentration of allochthonous hydrocarbons is found at the bottom boundary of MTD3. Moreover, the indigenous organic matter within MTDs can significantly influence various geochemical parameters. Therefore, when evaluating gas source kitchens using geochemical indicators, it is crucial to consider the influence of indigenous sediment within MTDs.

6. Conclusions

Based on a series of detailed geochemical analyses of hydrate-related gases and hydrate-bearing sediments within GMGS5-W08 in the QDNB, the source of the thermogenic gas and the influence of sediments on hydrate-related gas migration were investigated. The main conclusions are as follows:

1. The decrease in C_1 content and increase in $\delta^{13}C_1$ with depth suggest an increasing proportion of thermogenic gas, indicating that the MTDs act as cap rocks, impeding the migration of gas from deep source rocks.

2. The varying thermogenic gas contributions result in distinct changes in the distribution of sI and sII gas hydrates along the depth profile. It is concluded that MTDs significantly influence the distribution of sII gas hydrates, with important implications for gas hydrate resource assessment and exploration.
3. Hydrate-bearing sediments contain both indigenous organic matter and allochthonous hydrocarbons, with the allochthonous hydrocarbons originating from marine source rocks of low to moderate thermal maturity. It is crucial to distinguish the influence of indigenous organic matter when determining the origin of hydrate-related gas sources.
4. The vertical distribution patterns of source input, depositional environment, and organic matter maturity parameters within hydrate sediments exhibit distinct trends. These variations at the interfaces closely align with the boundaries of the MTDs, indicating that MTDs prevent upward migration of allochthonous hydrocarbons from depth and have a significant impact on the distribution of organic matter.

Author Contributions: H.L. (Huaxin Liu): Investigation, Methodology, Formal analysis, Interpretation, Writing—original draft; Writing—review & editing; M.L.: Project administration, Interpretation, Writing—review & editing; H.L. (Hongfei Lai): Funding acquisition, Data curation, Writing—review & editing; Y.F. (Ying Fu): Methodology, Data curation, Validation; Z.K. and Y.F. (Yunxin Fang): Data curation, Software. All authors have read and agreed to the published version of the manuscript.

Funding: This research was funded by [Guangdong Basic and Applied Basic Research] grant number [2023A1515010913, 2022A1515110569], [Guangzhou Science and Technology Project] grant number [202201011396], [Guangdong Major project of Basic and Applied Basic Research] grant number [2020B0301030003] and [China Geological Survey Project] grant number [DD20221705, DD20221700].

Data Availability Statement: The original contributions presented in the study are included in the article material, further inquiries can be directed to the corresponding authors.

Acknowledgments: The authors are grateful for all participants of the sixth China National Gas Hydrate Drilling Expeditions (GMGS5). We are grateful to the anonymous editors and reviewers for their insightful comments and suggestions, which significantly enhanced the quality of this manuscript.

Conflicts of Interest: The authors declare no conflict of interest.

References

1. Sloan, E.D. Gas Hydrates: Review of Physical/Chemical Properties. *Energy Fuels* **1998**, *2*, 191–196. [CrossRef]
2. Paull, C.K.; Dillon, W.P. *Natural Gas Hydrates: Occurrence, Distribution, and Detection*; American Geophysical Union Geophysical Monograph Series: Washington, DC, USA, 2001.
3. Sloan, E.D., Jr. Fundamental principles and applications of natural gas hydrates. *Nature* **2003**, *426*, 353–359. [CrossRef] [PubMed]
4. Tréhu, A.M.; Long, P.E.; Torres, M.E.; Bohrmann, G.; Rack, F.R.; Collett, T.S.; Goldberg, D.S.; Milkov, A.V.; Riedel, M.; Schultheiss, P.; et al. Three-dimensional distribution of gas hydrate beneath southern Hydrate Ridge: Constraints from ODP Leg 204. *Earth Planet. Sci. Lett.* **2004**, *222*, 845–862. [CrossRef]
5. Pytlak, L.; Gross, D.; Sachsenhofer, R.F.; Bechtel, A.; Gratzer, R.; Linzer, H.G. Generation, mixing and alteration of thermogenic and microbial gas in oil deposits: The case of the Alpine Foreland Basin (Austria). *Mar. Petrol. Geol.* **2016**, *78*, 575–592. [CrossRef]
6. Makogon, Y.F. Natural gas hydrates—A promising source of energy. *Nat. Gas Sci. Eng.* **2010**, *2*, 49–59. [CrossRef]
7. Wei, J.H.; Liang, J.; Lu, J.; Zhang, W.; He, Y.L. Characteristics and dynamics of gas hydrate systems in the northwestern South China Sea—Results of the fifth gas hydrate drilling expedition. *Mar. Pet. Geol.* **2019**, *110*, 287–298. [CrossRef]
8. Ye, J.; Wei, J.H.; Liang, J.; Lu, J.; Lu, H.; Zhang, W. Complex gas hydrate system in a gas chimney, South China Sea. *Mar. Pet. Geol.* **2019**, *104*, 29–39. [CrossRef]
9. Zhang, W.; Liang, J.; Wei, J.; Su, P.; Lin, L.; Huang, W. Origin of natural gases and associated gas hydrates in the Shenhu area, northern South China Sea: Results from the China gas hydrate drilling expeditions. *Asian Earth Sci.* **2019**, *183*, 103953. [CrossRef]
10. Zhang, W.; Liang, J.Q.; Lu, J.A.; Meng, M.M.; He, Y.L.; Deng, W.; Feng, J. Characteristics and controlling mechanism of typical leakage gas hydrate reservoir forming system in the Qiongdongnan Basin, northern South China Sea. *Nat. Gas Ind.* **2020**, *40*, 90–99. (In Chinese with English abstract).
11. Wei, J.; Wu, T.; Zhu, L.; Fang, Y.; Liang, J.; Lu, H.; Cai, W.; Xie, Z.; Lai, P.; Cao, J.; et al. Mixed gas sources induced co-existence of sI and sII gas hydrates in the Qiongdongnan Basin, South China Sea. *Mar. Pet. Geol.* **2021**, *128*, 105024. [CrossRef]

12. Liang, J.Q.; Zhang, W.; Lu, J.A.; Wei, J.H.; Kuang, Z.G.; He, Y.L. Geological occurrence and accumulation mechanism of natural gas hydrates in the eastern Qiongdongnan Basin of the South China Sea: Insights from site GMGS5-W9-2018. *Mar. Geol.* **2019**, *418*, 106042. [CrossRef]
13. Lai, H.F.; Fang, Y.X.; Kuang, Z.G.; Ren, J.; Liang, J.F.; Lu, J.A.; Wang, G.; Xing, C. Geochemistry, origin and accumulation of natural gas hydrates in the Qiongdongnan Basin, South China Sea: Implications from site GMGS5-W08. *Mar. Pet. Geol.* **2021**, *123*, 104774. [CrossRef]
14. Lai, H.F.; Qiu, H.J.; Liang, J.Q.; Kuang, Z.G.; Fang, Y.X.; Ren, J.F.; Lu, J.A. Geochemical Characteristics and Gas-to-Gas Correlation of Two Leakage-type Gas Hydrate Accumulations in the Western Qiongdongnan Basin, South China Sea. *Acta Geol. Sin. (Engl. Ed.)* **2022**, *2*, 680–690. [CrossRef]
15. Lai, H.F.; Deng, Y.N.; Yang, L.; Liang, J.Q.; Dai, L.R.; Li, L.; Fang, Y.X.; Liu, L.Y.; Kuang, Z.G. Origin of natural gas within the deep-sea uncompacted sediments of the Shenhu area, northern South China Sea: Geochemical and methanogenic cultivation results. *Mar. Pet. Geol.* **2023**, *147*, 106015. [CrossRef]
16. Ding, W.W.; Wang, Y.M.; Chen, H.L.; Yang, S.F.; Wu, N.Y. Deformation characters and its tectonic evolution of the Southwest Taiwan Basin. *J.-Zhejiang Univ.-Sci. Ed.* **2004**, *31*, 216–220.
17. Suess, E.; Huang, Y.; Wu, N.; Han, X.; Su, X. South China Sea continental margin: Geological methane budget and environmental effects of methane emissions and gas hydrates. In *Ifm-Geomar, Kiel. Rv Sonne Cruise Report So*; GEOMAR: Kiel, Germany, 2005; Volume 177.
18. Tong, H.P.; Feng, D.; Cheng, H.; Yang, S.X.; Wang, H.B.; Min, A.G.; Edwards, R.L.; Chen, Z.; Chen, D. Authigenic carbonates from seeps on the northern continental slope of the South China Sea: New insights into fluid sources and geochronology. *Mar. Pet. Geol.* **2013**, *43*, 260–271. [CrossRef]
19. He, Y.L.; Liang, J.Q.; Shi, W.Z.; Kuang, Z.; Deng, W.; Wang, R.; Xu, L.; Du, H. Influencing factors and accumulation modes of gas hydrate in the South low uplift and its surrounding area of Qiongdongnan Basin. *Earth Sci.* **2022**, *47*, 1711–1727. (In Chinese with English abstract).
20. Fang, Y.X.; Wei, J.H.; Lu, H.L.; Liang, J.Q.; Lu, J.A.; Fu, J.; Cao, J. Chemical and structural characteristics of gas hydrates from the Haima cold seeps in the Qiongdongnan Basin of the South China Sea. *Asian Earth Sci.* **2019**, *182*, 103924. [CrossRef]
21. He, Y.L.; Liang, J.Q.; Kuang, Z.G.; Deng, W.; Ren, J.F.; Lai, H.F.; Meng, M.M.; Zhang, W. Migration and accumulation characteristics of natural gas hydrates in the uplifts and their slope zones in the Qiongdongnan Basin, China. *China Geol.* **2022**, *5*, 234–250. [CrossRef]
22. Meng, M.M.; Liang, J.Q.; Lu, J.A.; Zhang, W.; Kuang, Z.G.; Fang, Y.X.; He, Y.L.; Deng, W.; Wei, H. Quaternary deep-water sedimentary characteristics and their relationship with the gas hydrate accumulations in the Qiongdongnan Basin, Northwest South China Sea. *Deep. Sea Res. Part I Oceanogr. Res. Pap.* **2021**, *177*, 103628. [CrossRef]
23. Zhu, W.L.; Huang, B.J.; Mi, L.J.; Wilkins, R.W.T.; Fu, N.; Xiao, X.M. Geochemistry, origin, and deep-water exploration potential of natural gases in the Pearl River Mouth and Qiongdongnan basins, South China Sea. *AAPG Bull.* **2009**, *93*, 741–761. [CrossRef]
24. Zhang, Y.D.; Sun, Y.G.; Liu, Q. Distribution and carbon isotope composition of pregnane in carbonate-evaporitic rocks from the Bonan Sag, Bohai Bay Basin, Eastern China: Insights into sources and associated lake environments. *Org. Geochem.* **2021**, *151*, 104127. [CrossRef]
25. Lei, C.; Ren, J.Y.; Pei, J.X.; Lin, H.T.; Yin, X.Y.; Tong, D.J. Tectonic Framework and Multiple Episode Tectonic Evolution in Deepwater Area of Qiongdongnan Basin, Northern Continental Margin of South China Sea. *Earth Sci.* **2011**, *36*, 151–162. (In Chinese with English abstract). [CrossRef]
26. Huang, B.J.; Tian, H.; Li, X.S.; Wang, Z.F.; Xiao, X.M. Geochemistry, origin and accumulation of natural gases in the deepwater area of the Qiongdongnan Basin, South China Sea. *Mar. Pet. Geol.* **2016**, *72*, 254–267. [CrossRef]
27. Su, A.; Chen, H.T.; Chen, X.; Liu, H.P.; Liu, Y.; Lei, M.Z. New insight into origin, accumulation and escape of natural gas in the Songdong and Baodao regions in the eastern Qiongdongnan basin, South China Sea. *Nat. Gas Sci. Eng.* **2018**, *52*, 467–483. [CrossRef]
28. Liang, G.; Gan, J.; Li, X. Genetic types and origin of natural gas in Lingshui sag, Qiongdongnan Basin. *China Offshore Oil Gas* **2015**, *27*, 47–53.
29. Xu, X.D.; Zhang, Y.Z.; Liang, G.; Xiong, X.F.; Li, X.; Guo, X.X.; Liu, H.Y. Hydrocarbon source condition and accumulation mechanism of natural gas in deepwater area of Qiongdongnan Basin, northern South China Sea. *Nat. Gas Geosci.* **2016**, *27*, 1985–1992.
30. Huang, H.T.; Huang, B.J.; Huang, Y.W.; Li, X.; Tian, H. Condensate origin and hydrocarbon accumulation mechanism of the deepwater giant gas field in western South China Sea: A case study of Lingshui 17-2 gas field in Qiongdongnan Basin. *Pet. Explor. Dev.* **2017**, *44*, 409–417. [CrossRef]
31. Zhu, Y.M.; Sun, L.T.; Hao, F.; Tuo, L. Geochemical composition and origin of Tertiary oils in the Yinggehai and Qiongdongnan Basins, offshore South China Sea. *Mar. Pet. Geol.* **2018**, *96*, 139–153. [CrossRef]
32. Gan, J.; Zhang, Y.Z.; Liang, G.; Yang, X.B.; Li, X.; Yang, J.H.; Guo, X.X. On accumulation process and dynamic mechanism of natural gas in the deep water area of central canyon, Qiongdongnan Basin. *Acta Geol. Sin.* **2018**, *92*, 2359–2367. (In Chinese with English abstract).
33. Xie, Y.H.; Zhang, G.C.; Sun, Z.P.; Zeng, Q.B.; Zhao, Z.; Guo, S.A. Reservoir forming conditions and key exploration technologies of Lingshui 17-2 giant gas field in deepwater area of Qiongdongnan Basin. *Pet. Res.* **2019**, *4*, 1–18. [CrossRef]

34. Su, M.; Xie, X.N.; Wang, Z.F.; Jiang, T.; Zhang, C.; He, Y.L. Sedimentary evolution of the central canyon system in Qiongdongnan Basin, northern South China Sea. *Acta Pet. Sin.* **2013**, *34*, 467–478. (In Chinese with English abstract). [CrossRef]
35. Wang, Z.F.; Sun, Z.P.; Zhang, Y.Z.; Guo, M.G.; Zhu, J.T.; Huang, B.J.; Zhang, D.J.; Jiang, R.F.; Man, X.; Zhang, H.Y. Distribution and hydrocarbon accumulation mechanism of the giant deepwater Central Canyon gas field in Qiongdongnan Basin, northern South China Sea. *China Pet. Explor.* **2016**, *21*, 54–64. (In Chinese with English abstract).
36. Hao, S.S.; Huang, Z.L.; Liu, G.D.; Zheng, Y.L. Geophysical properties of cap rocks in Qiongdongnan Basin, South China Sea. *Mar. Pet. Geol.* **2000**, *17*, 547–555. [CrossRef]
37. Espitalié, J.; Deroo, G.; Marquis, F. La pyrolyse Rock-Evaletses applications. Deuxième partie. *Rev. L'institut Français Pétrole* **1985**, *40*, 755–784. [CrossRef]
38. Marzi, R.; Torkelson, B.E.; Olson, R.K. A revised carbon preference index. *Org. Geochem.* **1993**, *20*, 1303–1306. [CrossRef]
39. Bourbonniere, R.A.; Meyers, P.A. Sedimentary geolipid records of historical changes in the watersheds and productivities of Lakes Ontario and Erie. *Limnol. Oceanogr.* **1996**, *41*, 352–359. [CrossRef]
40. Radke, M.; Welte, D.H.; Willsch, H. Geochemical study on a well in the Western Canada Basin: Relation of the aromatic distribution pattern to maturity of organic matter. *Geochim. Cosmochim. Acta* **1982**, *46*, 1–10. [CrossRef]
41. Bao, J.P.; Wang, T.G.; Zhou, Y.Q.; Yu, F.X.; Wang, J.Y.; Zhou, Q.L.; Chen, F.J. The relationship between methyl phenanthrene ratios and the evolution of organic matter. *J. Jianghan Pet. Inst.* **1992**, *14*, 8–13. (In Chinese with English abstract).
42. Radke, M.; Rullkötter, J.; Vriend, S.P. Distribution of naphthalenes in crude oils from the Java Sea: Source and maturation effects. *Geochim. Cosmochim. Acta* **1994**, *58*, 3675–3689. [CrossRef]
43. Milkov, A.V.; Etiope, G. Revised genetic diagrams for natural gases based on a global dataset of >20,000 samples. *Org. Geochem.* **2018**, *125*, 109–120. [CrossRef]
44. Xu, Y.; Shen, P. A Study of Natural Gas Origins in China. *AAPG Bull.* **1996**, *80*, 1604–1614.
45. Klauda, J.B.; Sandler, S.I. Global Distribution of Methane Hydrate in Ocean Sediment. *Energy Fuels* **2005**, *19*, 459–470. [CrossRef]
46. Sassen, R.; Milkov, A.V.; Ozgul, E.; Roberts, H.H.; Hunt, J.L.; Beeunas, M.A.; Chanton, J.P.; Defreitas, D.A.; Sweet, S.T. Gas venting and subsurface charge in the Green Canyon area, Gulf of Mexico continental slope. *Org. Geochem.* **2003**, *34*, 1455–1464. [CrossRef]
47. Chung, H.M.; Gormly, J.R.; Squires, R.M. Origin of gaseous hydrocarbons in subsurface environments: Theoretical considerations of carbon isotope distribution. *Chem. Geol.* **1988**, *71*, 97–104. [CrossRef]
48. Kida, M.; Khlystov, O.; Zemskaya, T.; Takahashi, N.; Minami, H.; Sakagami, H.; Krylov, A.; Hachikubo, A.; Yamashita, S.; Shoji, H. Coexistence of structure I and II gas hydrates in Lake Baikal suggesting gas sources from microbial and thermogenic origin. *Geophys. Res. Lett.* **2006**, *33*. [CrossRef]
49. Klapp, S.A.; Murshed, M.M.; Pape, T.; Klein, H.; Bohrmann, G.; Brewer, P.G.; Kuhs, W.F. Mixed gas hydrate structures at the Chapopote Knoll, southern Gulf of Mexico. *Earth Planet. Sci. Lett.* **2010**, *299*, 207–217. [CrossRef]
50. Bray, E.E.; Evans, E.D. Distribution of n-paraffins as a clue to recognition of source beds. *Geochim. Cosmochim. Acta* **1961**, *22*, 2–15. [CrossRef]
51. Eglinton, G.; Hamilton, R.J. Leaf Epicuticular Waxes: The waxy outer surfaces of most plants display a wide diversity of fine structure and chemical constituents. *Science* **1967**, *156*, 1322–1335. [CrossRef]
52. Peters, K.E.; Walters, C.C.; Moldowan, J.M. *The Biomarker Guide: Biomarkers and Isotopes in Petroleum Exploration and Earth History*; The Press Syndicate of the University of Cambridge: Cambridge, MA, USA, 2005; Volume 2.
53. Peters, K.E.; Peters, K.E.; Walters, C.C.; Moldowan, J.M. *The Biomarker Guide*; Cambridge University Press: Cambridge, MA, USA, 2005; Volume 1.
54. Xiao, H.; Wang, T.G.; Li, M.J.; Cheng, D.; Yang, Z. Organic geochemical heterogeneity of the Cretaceous Abu Gabra Formation and reassessment of oil sources in the Sufyan sub-basin, Sudan. *Org. Geochem.* **2021**, *162*, 104301. [CrossRef]
55. Shanmugam, G. Significance of coniferous rain forests and related organic matter in generating commercial quantities of oil, Gippsland Basin, Australia. *AAPG Bull.* **1985**, *69*, 1241–1254. [CrossRef]
56. Huang, W.Y.; Meinschein, W.G. Sterols as ecological indicators. *Geochim. Cosmochim. Acta* **1979**, *43*, 739–745. [CrossRef]
57. Ekweozor, C.M.; Strausz, O.P. Tricyclic terpanes in the Athabasca oil sands: Their geochemistry. *Adv. Org. Geochem.* **1981**, *1983*, 746–766.
58. Xiao, H.; Wang, T.G.; Li, M.J.; Fu, J.; Tang, Y.J.; Shi, S.B.; Yang, Z.; Lu, X.L. Occurrence and Distribution of Unusual Tri- and Tetracyclic Terpanes and Their Geochemical Significance in Some Paleogene Oils from China. *Energy Fuels* **2018**, *32*, 7393–7403. [CrossRef]
59. Xiao, H.; Li, M.J.; Yang, Z.; Zhu, Z.L. Distribution patterns and geochemical implications of C_{19}–C_{23} tricyclic terpanes in source rocks and crude oils occurring in various depositional environments. *Geochimica* **2019**, *48*, 161–170. (In Chinese with English abstract).
60. Bohacs, K.M.; Carroll, A.R.; Neal, J.E.; Mankiewicz, P.J. Lake-basin type, source potential, and hydrocarbon character: An integrated Sequence-Stratigraphic-Geochemical framework. *AAPG Stud. Geol.* **2000**, *46*, 3–34.

Journal of
Marine Science and Engineering

Article

Numerical Simulation of Production Behavior with Different Complex Structure Well Types in Class 1-Type Hydrate Reservoir

Tinghui Wan [1,2], Zhanzhao Li [1,2,*], Mingming Wen [1,2], Zongheng Chen [1,2], Lieyu Tian [1,2], Qi Li [1,2], Jia Qu [1,2] and Jingli Wang [1,2,*]

1 Guangzhou Marine Geology Survey, China Geological Survey, Ministry of Natural Resources, Guangzhou 511458, China; atomion@126.com (T.W.); wmingming@mail.cgs.gov.cn (M.W.); czhgs@126.com (Z.C.); tianlieyu23@163.com (L.T.); liqi3412@163.com (Q.L.); qujia2261520@163.com (J.Q.)

2 National Engineering Research Center for Gas Hydrate Exploration and Development, Guangzhou 511458, China

* Correspondence: 13650780173@163.com (Z.L.); wjl06012527@126.com (J.W.)

Abstract: Enhancing the production capacity of natural gas hydrates (NGHs) is critical for its commercial development. Complex structure wells may efficiently increase drainage areas while enhancing exploitation efficiency. Based on the field data of China's first offshore NGH test production, the numerical method was used to analyze the production performance of different complex structure well types by continuous depressurization production for 360 days under the preconditions of fixed effective completion length of 300 m and a pressure difference of 6 MPa. Results indicated that the complex structure well types deployed at the three-phase layer demonstrated superior production performance within 240 days of production; the DLW2 and HW2 well types stood out, with an average gas production rate Q_g reaching 43,333 m^3/d and a specific production index J of 24.1. After 360 days of production, benefiting from multi-layer combined production, the Cluster vertical well deployed at the multi-layer had the best production performance, with an average Q_g of 34,444 m^3/d and a J-index of 19.1. The research results provided insights into the complex structure well-type selection strategy for NGH depressurization in this sea area.

Keywords: natural gas hydrate; Shenhu Sea area; complex structure well; numerical simulation; TOUGH+HYDRATE

Citation: Wan, T.; Li, Z.; Wen, M.; Chen, Z.; Tian, L.; Li, Q.; Qu, J.; Wang, J. Numerical Simulation of Production Behavior with Different Complex Structure Well Types in Class 1-Type Hydrate Reservoir. *J. Mar. Sci. Eng.* **2024**, *12*, 508. https://doi.org/10.3390/jmse12030508

Academic Editor: Atilla Incecik

Received: 4 February 2024
Revised: 8 March 2024
Accepted: 15 March 2024
Published: 19 March 2024

1. Introduction

Natural gas hydrates (NGHs) are widely present in permafrost regions and deep-sea sediments. It is estimated that there is approximately 10^{15}–10^{18} m^3 CH_4 in the global NGH resources, which is considered to be enough to have the potential to replace fossil fuels [1–5]. The first NGH field test production was completed in 2002 at the Mallik facility in Canada [6]. Recent offshore NGH field test production at Japan's Nankai Trough and China's Shenhu Sea area have demonstrated the feasibility and superiority of the depressurization method [7–10]. However, the offshore field test production capacity remains significantly below the commercialization standard of 50×10^4 m^3/d [1]. As a result, improving gas recovery efficiency has become a critical challenge for the industrial exploitation of hydrates. Through the comprehensive analysis of various extraction methods and technologies for NGHs, complex structure well types represented by a horizontal and multilateral well are considered as the primary approaches to increasing NGH productivity [11]. Complex structure wells include all well types except that the traditional vertical well, such as horizontal well, extended reach well, and multilateral well [12].

Therefore, extracting hydrates from complex structured wells has become a research hotspot. For example, Xin X. et al. found that the deployment depth of the lateral branches in a multilateral well is the key factor affecting production capacity [13]. Mao P. et al.

evaluated the influence of different parameters related to helical multilateral well on production capacity and believed that it had the potential to achieve commercial development of natural gas hydrates [14]. Ye H. et al. discovered that a directional well and a multilateral well can greatly improve productivity, particularly in a cluster well, which can enhance gas productivity by around 2.2 times compared to a single well [15]. Mahmood M. et al. compared the production capacity of a radial lateral well (RLW) and a horizontal snake well (HSW) in extracting gas hydrates and found that RLW's productivity is directly related to the number, length, and radius of laterals; two key parameters affect HSW production capacity: one is wellbore length, and the other is radius [16]. Jin G. et al. found that under the same completion length, a multilateral well and a single well produced almost equivalent amounts of gas and water. Interference near branch intersections improves the hydrate dissociation [17]. Ye H. et al. evaluated the impact of different parameter settings on the productivity of different well types, and the results showed that branch parameters have the most significant impact on the productivity of a cluster horizontal well [18]. Hao Y. et al. found that a cluster multilateral well and a fishbone well are the optimal well types for the short- and long-term development of hydrates, respectively [19]. Ye H. et al. found that enlarging the wellbore diameter of a complex structure well can significantly increase production capacity in short-term production [20]. Cao X. et al. found that well interference occurs between lateral branches of a multilateral well, which is unfavorable for gas production and becomes worse as the number of branches increases [21]. Jin G. et al. estimated that 22 sets of multilateral wells each with a completion length of 1000 m are sufficient to achieve commercial development of NGHs in China's Shenhu Sea area [22]. He J. et al. simulated the productivity of a fishbone well with different lateral branch numbers and found that the production capacity of the six-branch fishbone well was about 59.3% higher than that of the single horizontal well [23].

However, the previous research on complex structure wells mentioned above generally targets 1~4 types of complex structure wells, such as cluster vertical well, cluster horizontal well, herringbone lateral well, and radial lateral well. As of today, both single vertical and single horizontal well have been validated for their technical feasibility through hydrate production testing, and these two types of wells can significantly improve the fault tolerance of the production system. The herringbone lateral well and radial lateral well have not undergone actual application of hydrate production testing, and still face technical challenges in the future. Previous research mainly focuses on the effects of different well deployment or branch lengths, angles, and other parameters of complex structure well on production capacity in the Class 3-type hydrate reservoir. Meanwhile, it is still unclear which complex structure well type has the best production capacity and cost advantage under the different parameter setting in the Class 1-type hydrate reservoir, which consists of an upper hydrate layer and a lower free gas layer (where gas and water can flow freely). The pressure and temperature distribution of the entire reservoir enables the stable existence of hydrates. Therefore, based on the field data of China's first offshore NGH test production, we take up to eight types of complex structure wells as the research object and use the numerical method to evaluate the production capacity of these well types under the precondition of the total completion length of 300 m and a fixed pressure difference of 6 MPa.

2. Methodology

2.1. Geological Background

China's first offshore NGH test site is located between the Xisha and Dongsha sea areas (Figure 1) [19]. Due to the exceptional geological and engineering conditions, this site is assessed to be the best location for field trials. The water depth at this site is about 1266 m and the seafloor temperature is around 3 °C with a geothermal gradient of 43.653 °C/km [24,25]. The reservoir is mainly composed of three layers: the free gas layer (FGL, 251–278 mbsf, which is composed of low saturation free gas and water), the three-phase layer (TPL, 236–251 mbsf, which contains hydrates, high saturation free gas, and water), and the gas hydrate-bearing layer (GHBL, 201–236 mbsf, which is rich in hydrate and water) [26,27].

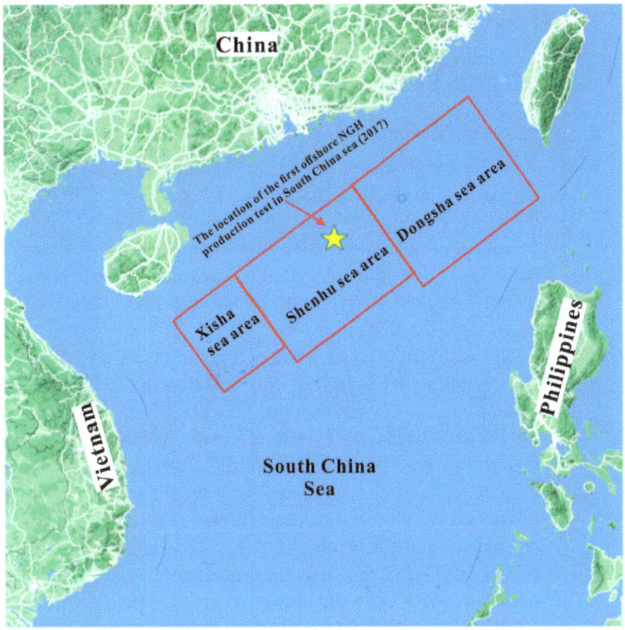

Figure 1. SHSC4 well location diagram (adapted from Ref. [19]).

2.2. Numerical Simulator

TOUGH+HYDRATE V1.0 is a well-known simulator widely used to predict NGH production [28–30]. To overcome the major issue in hydrate modeling, numerous grids are necessary for large-scale simulation [30]. Herein, we used the parallel version of this code and adopted the equilibrium model in this work [31,32]. But it can only simulate up to 50,000 grids, which is insufficient for predicting a large-scale model. Herein, we adopted the parallel version of this code (pT+H v1.0) and the equilibrium model in this work [31–33]. Furthermore, the user's manual for this code has a detailed description of the governing equations [27]. During the simulation process, we assumed that the Darcy's law is effective in the model and the wellbore is stable. The movement of geological media was neglected. The main governing equations were given as follows [27,34]:

1. Components and phases

Phase(β) = A, G, H, I is aqueous, gas, hydrate, and ice, respectively; Component(κ) = w, m, i, h is water, methane, salt, and hydrate, respectively.

2. Mass balance

The flow control equation for multi-component fluid mixtures follows mass conservation and is defined as follows:

$$\frac{d}{dt}\int_{V_n} M^{\kappa} dV = \int_{\Gamma_n} F^{\kappa} \cdot \mathrm{n} d\Gamma + \int_{V_n} q^{\kappa} dV \quad (1)$$

Here, M^{κ} is the mass accumulation, F^{κ} is the flux, and q^{κ} is the source/sink ratio.

3. Energy balance

The heat flow control equation follows energy conservation and is defined as follows:

$$\frac{d}{dt}\int_{V_n} M^{\theta} dV = \int_{\Gamma_n} F^{\theta} \cdot \mathrm{n} d\Gamma + \int_{V_n} q^{\theta} dV \tag{2}$$

Here, θ is the heat component, M^{θ}, F^{θ}, and q^{θ} are the heat accumulation, flux, and source/sink ratio, respectively.

2.3. Model Construction and Well Type Design

The logging curve of SHSC-4 was used to establish the reservoir's numerical model (Figure 2a), and the domain was 510 × 680 × 117 m in (x, y, z). The boundary layer with 20 m is enough to eliminate the boundary effects [35]. The reservoir can be divided into three layers based on the logging curve. The thicknesses of GHBL, TPL, and FGL were 35, 15, and 27 m, respectively [36]. To evaluate the gas production performance for different complex structure well types, a total of eight well types, including twenty-four simulation cases, were established in this work: (1) Cluster vertical well (CVW): CVW had four vertical wellbores with a wellbore spacing of 180 m, each vertical wellbore was an open hole completion with a length of 75 m and covered GHBL, TPL, and FGL (abbreviated as ML for multi-layer), as shown in Figure 2b. (2) Radial lateral well (RLW): RLW1–3 each had three lateral wellbores with a total open hole completion length of 300 m. RLW1's three lateral wellbores were located in the middle of the GHBL and TPL, as well as the upper part of the FGL. RLW2's three lateral wellbores were located at the top, middle, and lower part of the GHBL. RLW3's three lateral wellbores were located in the middle of the GHBL, as well as the upper and lower parts of the FGL, as shown in Figure 2c. (3) Cluster horizontal well (CHW): CHW1–3 each had three horizontal wells with a total open hole completion length of 300 m. CHW1–3 were located at the middle of GHBL, TPL, and FGL respectively, while CHW4 were located at the ML, as shown in Figure 2d. (4) Direction lateral well (DLW): DLW1–4 each contained 2 lateral wellbores with a total open hole completion length of 300 m; DLW1–4 were located at the GHBL, TPL, GHBL and ML, respectively, as shown in Figure 2e. (5) Herringbone lateral well (HLW): HLW1–3 had one main wellbore and two lateral wellbores each, with a total open hole completion length of 300 m; HLW1–3 were located at GHBL, TPL, and FGL respectively, as shown in Figure 2f. (6) Horizontal well (HW): HW1–3 had an open hole completion with a length of 300 m; HW1–3 were located at the middle of GHBL, TPL, and FGL, respectively, as shown in Figure 2g. (7) Horizontal snake well (HSW): HSW1–3 with a open hole completion length of 300 m and located in the middle of the GHBL, TPL, and FGL, respectively, as shown in Figure 2h. (8) Vertical lateral well (VLW): VLW1–3 had four lateral wellbores with a total open hole completion length of 300 m; VLW1–3 were located at the GHBL, TPL, and FGL, respectively, as shown in Figure 2i. The above simulation cases were divided into four groups, as listed in Table 1.

The discretization was carried out by the following steps to reduce the number of grids and decrease the computational resources: Different well type x-y plane domains were discretized separately and then extruded into 3D discretized grids along the z-axis. To depict the intricate multiphase flow near the wellbore, the minimum grid sizes in the x-, y- and z-axis directions were 2 m, 2 m, and 1 m, respectively. The x-y plane domain of the different well types, CVW, RLW, CHW, DLW, HLW, HW, HSW, and VLW were discretized into 13,923, 2166, 7469, 3040, 5341, 1740, 5478, and 7055 grids, respectively. The total grid was 1,127,763, 175,446, 604,989, 246,240, 432,621, 140,940, 443,718, and 571,455 respectively.

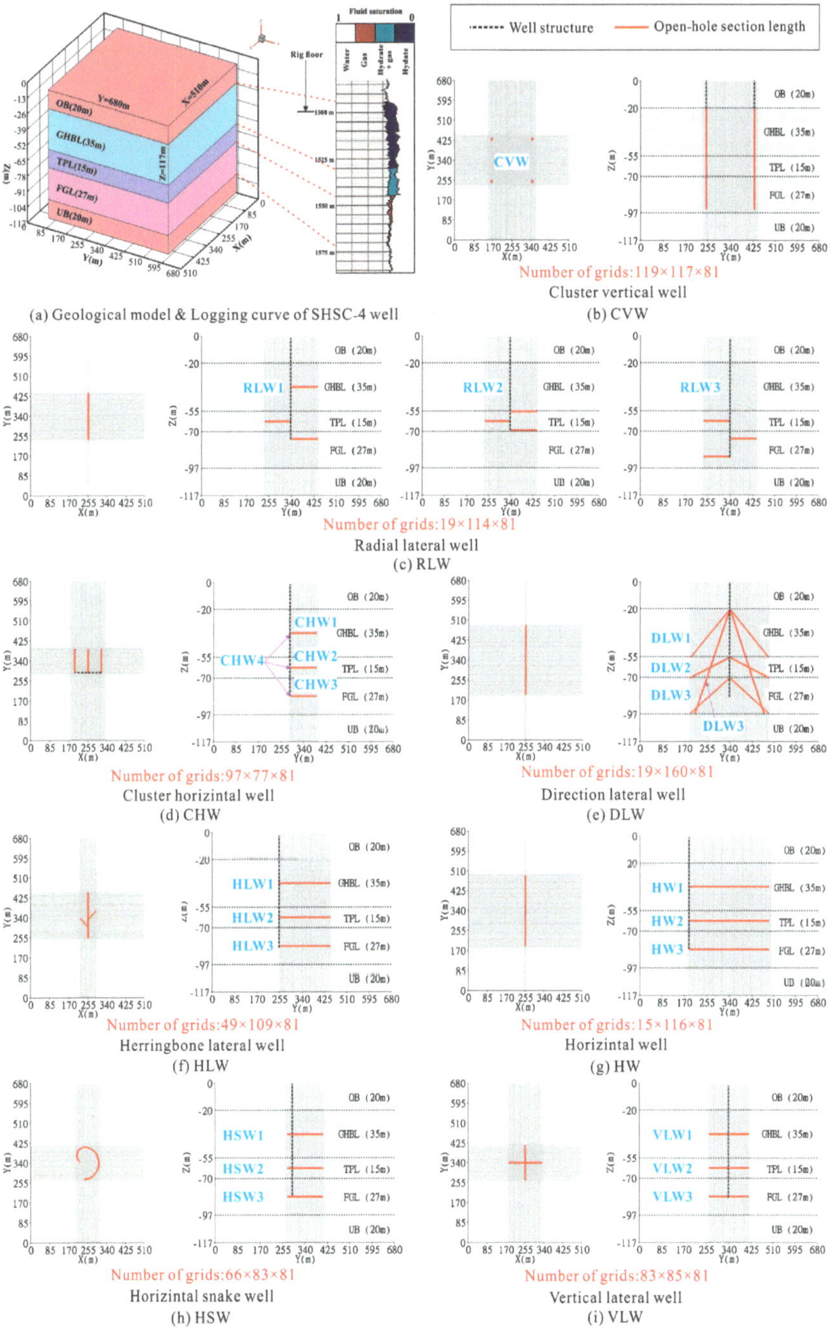

Figure 2. Model schematic diagram and well type design: (**a**) Geological model and Logging curve of SHSC-4 well. (**b**) Schematic diagram of CVW; (**c**) Schematic diagram of RLW; (**d**) Schematic diagram of CHW; (**e**) Schematic diagram of DLW; (**f**) Schematic diagram of HLW; (**g**) Schematic diagram of HW; (**h**) Schematic diagram of HSW; (**i**) Schematic diagram of VLW.

Table 1. The main parameters of simulation cases.

Groups	Case	Main Parameters			
		L/(m)	*l*/(m)	*n*	Location of Open Hole Section
A	CHW1	300	100	3	GHBL
	DLW1	300	150	2	
	HLW1	300	50	2	
	HW1	300	/	/	
	HSW1	300	/	/	
	VLW1	300	75	4	
B	CHW2	300	100	3	TPL
	DLW2	300	150	2	
	HLW2	300	50	2	
	HW2	300	/	/	
	HSW2	300	/	/	
	VLW2	300	75	4	
C	CHW3	300	100	3	FGL
	DLW3	300	150	2	
	HLW3	300	50	2	
	HW3	300	/	/	
	HSW3	300	/	/	
	VLW3	300	75	4	
D	CHW4	300	100	3	ML
	DLW4	300	150	2	
	CVW	300	75	4	
	RLW1	300	100	3	
	RLW2	300	100	3	
	RLW3	300	100	3	

Note: *n* is the quantity of lateral wellbore or single wellbore; ML is the abbreviation for multi-layer.

2.4. Initial and Boundary Conditions

The initial conditions of the model can be calculated by the code's self-balancing function, as shown in Figure 3 [36–39]. In order to establish the Dirichlet boundary, the top and bottom of the model were set to a fixed temperature and pressure [40]. The wellbore grids were given a fixed pressure difference of 6 MPa and pressure loss along the wellbore was ignored.

The reservoir characteristics (porosity, permeability, saturation) of each sublayer were according to the on-site measured data as follows: (a) The' porosity values of the three sublayers were 0.35, 0.33 and 0.32; (b) the permeability values of the three sublayers were 2.9, 1.5, and 7.4 mD; (c) the initial gas and hydrate saturations of the GHBL, TPL, and FGL were extracted from the logging curve [9]. It was assumed that the mean porosity of the OB and UB was 0.3 and the permeability was 2.0 mD. Table 2 lists the model's detailed parameter settings.

Table 2. Model's detailed parameter settings.

Parameter	Value and Unit
OB and UB's thickness [9,21,26,41]	20 m
GHBL's thickness [9,21,26,41]	35 m
TPL's thickness [9,21,26,41]	15 m
FGL's thickness [9,21,26,41]	27 m
OB and UB's permeability	2.0 mD
GHBL's permeability [9,21,26,41]	2.9 mD

Table 2. *Cont.*

Parameter	Value and Unit
TPL's permeability [9,21,26,41]	1.5 mD
FGL's permeability [9,21,26,41]	7.4 mD
Wellbore radius [9,21,26,41]	0.1 m
Salinity [9,21,26,41]	3.5%
GHBL and TPL's hydrate saturation [9,21,26,41]	Reference from logging curve (Figure 2a)
FGL's gas saturation [9,21,26,41]	Reference from logging curve (Figure 2a)
OB and UB's porosity	0.30
GHBL's porosity [9,21,26,41]	0.35
TPL's porosity [9,21,26,41]	0.33
FGL's porosity [9,21,26,41]	0.32
Grain density [21,26,41]	2600 kg/m^3
Geothermal gradient [21,26,41]	43.653 °C/km
Grain specific heat [21,26,41]	1000 $J \cdot kg^{-1} \cdot K^{-1}$
Gas composition [21,26,41]	100% CH_4
Dry thermal conductivity [21,26,41]	1.0 $W \cdot m^{-1} \cdot K^{-1}$
Wet thermal conductivity [21,26,41]	3.1 $W \cdot m^{-1} \cdot K^{-1}$
Capillary pressure model [21,26,41]	$P_{cap} = -P_0\left[(S^*)^{-1/\lambda} - 1\right]^{1-\lambda}$, $S^* = \frac{(S_A - S_{irA})}{(S_{mxA} - S_{irA})}$
Maximum reference aqueous saturation of capillary S_{mxA} [21,26,41]	1
Porosity distribution index λ [21,26,41]	0.45
Entry pressure P_0 [21,26,41]	10^4 Pa
Relative permeability model [21,26,41]	$K_{rA} = [(S_A - S_{irA})/(1 - S_{irA})]^{nA}$, $K_{rG} = [(S_G - S_{irG})/(1 - S_{irA})]^{nG}$
Permeability reduction exponent for aqueous phase n_A [21,26,41]	3.5
Permeability reduction exponent for gas phase n_G [21,26,41]	2.5
Irreducible saturation of gas phase S_{irG} [21,26,41]	0.03
Irreducible saturation of aqueous phase S_{irA} [21,26,41]	0.30

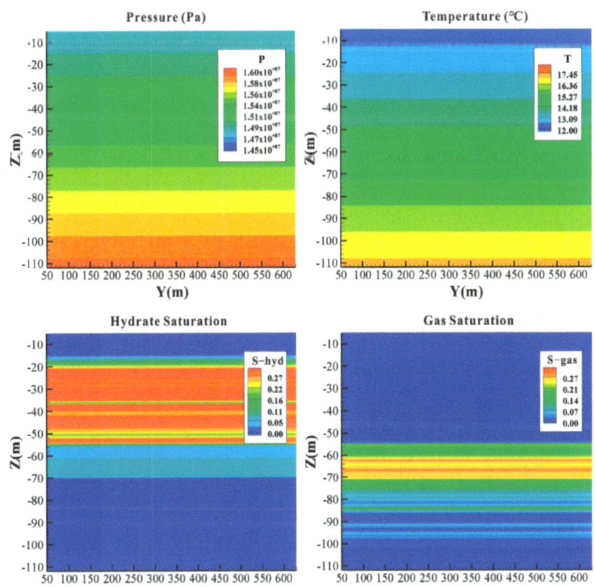

Figure 3. Initial conditions of the model.

2.5. Grid Independence Test

The numerical models established in this work were discretized separately in *x*-*y* plane domains, and the total number of grids between these models was different. To verify the errors of these models in predicting gas and water production within an acceptable range, we conducted the grid independence test. First, a vertical well with a completion section length of 70 m was set in the middle of these models, and then the same production pressure difference of 6 MPa was applied for continuous production for 360 days. Figure 4 shows the production results of these models within 360 days. The inconsistency number of grids among different models has almost no impact on the overall gas and water production. Therefore, these models can be used for subsequent comparison production performance of different complex structure well types.

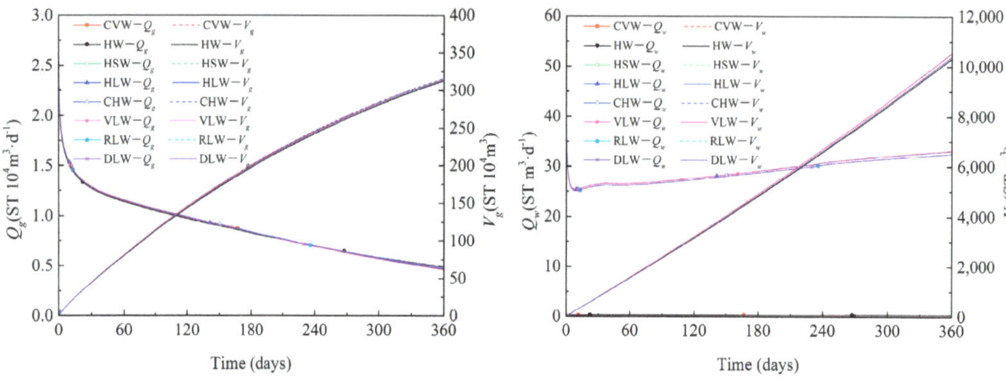

Figure 4. Grid independence test.

3. Results and Discussion

3.1. Well Types Deployed at GHBL

3.1.1. Gas and Water Characteristics

Figure 5a depicts the evolutions of the gas production rate (Q_g) and cumulative gas production (V_g) of different well types deployed in the middle of the GHBL within 360 days of production. In the initial stage (within 90 days), the Q_g of CHW1, HLW1, HW1, HSW1, and VLW1 were maintained relatively low. The presence of solid hydrates resulted in very low effective permeability of reservoirs. With the decomposition of solid hydrates, the seepage conditions near the wellbore gradually improved. After 90 days of depressurization, the Q_g and V_g curves began to show a sudden increase and enter the second stage, which was due to the free gas from the TPL beginning to reach and flow into the wellbores. In the later stage, it was gradually decreasing due to the effects of weak driving force. Unlike the other well types, DLW1 has distinct gas production trends that may likewise be separated into two stages. Q_g steadily increases and achieves a high value of 20,597 m^3/d in the first stage (within 180 days) and then decreases to 12,966 m^3/d by 360 days in the second stage. This is mainly because the lateral branches of DLW covered the entire GHBL and the root ends reached TPL. Due to the weakened of pressure driving force and the sustained Joule–Thomson effect, secondary hydrates were generated at the root ends of its lateral branches. As a result, the average Q_g is 16,666 m^3/d with a total V_g of 589 × 10^4 m^3 after 360 days of production, which is better than the values of the above well types. The evolution of the water production rate (Q_w) corresponds to the gas production behavior as shown in Figure 5b. The gas-to-water ratio R_{gw} (ST m^3 of CH_4/ST m^3 of water) is an important indicator for assessing gas production efficiency, and a high R_{gw} means higher recoverability. DLW1 has a higher R_{gw} of around 170 after 360 days of production, which indicates that it is more adaptable than the other well types during production. Table 3 presents the production capacity of the aforementioned well types.

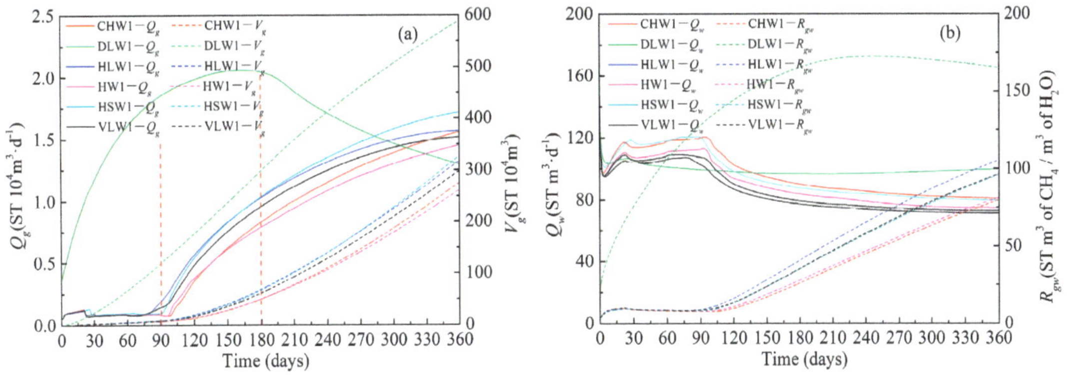

Figure 5. Production performances of different well types deployed at the middle of GHBL within 360 days: (**a**) Gas production rate Q_g and cumulative gas production V_g. (**b**) Water production rate Q_w and gas to water ratio R_{gw}.

Table 3. Productivity of different well types deployed at GHBL after 360 days of production.

Case	Average Q_g (10^4 m^3/d)	V_g (10^4 m^3)	Compared to the Reference Case
DLW1	1.64	589	224.81%
HSW1	0.91	327	124.81%
HLW1	0.88	317	120.99%
VLW1	0.83	298	113.74%
CHW1	0.77	276	105.34%
HW1 (ref)	0.73	262	100.00%

3.1.2. Characteristics of Reservoir Parameters

As shown in Figure 6a, due to the pressure drop superposition effect, the pressure drop areas between the main wellbore, and lateral branches of DLW1, HLW1, and VLW1 are larger. Similar to the pressure distribution diagram, in the temperature distribution diagram, Figure 6b, under the Joule–Thomson effect and the heat absorption of hydrate decomposition, the temperature near the wellbore of each well type significantly decreases and the low-temperature areas between the main wellbore and lateral branches of DLW1, HLW1 and VLW1 are larger. As shown in Figure 6c,d, after 360 days of production, the saturation of hydrates near the wellbore tends to be almost zero. The released gas accumulates near the wellbore, providing gas for long-term production.

3.2. Well Types Deployed at TPL

3.2.1. Gas and Water Characteristics

Figure 7a shows the evolutions of Q_g and V_g of different well types deployed in the middle of TPL within 360 days of production. When the different production well types were deployed in the middle of TPL, the high-saturation free gas of TPL and the hydrate decomposition gas of GHBL could be recovered at the same time, which greatly improved the overall productivity performance. The Q_g of all well types reached its peak in the initial stage, but slowly decreased in the later stage due to the weakening of the pressure driving force. As shown in Figure 8c, all well types generated a small amount of secondary hydrates at the root and toe ends of the wellbore after 360 days of production. Among all well types, the DLW2 and HW2 well types had the best productivity performance, with an average Q_g of 33,055 m^3/d and a total V_g of 1190 × 10^4 m^3 after 360 days of production, and the R_{gw} of all well types tended to be around 300~400 after 360 days of production. Table 4 presents the production capacity of the aforementioned well types.

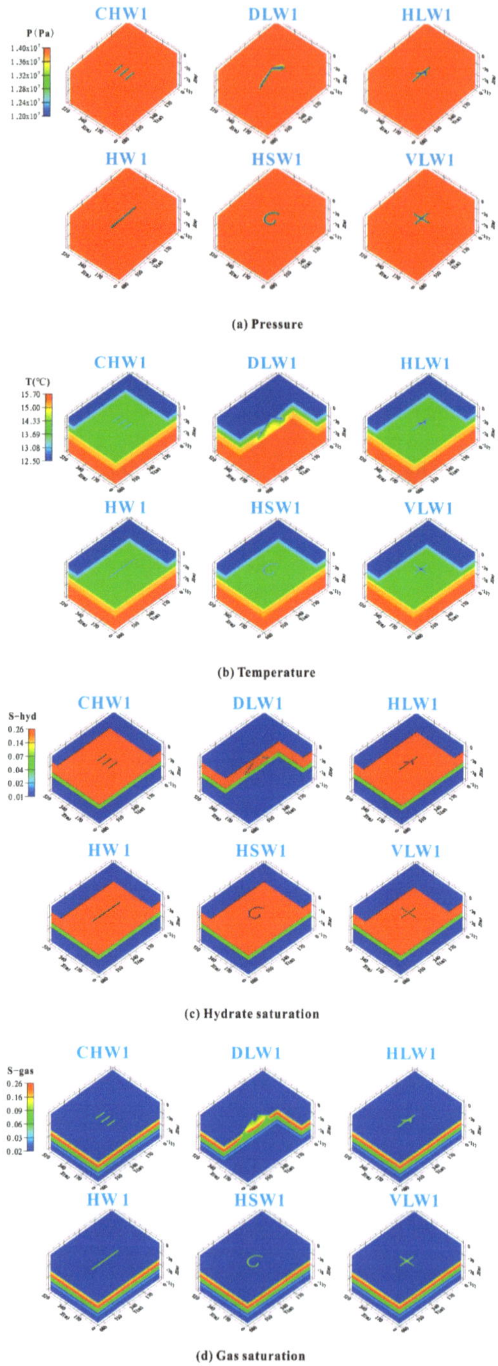

Figure 6. Spatial distribution diagram of physical properties on different well types deployed at the middle of GHBL, and t = 360 days.

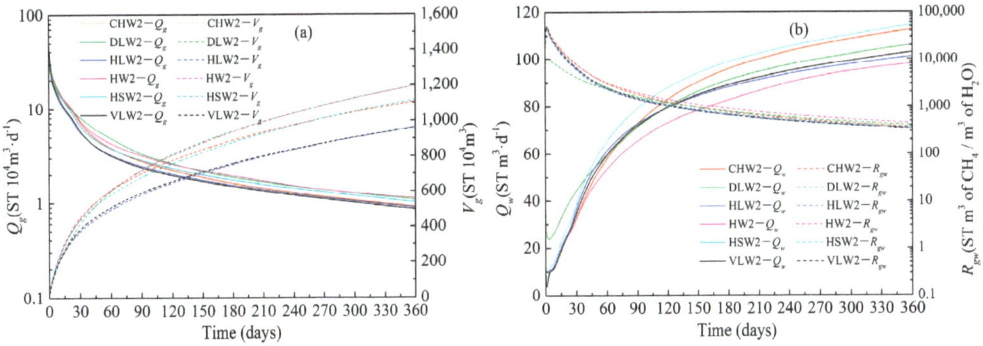

Figure 7. Production performances of different well types deployed in the middle of TPL within 360 days: (**a**) Gas production rate Q_g and cumulative gas production V_g. (**b**) Water production rate Q_w and gas to water ratio R_{gw}.

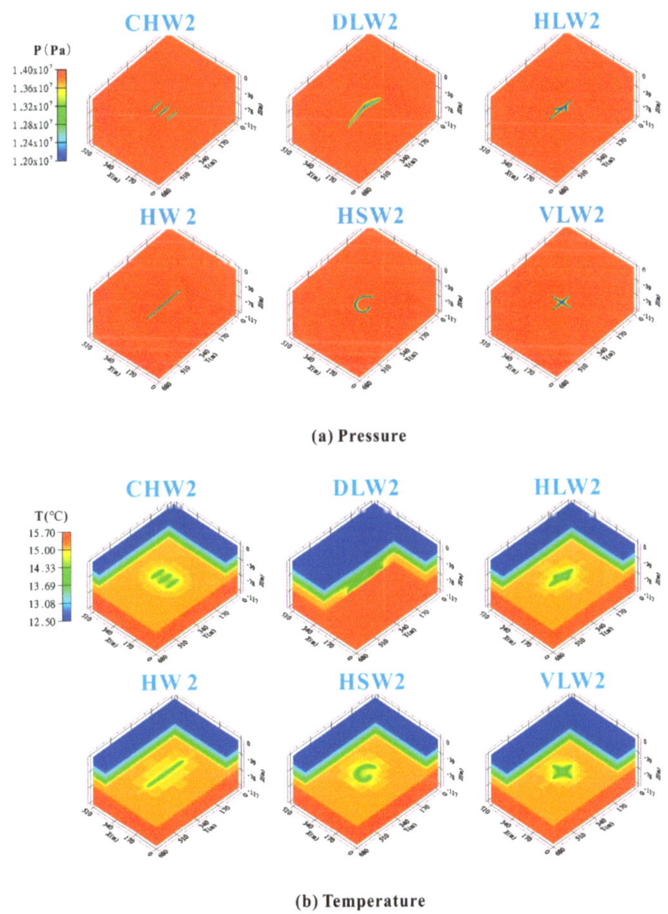

Figure 8. *Cont.*

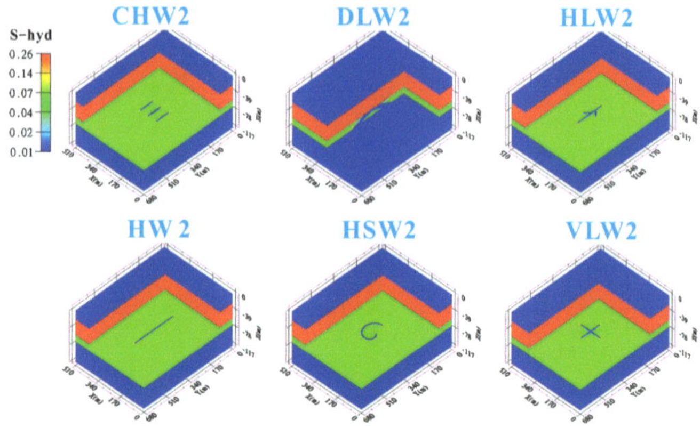

(c) Hydrate saturation

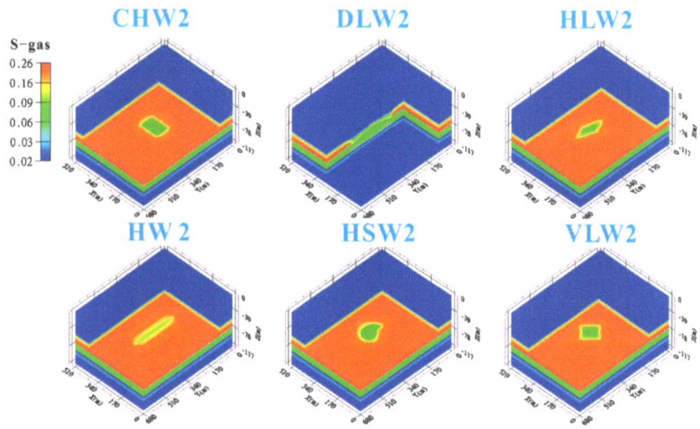

(d) Gas saturation

Figure 8. Spatial distribution diagram of physical properties on different well types deployed in the middle of TPL, and t = 360 days.

Table 4. Productivity of different well types deployed at the TPL after 360 days of production.

Case	Average Q_g (10^4 m^3/d)	V_g (10^4 m^3)	Compared to the Reference Case
DLW2	3.31	1190	123.96%
HW2	3.31	1190	123.96%
HSW2	3.08	1110	115.63%
CHW2	3.06	1100	114.58%
HLW2	2.67	960	100.00%
VLW2 (ref)	2.67	960	100.00%

3.2.2. Characteristics of Reservoir Parameters

As shown in Figure 8a, the pressure distribution of different well types deployed in the middle of the TPL is similar to that when deployed at the middle of the GHBL, with more obvious and larger areas of pressure drop between the main and lateral branches. It can be seen from Figure 8b that a large amount of gas from TPL is extracted in a short period, which leads to a strong Joule–Thomson effect and the low-temperature areas generated by all well types were significantly larger than those deployed at GHBL. As shown in

Figure 8c,d, after 360 days of production, different amounts of secondary hydrates were generated at the root and toe ends of all types of wellbores, and corresponding low gas saturation areas also formed near the wellbores.

3.3. Well Types Deployed at FGL

3.3.1. Gas and Water Characteristics

Figure 9a depicts the evolutions of Q_g and V_g of different well types deployed in the middle of FGL within 360 days of production. The Q_g curves of different well types show a sudden increase at 8 days. This is because the high-saturation free gas from TPL begins to reach and flow into the wellbores. Similarly, due to the effects of weak driving force, there is a gradual decrease trend in the later stage. The Q_g curves trend of DLW is inconsistent with that of other well types. This is because the lateral branches of DLW cover the entire FGL, and its Q_g is the highest among all well types in the initial stage. Among all well types deployed in the middle of the FGL, the HSW3 has the best gas performance, with an average Q_g of 24,222 m^3/d and a total V_g of 872 × 10^4 m^3 after 360 days of production. Water production behaviors are similar to gas production, as shown in Figure 9b; the R_{gw} of all well types remains steady around 135~180 after 360 days of production. Table 5 presents the production capacity of the aforementioned well types.

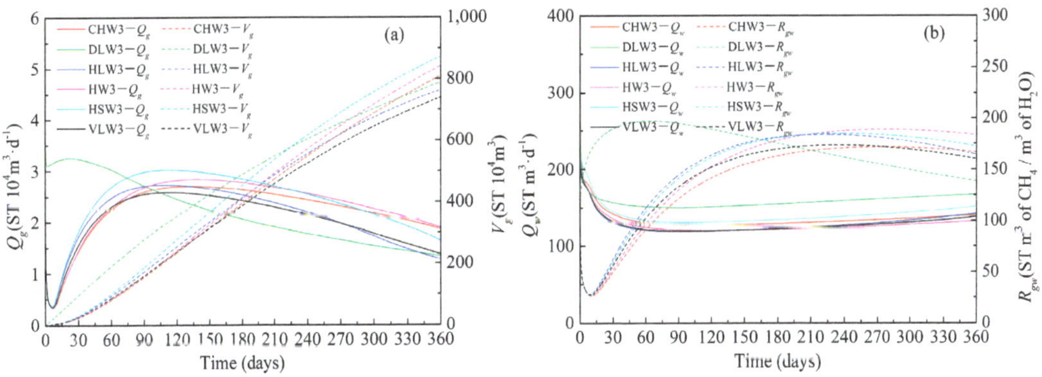

Figure 9. Production performances of different well types deployed in the middle of FGL within 360 days: (**a**) Gas production rate Q_g and cumulative gas production V_g. (**b**) Water production rate Q_w and gas to water ratio R_{gw}.

Table 5. Productivity of different well types deployed at the FGL after 360 days of production.

Case	Average Q_g (10^4 m^3/d)	V_g (10^4 m^3)	Compared to the Reference Case
HSW3	2.42	872	118.16%
HW3	2.34	842	114.09%
CHW3	2.25	810	109.76%
DLW3	2.19	789	106.91%
HLW3	2.12	763	103.39%
VLW3 (ref)	2.05	738	100.00%

3.3.2. Characteristics of Reservoir Parameters

As shown in Figure 12a, due to the presence of a gas expansion effect in the FGL, the pressure drop cannot be effectively transmitted to a distance; after 360 days of production, the pressure gradient around the DLW's wellbore is significantly lower than that of other well types. The areas of pressure drop between the main wellbore and lateral branches of other well types are bigger. As for the temperature distribution diagram (Figure 12b), due to the high initial sensible heat in the middle of the FGL, the low-temperature areas are not obvious after 360 days of production for all well types.

3.4. Well Types Deployed at ML

3.4.1. Gas and Water Characteristics

Figure 11a depicts the evolutions of Q_g and V_g of different well types deployed in the middle of ML within 360 days of production. Due to the advantages of multi-layer combined production, the Q_g curves peaked early and then gradually declined with the weakening of the driving force in the later stage. After 360 days of production, the CVW well type stood out with a wellbore spacing of 180 m and an average Q_g of 34,444 m^3/d with a total V_g of 1240 × 10^4 m^3. It is worth noting that starting from 40 days, the Q_g of the CVW well type began to surpass that of other well types. After 135 days of production, its V_g also began to surpass that of other well types. The RLW2 well had the highest productivity among the RLW well types because its three lateral branches were located at the upper, middle, and lower parts of the TPL, which means it could effectively recover free gas from the TPL and FGL, as well as hydrate decomposition gas from the GHBL, resulting in the average Q_g of 26,527 m^3/d with a total V_g of 955 × 10^4 m^3 after 360 days of production. The R_{gw} of all well types remained steady at around 200~300 after 360 days of production, as shown in Figure 11b. Table 6 presents the production capacity of the aforementioned well types.

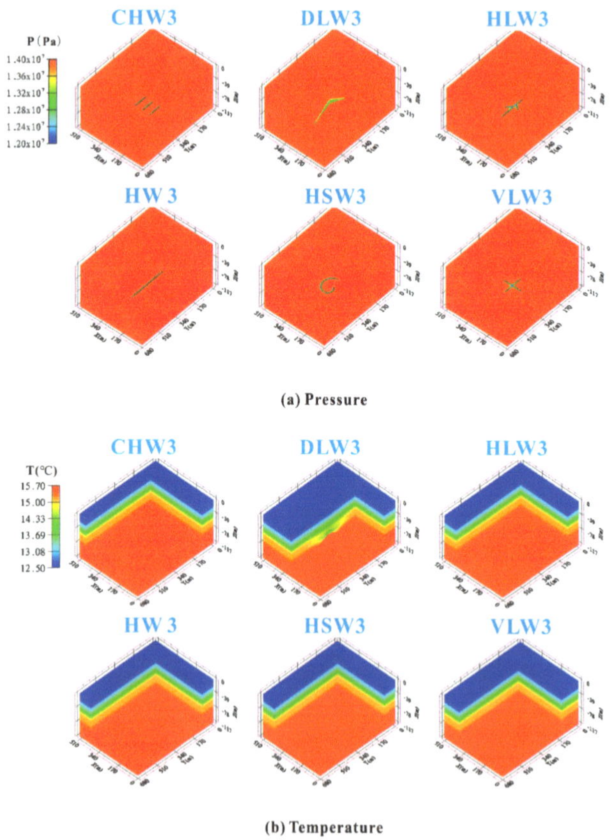

Figure 10. *Cont.*

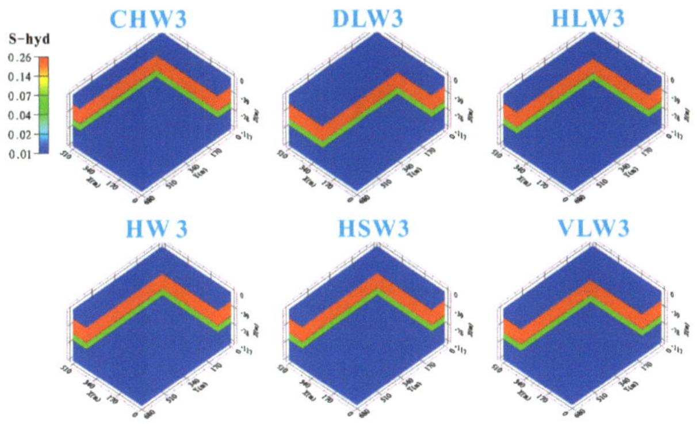

(c) Hydrate saturation

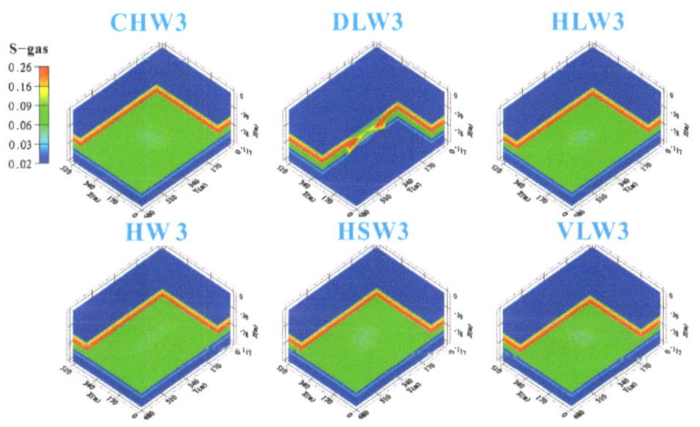

(d) Gas saturation

Figure 10. Spatial distribution diagram of physical properties on different well types deployed in the middle of FGL, and t = 360 days.

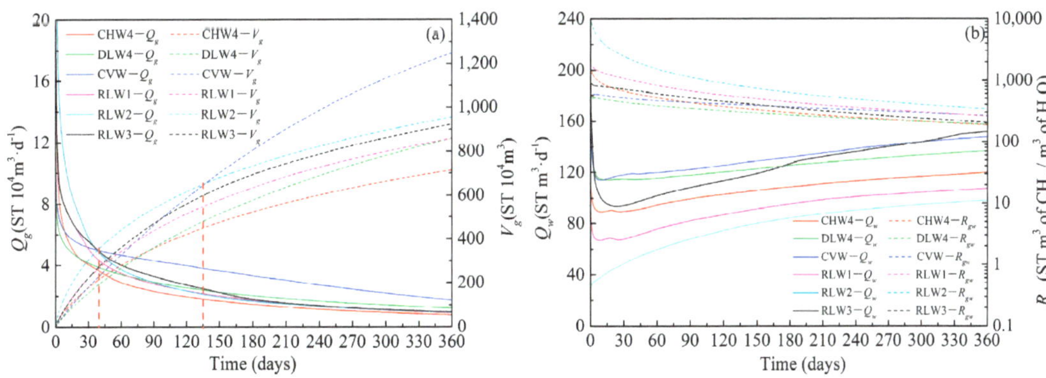

Figure 11. Production performances of different well types deployed at ML within 360 days: (**a**) Gas production rate Q_g and cumulative gas production V_g. (**b**) Water production rate Q_w and gas to water ratio R_{gw}.

Table 6. Productivity of different well types deployed at the ML after 360 days of production.

Case	Average Q_g (10^4 m^3/d)	V_g (10^4 m^3)	Compared to the Reference Case
CVW	3.44	1240	173.91%
RLW2	2.65	955	133.94%
RLW3	2.56	923	129.45%
RLW1	2.38	857	120.20%
DLW4	2.38	858	120.34%
CHW4 (ref)	1.98	713	100.00%

3.4.2. Characteristics of Reservoir Parameters

As shown in Figure 12a, it can be observed that the pressure gradient of the lower part of the wellbore structure of all well types was greater than the upper part. Especially for RLW3, the pressure gradient after 360 days of production were significantly lower than other well types. This is mainly because the gas expansion effect in the lower gas-bearing layer limited the pressure propagation. As for the temperature distribution diagram (Figure 12b), the wellbore structures located at the GHBL and TPL layers of all well types formed a noticeable low-temperature area around them. From Figure 12c,d, it can be seen that due to the high initial sensible heat of the FGL, there was no secondary hydrate formation near the wellbore structures located here. Secondary hydrates tend to generate near the wellbore structures located at TPL with high-saturation free gas here, such as the CVW well type where secondary hydrates are most prominent and the gas accumulates near the wellbore.

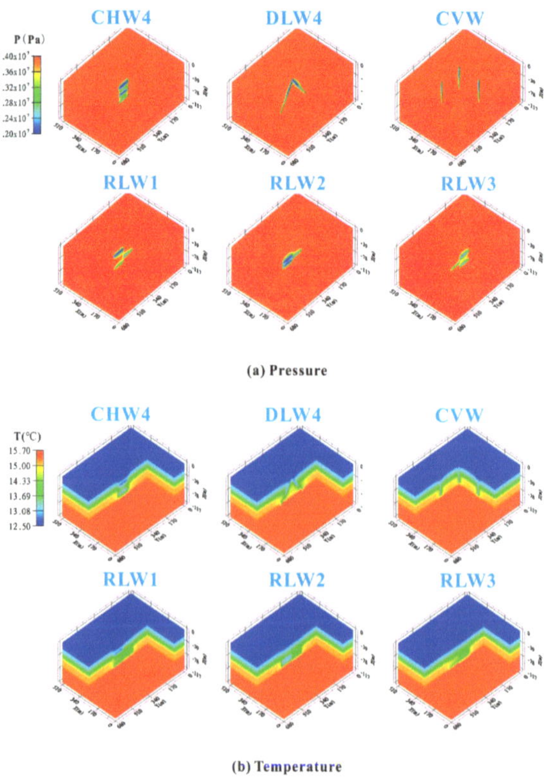

Figure 12. *Cont.*

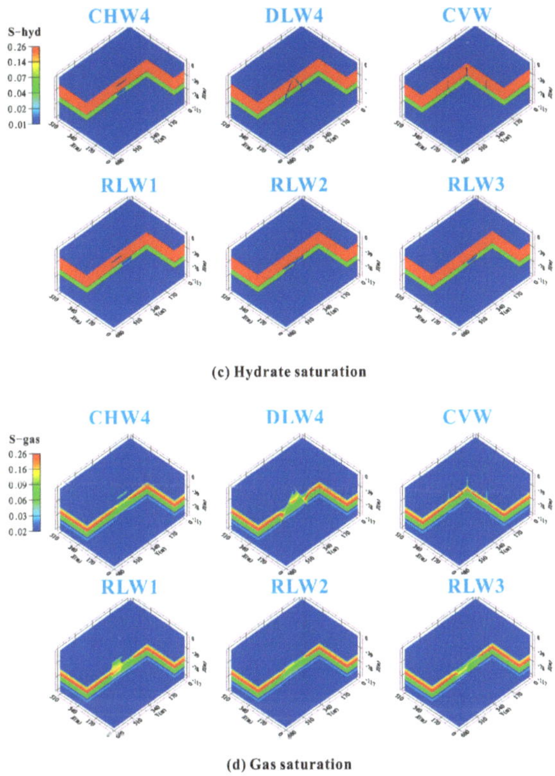

Figure 12. Spatial distribution diagram of physical properties on different well types deployed at ML, and t = 360 days.

3.5. Comparisons of Production Performances

In general, the average Q_g is the most direct indicator used to evaluate the yield-increasing effect, while relative criterion R_{gw} is used to evaluate hydrate production efficiency. The yield-increasing mechanism of complex structure wells mainly depends on enlarging the drainage area. However, the increase in production may not be entirely related to the length of completion of the well. Herein, specific index J is recommended as an additional indicator in this section to evaluate production capacity, which is mainly affected by the well type. The following is the definition of the index [10]:

$$J = Q_g / h\Delta P \tag{3}$$

h represents the completion length (m), and P represents the pressure difference (MPa). In this work, different complex well types were deployed at the GHBL, TPL, FGL, and ML, respectively. The average Q_g and the J index of these well types after 360 days production are shown in Figure 13. Only in terms of production capacity, during the 240-day production period, the production performance of different complex well types with different deployment were as follows: TPL > ML > FGL > GHBL, and the DLW2 and HW2 well types stood out, with an average Q_g of 43,333 m^3/d and a J-index of 24.1. After 360 days of production, the CVW well type deployed at ML had the best production capacity performance, with an average Q_g of 34,444 m^3/d and a J-index of 19.1. This is mainly due to the advantage of the multi-layer combined production, as well as the synergistic pressure reduction effect between wellbores. However, it is worth noting that no matter what kind of well type is adopted, one single well production mode cannot meet the commercial development of the offshore NGH. The "well factory" production mode such as a multi-complex well combined with depressurization is the best option for economic exploitation of NGH. Herein, we suggest adopting the CVW as the basic well type for the "well factory" production mode, not only because it has better production performance, but also because the technology threshold and cost involved are relatively low. CVW can improve system redundancy, and any single well can be easily converted into a monitoring or injection well during the production process, which is beneficial for adjusting production plans and requires further research.

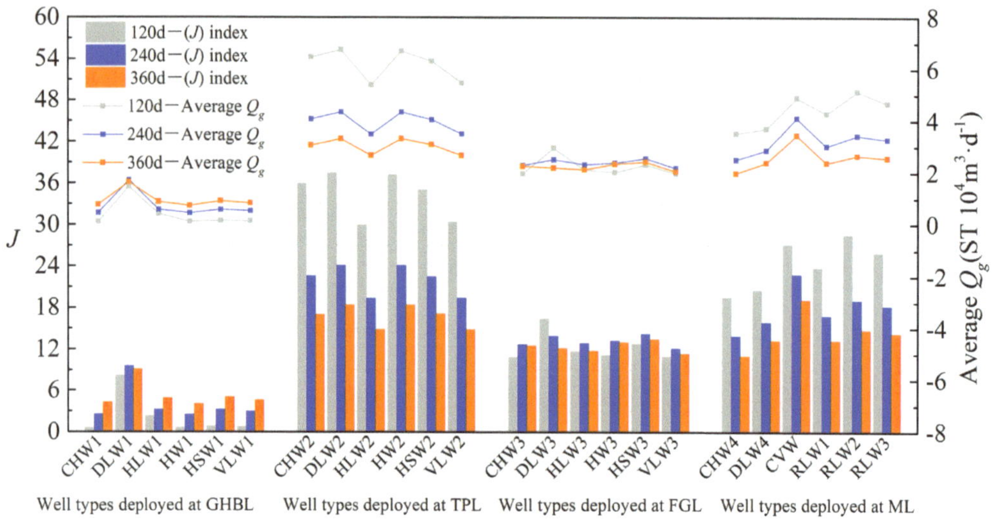

Figure 13. Histogram of J and average Q_g for different well types, and t = 120, 240 and 360 days.

4. Conclusions

Based on the field data of China's NGH test production in the Shenhu Sea area, the numerical method was used to evaluate the production capacity of different complex structure well types by continuous depressurization production for 360 days under the preconditions of fixed effective completion length of 300 m and a pressure difference of 6 MPa. The following results were obtained: during the 240-day production period, the complex structure well deployed at the TPL demonstrated higher production performance, and the DLW2 and HW2 well types stood out with an average Q_g of 43,333 m^3/d and a J-index of 24.1. After 360 days of production, the CVW deployed at the ML had the best production performance due to the advantage of the multi-layer combined production, as well as the synergistic pressure reduction production effect between wellbores, with an average Q_g of 34,444 m^3/d and a J-index of 19.1. According to the simulation results, CVW has excellent gas production performance, with the advantages of low technical threshold and cost. More importantly, CVW can improve the fault tolerance of the production system and can easily convert any single well into a monitoring well or injection well. Herein, we suggest adopting CVW as the basic well for the "well factory" production mode, which is worth further research.

Author Contributions: Conceptualization, T.W.; Methodology, T.W.; Software, T.W.; Formal analysis, M.W. and Z.C.; Investigation, M.W. and Z.C.; Resources, L.T.; Data curation, Q.L. and J.Q.; Writing—original draft, T.W.; Writing—review & editing, Z.L.; Visualization, Q.L. and J.Q.; Supervision, Z.L. and J.W.; Project administration, J.W. All authors have read and agreed to the published version of the manuscript.

Funding: This research was funded by the National Key Research and Development Program of China (No. 2021YFB3401405); National Key Research and Development Program of China (No. SQ2023YFC2800361); Guangdong Basic and Applied Basic Research Foundation (No. 2022A1515011902); Guangzhou Science and Technology Program (No. 202206050002); and the Director General's Scientific Research Fund of Guangzhou Marine Geological Survey, China (No. 2023GMGSJZJJ00027) for financial assistance to this research.

Data Availability Statement: The data presented in this study are available on request from the corresponding author.

Conflicts of Interest: The authors declare that they do not have any commercial or associative interests that represent conflicts of interest in connection with the submitted work.

Nomenclature

Symbols	
L	open hole section length of wellbore (m)
l	length of each lateral wellbore or single wellbore (m)
n	quantity of lateral wellbore or single wellbore
t	times (s)
x, y, z	cartesian coordinates (m)
Q_g	gas production rates at well (m^3/d)
Q_w	water production rates at well (m^3/d)
V_g	cumulative gas production at well (m^3/d)
R_{gw}	ratio of cumulative gas to cumulative gas(ST m^3 of CH_4/m^3 of H_2O)
J	specific production index
T	temperature (°C)
P_{cap}	capillary pressure (Pa)
P_0	entry pressure of capillary pressure model (Pa)
S^*	saturation for capillary pressure
S_{mxA}	maximum reference aqueous saturation of capillary
S_{irA}	irreducible saturation of aqueous phase
S_{irG}	irreducible saturation of gas phase
n_A	permeability reduction exponent for aqueous phase
n_G	permeability reduction exponent for gas phase
λ	porosity distribution index
k	permeability (m^2)
$k_{r\beta}$	relative permeability of phase β
φ	porosity
ρ_β	density of phase β
ρ_R	density of rock grain (kg/m^3)
M^κ	mass accumulation of component κ, (kg/m^3)
F^κ	mass flux of component κ, $kg/(m^2\cdot s)$
q^κ	sink/source of component κ, $kg/(m^3\cdot s)$
M^θ	energy accumulation (J/m^3)
F^θ	energy flux, $J/(m^2\cdot s)$
q^θ	sink/source of heat, $J/(m^3\cdot s)$
V	volume (m^3)
Γ	surface area (m^2)
β	phase, $\beta = A, G, H, I$ is aqueous, gas, hydrate, and ice, respectively
κ	component, $\kappa = w, m, i, h$ is water, methane, salt, and hydrate, respectively
Abbreviations	
OB	overburden layer
UB	underburden layer
GHBL	gas hydrate bearing layer
TPL	three phase layer
FGL	free gas layer
ML	multi-layer
NGH	natural gas hydrate

HW	horizontal well
CVW	cluster vertical well
CHW	cluster horizontal well
HLW	herringbone lateral well
HSW	horizontal snake well
VLW	vertical lateral well
RLW	radial lateral well
DLW	direction lateral well

References

1. Sloan, E. Fundamental principles and applications of natural gas hydrates. *Nature* **2003**, *426*, 353–359. [CrossRef] [PubMed]
2. Boswell, R. Is gas hydrate energy within reach? *Science* **2009**, *325*, 957–958. [CrossRef] [PubMed]
3. Chong, Z.; Yang, S.; Babu, P.; Linga, P.; Li, X. Review of natural gas hydrates as an energy resource: Prospects and challenges. *Appl. Energy* **2016**, *162*, 1633–1652. [CrossRef]
4. Boswell, R.; Collett, T.S. Current perspectives on gas hydrate resources. *Energy Environ. Sci.* **2011**, *4*, 1206–1215. [CrossRef]
5. Hassanpouryouzband, A.; Joonaki, E.; Farahani, M.; Takeya, S.; Ruppel, C.; Yang, J.; English, N.; Schicks, J.; Edlmann, K.; Mehrabian, H.; et al. Gas hydrates in sustainable chemistry. *Chem. Soc. Rev.* **2020**, *49*, 5225–5309. [CrossRef]
6. Moridis, G.J.; Collett, T.S.; Dallimore, S.R.; Satoh, T.; Hancock, S.; Weatherill, B. Numerical studies of gas production from several CH4 hydrate zones at the Mallik site, Mackenzie Delta, Canada. *J. Pet. Sci. Eng.* **2004**, *43*, 219–238. [CrossRef]
7. Yamamoto, K.; Terao, Y.; Fujii, T.; Ikawa, T.; Seki, M.; Matsuzawa, M.; Kanno, T. *Operational Overview of the First Offshore Production Test of Methane Hydrates in the Eastern Nankai Trough, Offshore Technology Conference*; OnePetro: Richardson, TX, USA, 2014.
8. Yamamoto, K.; Wang, X.X.; Tamaki, M.; Suzuki, K. The second offshore production of methane hydrate in the Nankai Trough and gas production behavior from a heterogeneous methane hydrate reservoir. *RSC Adv.* **2019**, *9*, 25987–26013. [CrossRef] [PubMed]
9. Li, J.; Ye, J.; Qin, X.; Qiu, H.; Wu, N.; Lu, H.; Xie, W.; Lu, J.; Peng, F.; Xu, Z.; et al. The first offshore natural gas hydrate production test in South China Sea. *China Geol.* **2018**, *1*, 5–16. [CrossRef]
10. Ye, J.; Qin, X.; Xie, W.; Lu, H.; Ma, B.; Qiu, H.; Liang, J.; Lu, J.; Kuang, Z.; Lu, C.; et al. The second natural gas hydrate production test in the South China Sea. *China Geol.* **2020**, *3*, 197–209. [CrossRef]
11. Wu, N.; Li, Y.; Wan, Y.; Sun, J.; Huang, L.; Mao, P. Prospect of marine natural gas hydrate stimulation theory and technology system. *Nat. Gas Ind. B* **2021**, *40*, 173–187. [CrossRef]
12. Gao, D. Discussin on development modes and engineering techniques for deepwater natural gas and its hydrates. *Nat. Gas Ind.* **2020**, *40*, 136–143.
13. Xin, X.; Li, S.; Xu, T.; Yuan, Y. Numerical investigation on gas production performance in methane hydrate of multilateral well under depressurization in Krishna-Godavari basin. *Geofluids* **2021**, *2021*, 9936872. [CrossRef]
14. Mao, P.; Wan, Y.; Sun, J.; Li, Y.; Hu, G.; Ning, F.; Wu, N. Numerical study of gas production from fine-grained hydrate reservoirs using a multilateral horizontal well system. *Appl. Energy* **2021**, *301*, 117450. [CrossRef]
15. Ye, H.; Wu, X.; Li, D. Numerical Simulation of Natural Gas Hydrate Exploitation in Complex Structure Wells: Productivity Improvement Analysis. *Mathematics* **2021**, *9*, 2184. [CrossRef]
16. Mahmood, M.; Guo, B. Productivity comparison of radial lateral wells and horizontal snake wells applied to marine gas hydrate reservoir development. *Petroleum* **2021**, *7*, 407–413. [CrossRef]
17. Jin, G.; Peng, Y.; Liu, L.; Su, Z.; Liu, J.; Li, T.; Wu, D. Enhancement of gas production from low-permeability hydrate by radially branched horizontal well: Shenhu Area, South China Sea. *Energy* **2022**, *253*, 124129. [CrossRef]
18. Ye, H.; Wu, X.; Li, D.; Jiang, Y. Numerical simulation of productivity improvement of natural gas hydrate with various well types: Influence of branch parameters. *J. Nat. Gas Sci. Eng.* **2022**, *103*, 104630. [CrossRef]
19. Hao, Y.; Yang, F.; Wang, J.; Fan, M.; Li, S.; Yang, S.; Wang, C.; Xiao, X. Dynamic analysis of exploitation of different types of multilateral wells of a hydrate reservoir in the South China sea. *Energy Fuels* **2022**, *36*, 6083–6095. [CrossRef]
20. Ye, H.; Wu, X.; Guo, G.; Huang, Q.; Chen, J.; Li, D. Application of the enlarged wellbore diameter to gas production enhancement from natural gas hydrates by complex structure well in the Shenhu sea area. *Energy* **2022**, *264*, 126025. [CrossRef]
21. Cao, X.; Sun, J.; Qin, F.; Ning, F.; Mao, P.; Gu, Y.; Li, Y.; Zhang, H.; Yu, Y.; Wu, N. Numerical analysis on gas production performance by using a multilateral well system at the first offshore hydrate production test site in the Shenhu area. *Energy* **2023**, *270*, 126690. [CrossRef]
22. Jin, G.; Su, Z.; Zhai, H.; Feng, C.; Liu, J.; Peng, Y.; Liu, L. Enhancement of gas production from hydrate reservoir using a novel deployment of multilateral horizontal well. *Energy* **2023**, *270*, 126867. [CrossRef]
23. He, J. Numerical simulation of a Class I gas hydrate reservoir depressurized by a fishbone well. *Processes* **2023**, *11*, 771. [CrossRef]
24. Qin, X.; Lu, J.; Lu, H.; Qiu, H.; Liang, J.; Kang, D.; Zhan, L.; Lu, H.; Kuang, Z. Coexistence of natural gas hydrate, free gas and water in the gas hydrate system in the Shenhu Area, South China Sea. *China Geol.* **2020**, *3*, 210–220. [CrossRef]
25. Zhang, W.; Liang, J.; Lu, J.; Wei, J.; Su, P.; Fang, Y.; Guo, Y.; Yang, S.; Zang, G. Accumulation features and mechanisms of high saturation natural gas hydrate in Shenhu area, northern south China sea. *Pet. Explor. Dev.* **2017**, *44*, 708–719. [CrossRef]
26. Sun, Y.; Ma, X.; Guo, W.; Jia, R.; Li, B. Numerical simulation of the short- and long-term production behavior of the first offshore gas hydrate production test in the South China Sea. *J. Pet. Sci. Eng.* **2019**, *181*, 106196. [CrossRef]

27. Qin, X.; Liang, Q.; Yang, L.; Qiu, H.; Xie, W.; Liang, J.; Lu, J.; Lu, C.; Lu, H.; Ma, B.; et al. The response of temperature and pressure of hydrate reservoirs in the first gas hydrate production test in South China Sea. *Appl. Energy* **2020**, *278*, 115649. [CrossRef]
28. Moridis, G.; Kowalsky, M.; Pruess, K. *TOUGH+ Hydrate V1.0 User's Manual*; Report LBNL-0149E; Lawrence Berkeley National Laboratory: Berkeley, CA, USA, 2008.
29. Sun, J.; Ning, F.; Lei, H.; Gai, X.; Sánchez, M.; Lu, J.; Li, Y.; Liu, L.; Liu, C.; Wu, N.; et al. Wellbore stability analysis during drilling through marine gas hydrate-bearing sediments in Shenhu area: A case study. *J. Pet. Sci. Eng.* **2018**, *170*, 345–367. [CrossRef]
30. Zhu, H.; Xu, T.; Yuan, Y.; Xia, Y.; Xin, X. Numerical investigation of the natural gas hydrate production tests in the Nankai Trough by incorporating sand migration. *Appl. Energy* **2020**, *275*, 115384. [CrossRef]
31. Reagan, M.; Moridis, G.; Johnson, J.; Pan, L.; Freeman, C.; Boyle, K.; Keen, N.; Husebo, J. Field-scale simulation of production from oceanic gas hydrate deposits. *Transp. Porous Media* **2015**, *108*, 151–169. [CrossRef]
32. Zhang, K.; Moridis, G.; Wu, Y.; Pruess, K. A domain decompositionapproach for large-scale simulations of flow processes in hydrate-bearing geologic media. In Proceedings of the 6th International Conference on Gas Hydrates, ICGH 2008, Vancouver, BC, Canada, 6–10 July 2008.
33. Yin, Z.; Chong, Z.R.; Tan, H.K.; Linga, P. Review of gas hydrate dissociation kinetic models for energy recovery. *J. Nat. Gas Sci. Eng.* **2016**, *35*, 1362–1387. [CrossRef]
34. Okwananke, A.; Hassanpouryouzband, A.; Farahani, M.; Yang, J.; Tohidi, B.; Chuvilin, E.; Istomin, V.; Bukhanov, B. Methane recovery from gas hydrate-bearing sediments: An experimental study on the gas permeation characteristics under varying pressure. *J. Pet. Sci. Eng.* **2019**, *180*, 435–444. [CrossRef]
35. Yuan, Y.; Xu, T.; Jin, C.; Zhu, H.; Gong, Y.; Wang, F. Multiphase flow and mechanical behaviors induced by gas production from clayey-silt hydrate reservoirs using horizontal well. *J. Clean. Prod.* **2021**, *328*, 129578. [CrossRef]
36. Yu, T.; Guan, G.; Wang, D.; Song, Y.; Abudula, A. Numerical investigation on the long-term gas production behavior at the 2017 Shenhu methane hydrate production-site. *Appl. Energy* **2021**, *285*, 116466. [CrossRef]
37. Sun, J.; Zhang, L.; Ning, F.; Lei, H.; Liu, T.; Hu, G.; Lu, H.; Lu, J.; Liu, C.; Jiang, G.; et al. Production potential and stability of hydrate-bearing sediments at the site GMGS3-W19 in the South China Sea: A preliminary feasibility study. *Mar. Pet. Geol.* **2017**, *86*, 447–473. [CrossRef]
38. Yuan, Y.; Xu, T.; Xin, X.; Xia, Y. Multiphase Flow Behavior of Layered Methane Hydrate Reservoir Induced by Gas Production. *Geofluids* **2017**, *2017*, 7851031. [CrossRef]
39. Sun, J.; Ning, F.; Li, S.; Zhang, K.; Liu, T.; Zhang, L.; Jiang, G.; Wu, N. Numerical simulation of gas production from hydrate-bearing sediments in the Shenhu area by depressurising: The effect of burden permeability. *J. Unconv. Oil Gas Resour.* **2015**, *12*, 23–33. [CrossRef]
40. Feng, Y.; Chen, L.; Suzuki, A.; Kogawa, T.; Okajima, J.; Komiya, A.; Maruyama, S. Enhancement of gas production from methane hydrate reservoirs by the combination of hydraulic fracturing and depressurization method. *Energy Convers. Manag.* **2019**, *184*, 194–204. [CrossRef]
41. Ma, X.; Sun, Y.; Liu, B.; Guo, W.; Jia, R.; Li, B.; Li, S. Numerical study of depressurization and hot water injection for gas hydrate production in China's first offshore test site. *J. Nat. Gas Sci. Eng.* **2020**, *83*, 103530. [CrossRef]

energies

MDPI

Article

Effect of Entrainment on the Liquid Film Behavior in Pipe Elbows

Zhenqiang Xie [1,2,*] and Xuewen Cao [2]

[1] CNOOC Research Institute, Beijing 100028, China
[2] College of Pipeline and Civil Engineering, China University of Petroleum (East China), Qingdao 266580, China; caoxw@upc.edu.cn
* Correspondence: xiezhq4@cnooc.com.cn

Abstract: Multiphase flow entrainment in natural gas engineering significantly influences the safety and efficiency of oil companies since it affects both the flow and the heat transfer process, but its mechanisms are not fully understood. Additionally, current computational fluid dynamics (CFD) methodologies seldom consider entrainment behavioral changes in pipe elbows. In this article, a verified CFD method is used to study the entrainment behavior, mechanism, and changes in an elbow. The results show that droplet diameter in a developed annular flow follows a negative skewness distribution; as the radial distance (from the wall) increases, the fluctuation in the droplets becomes stronger, and the velocity difference between the gas and the droplets increases linearly. Turbulence bursts and vortices sucking near the wall jointly contribute to droplet entrainment. As the annular flow enters the elbow, the secondary flow promotes the film expansion to the upper and lower parts of the pipe. Droplets re-occur near the elbow exit intrados, and their size is much smaller than those in the upstream pipe. Vortices sucking under low gas velocity play an important role in this process. These findings provide guidelines for safety and flow assurance issues in natural gas production and transportation and bridge the gap between multiphase flow theory and natural gas engineering.

Keywords: entrainment; liquid film; droplet; elbow

Citation: Xie, Z.; Cao, X. Effect of Entrainment on the Liquid Film Behavior in Pipe Elbows. *Energies* **2024**, *17*, 1983. https://doi.org/10.3390/en17081983

Academic Editor: Helena M. Ramos

Received: 9 March 2024
Revised: 8 April 2024
Accepted: 12 April 2024
Published: 22 April 2024

1. Introduction

Currently, gas-dominant multiphase flow (i.e., churn flow and annular flow) has many natural gas engineering applications. Typically, in annular flow, high-speed gas moves through the pipe center while a liquid film flows around the pipe wall [1–3]. In this process, the strong interaction between the gas core and liquid film shears part of the liquid roll-wave crests and leads to droplet entrainment [4]. Entrainment is critical in gas-dominant multiphase flow since it affects the mass and heat exchanges between phases as well as phase distribution and pressure drop. These changes destabilize the flow and make it hard to predict, which negatively affects the safety and flow assurance in natural gas production and transportation. Droplet entrainment characteristics are also important for developing an accurate multiphase flow model [5,6]. However, the entrainment mechanism is not fully understood, and the empirical models available are based on simplified assumptions in specific conditions that are not accurate in many cases.

Some researchers have argued that the interfacial wave instability between two phases affects droplet entrainment [7–9]. Depending on the liquid flow rate, two types of waves exist on the liquid film: ripple waves and disturbance waves. When the liquid flow rate is low, ripple waves (small-scale) occur, which exist for a short time and move slowly. Therefore, droplets cannot be extracted from the liquid film. For high gas and liquid flow rates, the disturbance waves (long-length) become dominant. Since disturbance waves move faster than ripple waves, their amplitude is much higher, and their lifespan is longer [10,11]. Usually, for pipelines with a small diameter, the disturbance waves occur coherently on the circumference [12]. According to the literature, high-speed gas shears off the disturbance wave crest to form droplets; then, the droplets enter the gas core [13,14]. In this process, the gas drag force, as well as gravity, deforms and stretches the crest against the retaining liquid surface tension. As a result, the liquid ligament breaks up, and the droplet enters the gas core. However, the liquid volume detached from the wave crest is not taken into consideration in this mechanism, which makes it impossible to calculate the entrainment rate.

There are many studies on liquid sampling, flow patterns, and factors that influence liquid distribution as well as erosion using both experimental and simulation methods. Some of these are summarized in Table 1.

Table 1. Literature review.

Authors	Year	Summary of Work	Output
Wicks et al. [15]	1960	Isokinetic sampling experimental study on droplet entrainment characteristics	A model to calculate the droplet mass flow rate
Paleev et al. [16]	1966	Liquid film removal experimental study on entrainment characteristics	An empirical correlation to calculate film flow rate
Xu et al. [17]	2013	Water displacement characteristics investigation in inclined pipe	Critical value of oil flow rate to flush out trapped water
Haung et al. [18]	2013	Flow pattern observation in uphill and downhill pipes using high-speed camera	Experimental program to judge flow pattern in hilly-terrain flow
Parsi et al. [19]	2015	Experimental study on churn flow using wire-mesh sensor	Interfacial structures analysis of liquid film
Magnini et al. [20]	2018	Numerical study on effects of oil velocity and wall-wetting conditions on flow pattern	A model for predicting the drift velocity along upward pipes
Farokhipour et al. [21]	2020	Factors affecting elbow erosion and film distribution using simulation	Higher liquid velocity enhances film cushion and weakens erosion
Pshenin et al. [22]	2022	Hydrocarbon evaporation losses during loading of tanker fleets	A model of hydrocarbon vapor displacement from the inner part of a tanker
Korshak et al. [23]	2023	Water slug behavior and removal based on pumping liquid simulation	Slug removal algorithms that agree with experiment
Zhao et al. [24]	2023	Liquid distribution and erosion pattern of elbow under upward annular flow	Wave undercut and roll-wave entrainment mechanism

Regarding the empirical modeling, for vertical flow, Oliemans et al. [25] developed a correlation based on their experimental database that covers a wide range of flow conditions. Later, Ishii and Mishima [26] derived a verified correlation based on the gas Weber number and the liquid Reynolds number. Researchers also studied droplet sizes entrained in the gas core and proposed correlations based on their experimental database. Most of the correlations [27–29] considered the Weber number (the ratio between inertia and surface forces) as a key parameter since it indicates the interaction between the external stress force and the surface forces acting on droplets, an interaction which is important for the droplet breakup mechanism and droplet size. Kolev [27] derived a correlation to calculate the droplet diameter from Kelvin–Helmholtz instability. Fore [30] put forward a modified expression to predict the volume median diameter in vertical and horizontal flows in a pipe. Ishii [31] proposed a semi-empirical correlation to estimate the average maximum droplet size for roll-wave entrainment.

Summarizing the entrainment models above, Wicks's model and Paleev's model are based on experimental data in horizontal annular flow, so they are unlikely to provide a precise prediction for vertical or inclined annular flow. In contrast, Oliemans' and Ishii's models can provide accurate entrainment in vertical annular flow (superficial gas velocities of 15 to 40 m/s, superficial liquid velocities of 0.06 to 0.2 m/s, and pipeline diameter of 6 to 32 mm) since they use a vertical annular flow database. Magrini's experiments [32] show that Paleev's model provides the best prediction for horizontal annular flow. For the entrained droplet size calculation, most of the correlations choose the Weber number as the key factor. Additionally, these semi-empirical correlations are based on experimental data in specific conditions, which limits their applications.

In brief, the entrainment mechanism in annular flow is not fully understood; related models and correlations are empirical and inaccurate, which limits their application in practice. Many details on droplet entrainment are not included in the current CFD methodologies available in the literature, and no studies on how entrainment affects liquid film behavior in the elbow were found.

In this article, a verified CFD method is used to study entrainment behavior in pipelines. Firstly, entrainment characteristics like droplet diameter and velocity distribution are analyzed in detail. Then, the entrainment mechanism is interpreted based on a vortex theory. Lastly, entrainment changes in elbows are studied. These detailed analyses of liquid film and droplet entrainment help to evaluate elbow safety, with elbows being sensitive to erosion as well as corrosion, and solve flow assurance problems in natural gas production and transportation.

2. Theories and Modeling

2.1. Entrainment Modeling

The correlation established by Wicks and Dukler [15] is shown below:

$$R = \frac{q_L}{q_G} \cdot \frac{N_{We_{cr}} \cdot W_{LE}}{(dp/dL)_G} \tag{1}$$

where R is the entrainment group (unit ft^3/lbf-hr); q_L and q_G are the mass flow rates of liquid and gas, respectively; N_{Wecr} is the critical Weber number ranging from 13 to 22; W_{LE} is the droplet mass flow rate; and dp/dL is the pressure gradient.

The correlation developed by Paleev [16] is listed below:

$$F_E = 0.015 + 0.44 log\left[\left(\frac{\rho_C}{\rho_L}\right) \cdot \left(\frac{\mu_L v_{SG}}{\sigma}\right)^2 \times 10^4\right] \tag{2}$$

where ρ_C is the mixture density of the core defined as

$$\rho_C = \rho_G\left[1 + F_E\left(\frac{v_{SL}}{v_{SG}}\right)\left(\frac{\rho_L}{\rho_G}\right)\right] \tag{3}$$

where F_E is the entrainment fraction; ρ_L and ρ_G are the density of the liquid and gas, respectively; μ_L is the dynamic viscosity of the liquid; v_{SL} and v_{SG} are the superficial liquid velocity and superficial gas velocity, respectively; and σ is the surface tension.

Oliemans' model [25] is given below:

$$\frac{F_E}{(1 - F_E)} = 10^{-2.52} \rho_L^{1.08} \rho_G^{1.08} \mu_L^{0.27} \mu_G^{0.28} \sigma^{-1.8} D^{1.72} v_{SL}^{0.7} v_{SG}^{1.44} g^{0.46} \tag{4}$$

Ishii's model [26] is represented below:

$$F_E = tanh\left(7.25 \times 10^{-7} N_{We_{SG}}^{1.25} N_{Re_{SL}}^{0.25}\right) \tag{5}$$

where

$$N_{We_{SG}} = \frac{\rho_G v_{SG}^2 D}{\sigma\left[\frac{(\rho_L - \rho_G)}{\rho_G}\right]^{1/3}} \tag{6}$$

$$N_{Re_{SL}} = \frac{\rho_L v_{SL} D}{\mu_L} \tag{7}$$

2.2. Droplet Size

The critical Weber number is the most widely used criterion to estimate the droplet size; for the condition of high density difference (droplet in gas core), the empirical value of the Weber number is 12. The expression is as follows [27]:

$$\mathrm{We}_{crit} = \frac{\rho_G u_g^2 \varnothing_d}{\sigma} = 12 \tag{8}$$

where We_{crit} is the critical Weber number, u_g is the gas velocity, and $\varnothing_d$ is the droplet diameter.

For annular flow, the volume median diameter of the droplet derived by Fore [30] can be calculated using the following equation:

$$\frac{\varnothing_{vm}}{D} = 0.106 \cdot We_g^{-\frac{1}{2}} Re_g^{\frac{1}{10}} \tag{9}$$

where $\varnothing_{vm}$ is the droplet volume median diameter, and We_g and Re_g are the gas Weber number and the gas Reynolds number, respectively.

$$We_g = \frac{\rho_g v_{SG}^2 D}{\sigma} \tag{10}$$

$$Re_g = \rho_g v_{SG} D / \mu_g \tag{11}$$

This correlation is simple and accurate enough to predict the droplet volume median diameter in horizontal and vertical pipelines.

The semi-empirical correlation to estimate the average maximum droplet size proposed by Ishii [26] is as follows:

$$\varnothing_{max} = 0.031 \frac{\sigma}{\rho_g v_{SG}^2} Re_g^{\frac{2}{3}} \left(\frac{\rho_G}{\rho_L}\right)^{-\frac{1}{3}} \left(\frac{\mu_G}{\mu_L}\right)^{\frac{2}{3}} \tag{12}$$

where $\varnothing_{max}$ is the average maximum droplet size.

3. Experimental Background

The experimental data used for liquid entrainment verification in this study were taken from Magrini [32]. In the experiment, air from the compressor and water from the tank were measured separately and then mixed to develop a horizontal annular flow in a 76.2 mm ID pipeline. Then, the fluid entered the test section after it was fully developed. The test section included two parts: the isokinetic sampling section and the film removal section. The isokinetic sampling section measured the liquid entrainment in the gas, and the film removal section determined the liquid film flow rate. Finally, the fluid flowed out of the system from the outlet.

The isokinetic sampling section included an L-shape isokinetic probe, two valves, a container, and a gas flowmeter. The probe inlet was placed against the gas flow direction so that the droplets in the gas core could enter the probe easily. The probe could be located at different positions of the pipe cross-section. Valves were applied to control the sampling rate and keep the gas velocity in both the probe and the pipeline the same. The sampled droplets were collected in the container, and the sampled gas was vented out through the flowmeter. Then, the container was weighed to calculate the entrainment.

The film removal section includes an inserted porous sleeve that allows the liquid film to pass through, and the cavity between the sleeve and outer shell is used to store the sampled liquid. When passing the film removal section, liquid film passes through the porous section and enters the cavity. The entrained droplets in the pipe center move out directly without being removed. The collected liquid is used to calculate the film flow rate.

In these experiments, the superficial gas velocity ranged from 40 m/s to 80 m/s, and the superficial liquid velocity ranged from 0.0035 m/s to 0.04 m/s. Each test lasted for 5 min. The entrainment flux at each location can be calculated as

$$E_x = \frac{V_E}{A_{probe} t_s} \tag{13}$$

The entrainment fraction can be calculated as

$$F_E = \frac{\int E_x \partial A}{A_{pipe} v_{SL}} \tag{14}$$

where E_X is the entrainment flux, V_E is the collected liquid volume, t_S is the sample time, and A is the cross-sectional area.

4. CFD Simulation

It is difficult to acquire details of the entrainment process since it is random and changes fast. Fortunately, CFD can provide abundant information about the whole domain at any specific time. For liquid entrainment simulation, a volume of fluid (VOF) model based on the Euler method is adopted, in which the interface between immiscible fluids can be obtained. The main steps in the liquid entrainment simulation are as follows: (1) solve the single set of momentum equations shared by all the phases; (2) track the volume fraction in each computational cell through the domain; (3) solve the volume fraction equation through implicit time discretization. A flow chart representing the methodology is shown in Figure 1.

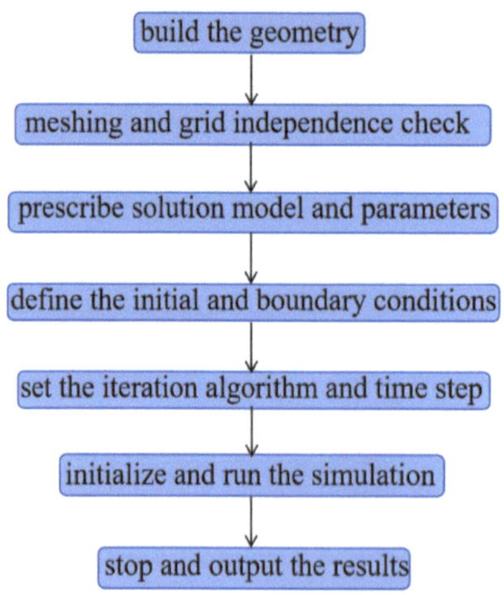

Figure 1. Flow chart for the methodology.

4.1. Geometry

The geometry used in this investigation is shown in Figure 2. The upstream length, L, is $10D$, the curvature radius of the elbow is $1.5D$, and the downstream length is $3D$.

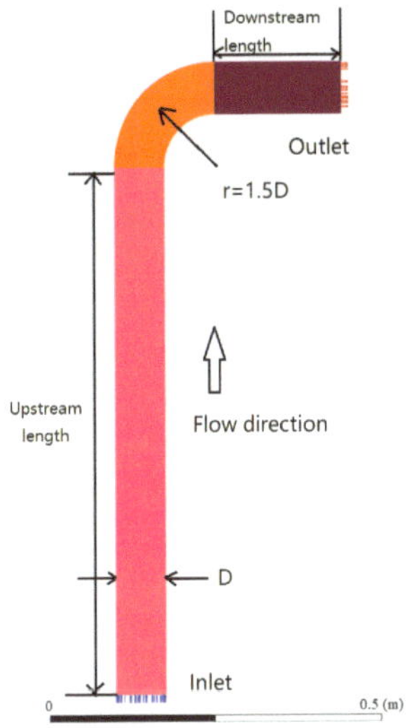

Figure 2. Geometry configuration of the computational domain.

4.2. Meshing

A high-quality mesh is critical for a successful simulation. Since the entrainment process occurs at the interface of the two phases near the wall, elements in this region were refined to obtain the details. Figure 3 provides the overview and cross-section details of the mesh.

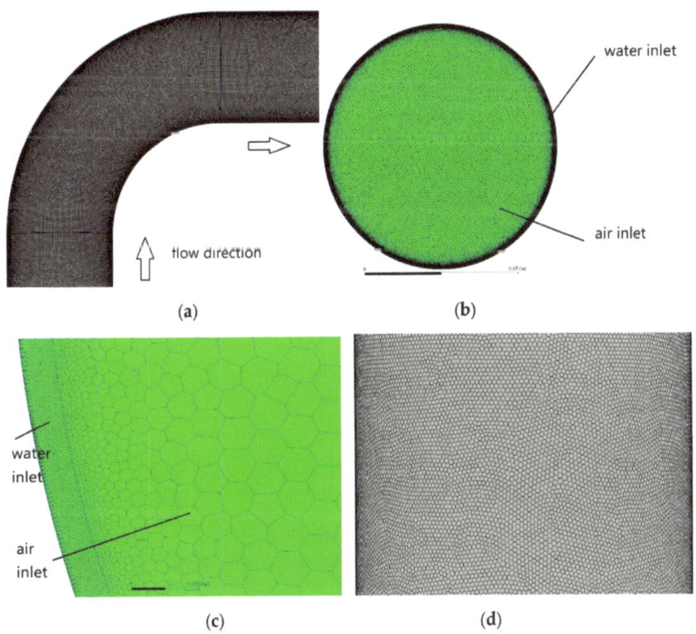

Figure 3. Mesh configuration of the domain: (**a**) overview, (**b**) inlet section, (**c**) near wall, (**d**) outer wall).

It can be seen that the domain is meshed by polygonal cells. Polygonal cells can reduce the skewness, improve the parallel solver load balancing, and minimize communication by decreasing the number of partition interface cells [33]. Figure 3c shows an enlarged view of the near-wall region, where the first-layer thickness of the mesh is 20 μm and the maximum cell size is 1 mm, the growth ratio is 1.2, and the row number of the boundary is 5. The length of the cells in both circumferential and axial directions is 1 mm, and the total number of elements is 11,338,564. Since the maximum cell size is in the same order as the droplet size, this mesh can reveal detailed information during the entrainment process.

4.3. Simulation Settings and Boundary Conditions

In this study, the annular flow simulation is performed first to analyze the entrainment process in the pipeline. The settings of the simulation are shown in Table 2.

Table 2. Settings of the flow simulation.

Items	Settings
Multiphase model	Volume of fluid
Turbulence model	k-e Realizable
Pressure–velocity coupling	SIMPLE
Interface modeling	Sharp
Spatial discretization	Second-order upwind
Primary phase	Air
Secondary phase	Water
Phase interaction	Surface tension force modeling
Surface tension coefficient	0.072 N/m
Near-wall treatment	Enhanced wall treatment

The boundary conditions of the flow simulation are shown in Table 2. Water (liquid phase) enters the pipe circumferentially along the pipe wall, and the water entrance is ring-shaped. Air (gas phase) enters the pipe in the center of the cross-section, and the shape of the air entrance is a circle. The superficial gas velocity is 24.6 m/s, and the superficial liquid velocity is 0.1 m/s; the initial values of the corresponding air and water mass flow rates are listed in Table 3, and the time step is 0.000005 s.

Table 3. Boundary conditions of the flow simulation.

Items	Settings
Air inlet (mass flow inlet)	0.145 kg/s
Water inlet (mass flow inlet)	0.456 kg/s
Outlet (pressure outlet)	Atmosphere
Turbulence intensity	5%
Turbulence viscosity ratio	10
Wall	No slip
Wall contact angle	90°

Once the mass flow rate difference between the inlet and outlet decreases below 0.01 kg/s and the flow state in the pipe becomes steady, the simulation can be stopped.

5. Results and Discussion

5.1. Entrainment Fraction Validation

In this section, experimental data from the literature [32] is adopted to verify the entrainment fraction in a CFD simulation. The working conditions in both experiments and simulations are the same. The superficial gas and liquid velocities are 40 m/s and 0.01 m/s, respectively. The entrainment fractions measured by isokinetic sampling and film removal methods in a vertical orientation are 0.42 and 0.51, respectively.

Based on Equations (18)–(20), the entrainment fractions calculated by Oliemans' and Ishii's models are 0.53 and 0.55, respectively. As for the simulation, at the position $1D$ before the elbow entrance, the area-weighted average void fraction of the gas core and pipe cross-section are extracted and converted to the average entrainment fraction using the following equations:

$$A_d = h_{gc} A_c \tag{15}$$

$$A_f = h_P A_P - A_d \tag{16}$$

$$F_{ECFD} = A_d / \left(A_d + A_f \right) \tag{17}$$

where A_d and A_f are the area of droplets and liquid film in the cross-section, respectively; A_c is the area of the gas core; h_{gc} and h_P are the liquid holdup of the gas core and pipe cross-section, respectively; and F_{ECFD} is the entrainment fraction calculated using CFD. The entrainment fractions calculated above use Equations (4)–(7) and are summarized in Table 4.

Table 4. Entrainment fraction obtained using different methods.

$N_{Re_{SL}}$	$N_{We_{SG}}$	Experimental Data	Oliemans' Model	Ishii's Model	CFD Data
1520	2376	0.4	0.53	0.55	0.47

As shown in Table 4, the entrainment fraction measured using isokinetic sampling is smaller than that of the film removal method. Droplets adhering to the small diameter tube surface may cause this difference. As for the film removal method, the liquid collected may contain the redeposited droplets, which makes the measured value larger. Thus, the real value should be between 0.42 and 0.51. Both Oliemans' model and Ishii's model give larger values than the experiments, indicating that the two models are not accurate enough. The entrainment fraction predicted using CFD is 0.47, just within the range (0.42 to 0.51), so the entrainment fraction predicted with CFD is reasonable.

5.2. Entrainment Behavior in the Pipeline

After the flow simulation stops, the phase contours as well as the velocity vectors of the pipeline longitudinal section and representative cross-sections are extracted to analyze the entrainment behavior in the pipeline.

5.2.1. Entrainment Distribution

Figure 4 shows the phase distribution and velocity vector in the pipeline. It can be seen that as the upstream length increases, the gas in the pipe center moves much faster than that near the wall, the gas velocity fluctuation in the pipe becomes obvious, and the uniform liquid film starts waving and then breaks up into droplets (see Figure 4b). Then, the droplets are gradually entrained into the pipe center by the high-speed gas. When the upstream length reaches $6D$, the droplet distribution seldom changes, and the entrainment process reaches dynamic equilibrium. After entering the elbow, most of the droplets hit the elbow extrados and coalesce into liquid film. Then, the film flows along the extrados side to the downstream pipe (see Figure 4c).

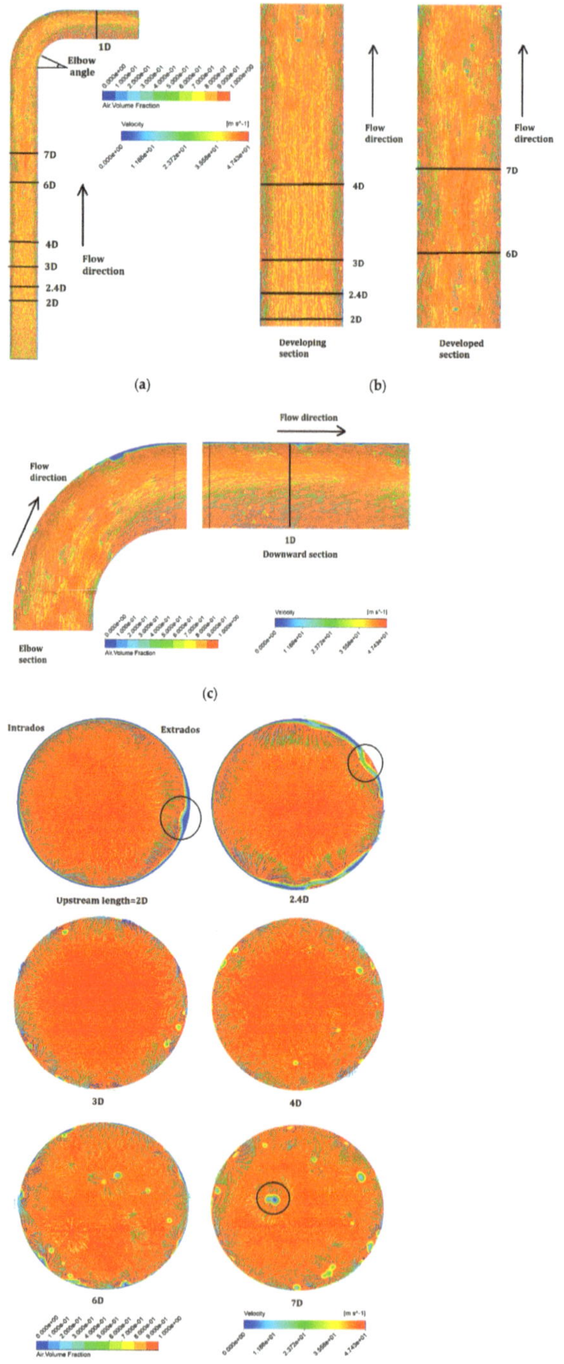

Figure 4. Phase distribution and velocity vector of the pipeline: (**a**) overview; (**b**) enlarged view of the upstream; (**c**) enlarged view of elbow and downstream; (**d**) representative cross-sections.

Representative cross-sections of the upstream pipe are extracted to reveal more information (see Figure 4d). At an upstream length of 2*D*, there is a film wave at the right side of the cross-section (circled). The corresponding velocity vectors indicate that a turbulence burst here contributes to the wave. When the upstream length reaches 2.4*D*, the film becomes discontinuous, and "liquid bridges" (circled) occur, which are also related to the turbulence bursts near the wall. As the flow develops, turbulence bursts become stronger and break the liquid bridges into droplets, and then, the droplets are entrained into the pipe center by the gas. At an upstream length beyond 6*D*, the droplets are distributed uniformly in the pipe cross-section. Droplet coalescence (circled) is also found in the flow at an upstream length of 7*D*.

Other researchers like Zahedi [34] and Farokhipour [21] also studied liquid distribution in annular flow pipelines. Zahedi [34] only reported the liquid distribution in the middle of the elbow (elbow angle = 45°), and no liquid was found at the elbow extrados; however, this seems inaccurate, because droplets carried by the gas core will hit the extrados and form a liquid film. Farokhipour's simulation [21] showed the liquid film distribution in the elbow well, but no droplets can be observed in the figures. Compared with their studies, the simulation in this study provides more details of droplet distribution, like droplet coalescence, which can be seen as a remedy for entrainment simulation. Another difference is the grid. Other researchers use hexahedral cells in their simulations, but polygonal cells are adopted in this study to reduce skewness to obtain more accurate results. Additionally, droplet distribution and entrainment changes in the elbow are also investigated, which have not featured in other simulations.

5.2.2. Entrainment Characteristics

Further, the void fraction (area-weighted average) of the cross-sections (every 0.25*D*) in the upstream section is extracted and summarized in Figure 5.

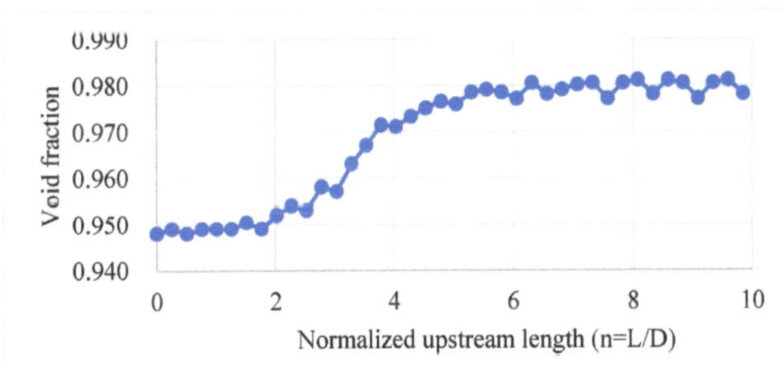

Figure 5. Relationship between upstream length and area-weighted average void fraction of the pipe cross-sections.

It can be seen that for the upstream length of 2*D*, the void fraction is steady but fluctuates slightly around 0.95. As the upstream length increases to 6*D*, the void fraction rises to 0.978. Meanwhile, its fluctuation becomes stronger, and droplets start to take shape and enter the gas core. Then, after 6*D*, the void fraction becomes steady again, with its value fluctuating more vigorously between 0.978 and 0.981. It seems that for the annular liquid inlet, the liquid droplet entrainment and re-deposition process reach dynamic equilibrium when the upstream length is beyond 6*D*.

According to Ishii and Kataoka [5,35], for the case where the liquid is injected smoothly as a film at the inlet, the entrained fraction expression takes the form of exponential relaxation. For the case in this study, the trend of the curve in Figure 5 is also exponential relaxation, which means the CFD results here are reasonable.

Then, the droplet information (droplet number, droplet diameter, droplet velocity, and droplet position (radial distance from the wall)) in the pipeline's longitudinal section is collected to further analyze the entrainment. Droplet diameters and positions are measured by the scale of the phase distribution contour; droplet velocity is determined by the velocity vector in the center of the droplet.

Figure 6 shows a histogram of the relationship between droplet diameter and droplet number distribution in the longitudinal section in the developed annular flow. It can be seen that the droplet number roughly follows a negative skew distribution. Most droplet diameters are distributed in a 1 mm to 2.5 mm range, and the droplet number in the range of 2 mm to 2.5 mm is the largest, while in the range of 0 to 0.5 mm, it is the smallest.

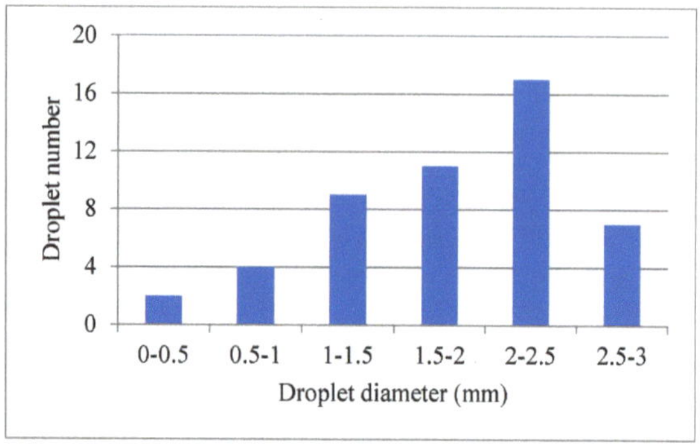

Figure 6. Relationship between droplet diameter and droplet number distribution.

In contrast to one study [36], for the conditions of low-pressure (below 3 atm) and low-viscosity fluid (water), the droplet diameter distribution in this study is similar to the upper limit log-normal (ULLN) distribution given by Mugele and Evans [37]. The maximum size of the droplet can be calculated based on the Kelvin–Helmholtz instability and critical Weber number.

$$\varnothing_{dmax} = \frac{We_{crit}\sigma}{\rho_G(v_G - v_L)^2} \tag{18}$$

where $\varnothing_{dmax}$ is the maximum diameter of the droplet, and We_{crit} is the critical Weber number, which equals 12. For this study, the maximum diameter of the droplet calculated using this equation is 2.7 mm, which is represented by the last bar in Figure 6.

For this study, the calculated volume median diameter of the droplet is 1.1 mm, while the volume median droplet diameter predicted using CFD simulation is 1.7 mm (54.5% higher than the calculated value). Further improvement may be achieved by refining the mesh, which is limited by the current computational resources.

Figure 7 shows the relationship between the radial distance (from the wall) and the droplet diameter in a developed annular flow. As shown in this figure, the average value of all the droplet diameters is 1.81 mm. Droplet diameter fluctuates around the average value, and its amplitude increases with the radial distance.

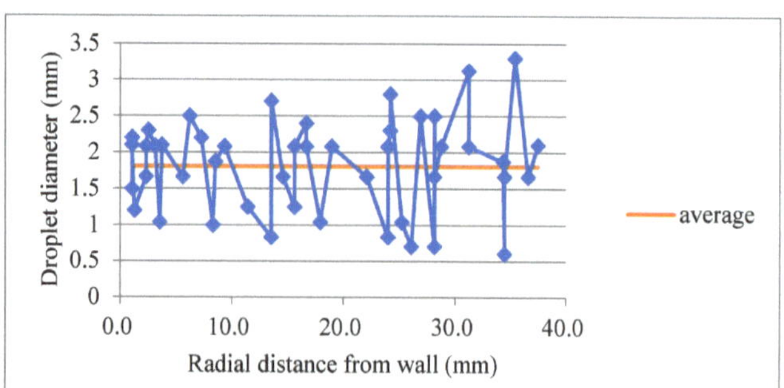

Figure 7. Relationship between radial distance (from the wall) and droplet diameter.

Table 5 presents the variance distribution of droplet diameters. It can be seen that as the radial distance (from the wall) increases, the expectation of the droplet diameter varies very slightly around 1.81 mm, by less than ±0.05 mm. However, for the variance, it increases from 0.19 to 0.63. This means that the change in droplet diameter becomes increasingly unstable the closer the droplet is to the gas core. This may be because droplets closer to the pipe center are more likely to be broken up or coalesced by the gas core turbulence, which strengthens the fluctuation.

Table 5. Variance distribution of droplet diameters.

Radial Distance from Wall (mm)	Expectation (mm)	Variance
0~10	1.85	0.19
10~20	1.76	0.36
20~30	1.76	0.51
30~40	1.86	0.63

According to Kolev [27], the droplet breakup mechanism can be expressed as a balance between the external stress force and the surface force. During the breakup, external stress force tries to disrupt the droplets, while surface tension force tries to avoid droplet deformation. From the critical Weber number, Equation (11), it can be seen that the droplet diameter varies with the inverse square of the gas velocity. As the radial distance from the wall increases, the gas velocity becomes higher, so the minimum value of droplet diameter decreases. On the other hand, according to the Kelvin–Helmholtz instability, as the gas velocity increases, the gas–liquid interface becomes more unstable. Additionally, droplet diffusion becomes stronger, so droplets are more likely to coalesce, which increases the maximum droplet diameter.

Figure 8 shows velocity distribution (both gas and droplet) according to radial distance (from the wall). It can be seen that both gas and droplet velocities rise and fluctuate as radial distance increases, while the velocity difference between gas and droplets also increases, which means that the slippage becomes more obvious. Moreover, the fluctuations in both gas and droplets become stronger.

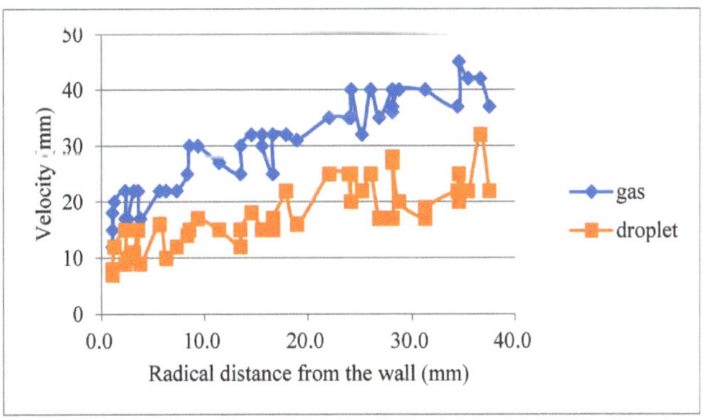

Figure 8. Velocity distribution according to radial distance (from the wall).

In order to clearly express the droplet slippage, the droplet slippage ratio can be calculated with the following equation.

$$S_d = \frac{u_g - u_d}{u_g} \times 100\% \tag{19}$$

where S_d is the droplet slippage ratio, and u_d is the droplet velocity. It can be found that the slippage ratio near the wall (27.3%) is smaller than that in the pipeline center (33.8%).

5.2.3. Entrainment Mechanism

In this subsection, the mechanism of the entrainment process in annular flow is interpreted gradually, as shown in Figure 9. In the initial state (see Figure 9a), the liquid film and gas are uniform and there is no fluctuation. Since there is a velocity difference between gas and liquid film, vortices occur at the two-phase interface under the effect of shear and friction, which produce waves in the film and spread downward (see Figure 9b). As the flow develops, the crest stretched by the vortices becomes higher, and its bottom is necking under the effect of surface tension (see Figure 9c,d). Then, the necking breaks, the liquid entering the gas becomes droplets, and the rest of the liquid becomes film (see Figure 9e). The high-speed gas moving in the pipe center lowers the dynamic pressure here, so the droplets can be entrained under the pressure difference (see Figure 9f). This process reduces the film volume and makes the film discontinuous.

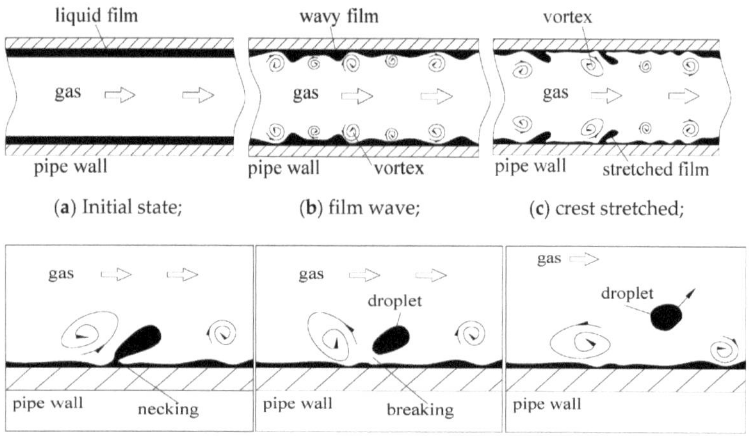

(**a**) Initial state; (**b**) film wave; (**c**) crest stretched;

(**d**) bottom necking; (**e**) bottom breaking; (**f**) droplet entrainment.

Figure 9. Mechanism of the entrainment process in annular flow.

The Kelvin–Helmholtz instability is mainly associated with flows that have tangential variation in the velocity field [38–40]. This instability is caused by the hydrodynamic amplification of perturbations that arise at the gas–liquid interface with a discontinuity in the velocity field. The entrained droplet size is approximately equal to the height of the most unstable wavelength.

$$\Delta h_{w,K-H} = 3\pi \frac{\left(1 + \frac{\rho_G}{\rho_L}\right)\sigma}{\rho_G\left(u_g - u_d\right)^2} \tag{20}$$

$$\phi_d \approx 1.5\Delta h_{w,K-H} \tag{21}$$

where $\Delta h_{w,K-H}$ is the height of the wavelength. For this study, the calculated droplet diameter is 1.8 mm, which is close to the average droplet diameter in Figure 6. This indicates that the simulation results in this study fit the Kelvin–Helmholtz instability theory and are reasonable.

As the flow develops downstream, the entrained droplets travel to the wall and are deposited into the liquid film again. As the entrainment and deposition reach dynamic equilibrium, the flow is fully developed. The most important factor that affects droplet entrainment is the vortices near the gas–liquid interface.

5.3. Entrainment Changes in the Elbow

The entrainment changes when the fully developed flow enters the elbow. Figure 10 shows the phase distribution and velocity vectors of the flow in the elbow: the flow is restricted by the geometry, and the gas moves toward the elbow intrados as the flow first enters the elbow. Droplets are uniformly distributed in the pipe cross-section (elbow angle = 0°).

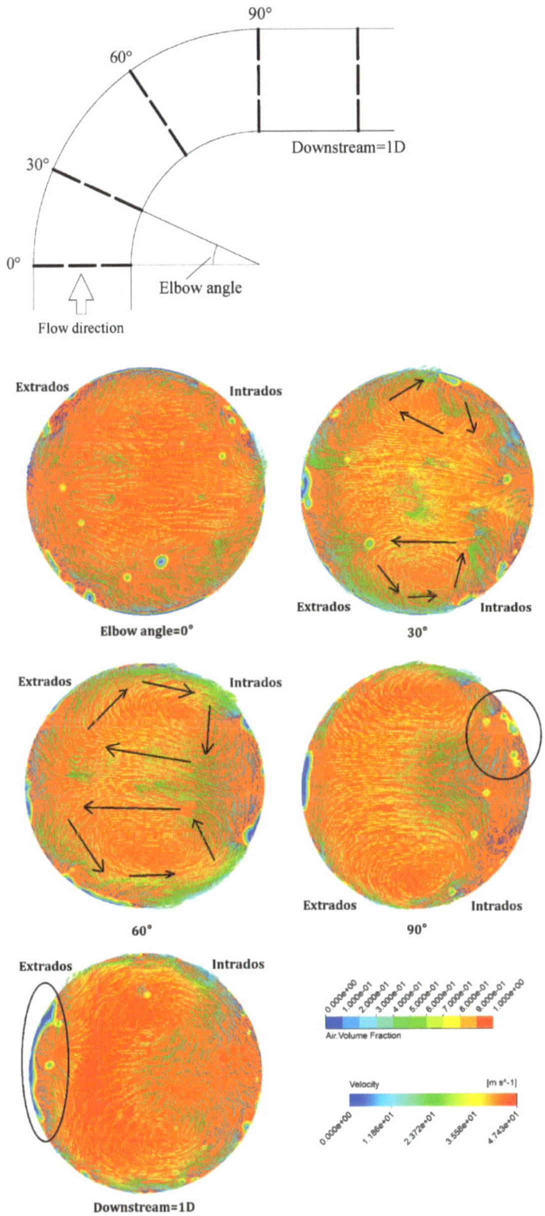

Figure 10. Phases and velocity distribution for different elbow angles.

When the flow reaches an elbow angle of 30°, more droplets are deposited on the extrados, and the liquid film starts to take shape. The flow near the wall moves toward the intrados and forms the secondary flow. For an elbow angle of 60°, most droplets are deposited on the extrados, and droplets in the gas become less frequent, while the film becomes thicker. Two counter-rotated vortices drive the film toward the extrados. For an elbow angle of 90°, the film at the extrados thickens. Droplets occur again at the intrados under the strong effect of the vortices. After entering the downstream pipe, the film thickness at the extrados increases, and the vortices become weaker.

Unlike the roll-wave mechanism proposed by Ishii [26], which is reasonable for high gas velocity in a straight pipe, droplets re-entrained at the intrados seem to be caused by the mechanism of vortices sucking, since the gas velocity here is relatively low. As shown in Figure 4c, it is also found that the re-entrained droplet size at the intrados is smaller (1 mm–1.5 mm). This may be because vortices at the intrados provide more sucking and external stress force on the droplets for their breakup.

6. Conclusions

In this study, a verified CFD method is used to investigate entrainment behavior in a pipeline. Firstly, entrainment characteristics like droplet diameter and velocity distribution are analyzed in detail. Then, the entrainment mechanism is interpreted based on a vortex theory. Lastly, entrainment changes in the elbow are studied. The results are valuable and provide guidance for entrainment study in multiphase flow. The main conclusions are summarized below:

(1) Under the annular inlet condition, the entrainment distribution along the upstream developing length from the entrance is close to a form of exponential relaxation. The droplet diameter in a developed annular flow follows a negative skewness distribution, with most droplet diameters ranging from 1.5 mm to 2 mm. As the radial distance (from the wall) increases, the fluctuation of the droplets becomes stronger, and the velocity difference between gas and droplets increases linearly.

(2) Turbulence bursts and vortices near the wall jointly deform the liquid film and crest so that they begin to take shape; then, the external stress force acting on the liquid crest overcomes the surface force, and droplets form. Finally, vortices sucking transforms the droplets into gas, and the entrained droplet size is approximately equal to the height of the most unstable wavelength.

(3) As the annular flow enters the elbow, droplets hit the extrados and form a liquid film, the droplet number in gas decreases, and at an elbow angle of 60°, there are no droplets in the pipe. Secondary flow promotes the film formation and expansion of the film to the upper and lower parts of the pipe. However, for an elbow angle of 90°, droplets re-occur near the elbow intrados, and the re-entrained droplet size is much smaller than that in the straight upstream pipe. Vortices sucking at low gas velocity play an important role in this process.

Unique work description

In this study, a more accurate simulation using polygonal cells instead of hexahedral cells was carried out to investigate details of droplet distribution and entrainment changes in the elbow, which is unique and different from other simulations.

Recommendations and future work

Although polygonal cells can provide more accurate results, they require more time and more computational resources, especially for finer grids. Grid optimization is a promising topic. In the future, AI technology like cloud computing and large data can be integrated into CFD simulations to enhance efficiency and consume fewer resources. Combined with pipeline-monitoring technologies, CFD simulation can make the detection system's response faster and smarter.

Author Contributions: Z.X.: conceptualization, formal analysis, investigation, writing—original draft. X.C.: conceptualization, supervision. All authors have read and agreed to the published version of the manuscript.

Funding: This research was supported by the National Natural Science Foundation of China (Funding number: 51874340, Funder: Xuewen Cao).

Institutional Review Board Statement: This article does not contain any studies with human participants or animals performed by any of the authors.

Informed Consent Statement: Informed consent was obtained from all the individual participants included in the study.

Data Availability Statement: The original contributions presented in the study are included in the article, further inquiries can be directed to the corresponding authors.

Acknowledgments: The authors would like to thank Jun Zhang and the help from E/CRC at the University of Tulsa.

Conflicts of Interest: Author Zhenqiang Xie was employed by the CNOOC Research Institute, and author Xuewen Cao was employed by China University of Petroleum (East China). The remaining authors declare that the research was conducted in the absence of any commercial or financial relationships that could be construed as a potential conflict of interest.

Nomenclature

Symbol	Description	Unit
A	Cross sectional area	m^2
A_c	Area of the gas core	m^2
A_d	Area of droplets on the cross-section	m^2
A_f	Area of liquid film on the cross-section	m^2
D	Pipe diameter	m
dp/dL	Pressure gradient	MPa/m
E_X	Entrainment flux	m/s
F_E	Entrainment fraction	/
F_{ECFD}	Entrainment fraction calculated by simulation	/
g	Gravitational acceleration	m/s^2
h_{gc}	Liquid holdup of gas core	/
h_P	Liquid holdup of pipe cross-section	/
$\Delta h_{w,K-H}$	Height of the wavelength	m
L	Distance from the pipe inlet	m
N_{Wecr}	Critical Weber number	/
q_L	Liquid mass flow rate	kg/h
q_G	Gas mass flow rate	kg/h
R	Entrainment group	ft^3/lbf-hr
Re_g	Gas Reynolds number	/
Re_{LF}	Reynolds number of the liquid film	/
S_d	Droplet slippage ratio	/
t_S	Sample time	s
u_g	Gas velocity	m/s
u_d	Droplet velocity	m/s
V_E	Collected liquid volume	m^3
We_{crit}	Critical Weber number	/
We_g	Gas Weber number	/
W_{LE}	Droplet mass flow rate	kg/h
ρ_C	Mixture density of the core	kg/m^3
ρ_G	Gas density	kg/m^3
ρ_L	Liquid density	kg/m^3
μ_L	Dynamic viscosity of liquid	Pa·s
μ_G	Dynamic viscosity of gas	Pa·s
v_{SL}	Superficial liquid velocity	m/s
v_{SG}	Superficial gas velocity	m/s

σ	Surface tension	N·m
$\varnothing_d$	Droplet diameter	m
$\varnothing_{max}$	Average maximum droplet size	m
$\varnothing_{vm}$	Droplet volume median diameter	m

References

1. Mishima, K.; Ishii, M. Flow regime transition criteria for upward two-phase flow in vertical tubes. *Int. J. Heat Mass Transf.* **1984**, *27*, 723–737.
2. Wu, B.; Firouzi, M.; Mitchell, T.; Rufford, T.E.; Leonardi, C.; Towler, B. A critical review of flow maps for gas-liquid flows in vertical pipes and annuli. *Chem. Eng. J.* **2017**, *326*, 350–377. [CrossRef]
3. Dasgupta, A.; Chandraker, D.K.; Kshirasagar, S.; Reddy, B.R.; Rajalakshmi, R.; Nayak, A.K.; Walker, S.P.; Vijayan, P.K.; Hewitt, G.F. Experimental investigation on dominant waves in upward air-water two-phase flow in churn and annular regime. *Exp. Therm. Fluid Sci.* **2017**, *81*, 147–163. [CrossRef]
4. Ryu, S.H.; Park, G.C. A droplet entrainment model based on the force balance of an interfacial wave in two-phase annular flow. *Nucl. Eng. Des.* **2011**, *241*, 3890–3897. [CrossRef]
5. Kataoka, I.; Ishii, M.; Mishima, K. Generation and size distribution of droplet in annular two-phase flow. *J. Fluids Eng.* **1983**, *105*, 230–238. [CrossRef]
6. Liu, L.; Bai, B. Generalization of droplet entrainment rate correlation for annular flow considering disturbance wave properties. *Chem. Eng. Sci.* **2017**, *164*, 279–291. [CrossRef]
7. Berna, C.; Escriva, A.; Munoz-Cobo, J.L.; Herranz, L.E. Review of droplet entrainment in annular flow: Interfacial waves and onset of entrainment. *Prog. Nucl. Energy* **2014**, *74*, 14–43. [CrossRef]
8. Pan, L.; Hanratty, T.J. Correlation of entrainment for annular flow in horizontal pipes. *Int. J. Multiph. Flow* **2002**, *32*, 385–408. [CrossRef]
9. Okawa, T.; Kotani, A.; Kataoka, I.; Natio, M. Prediction of critical heat flux in annular flow using a film flow model. *J. Nucl. Sci. Technol.* **2003**, *40*, 388–396. [CrossRef]
10. Alekseenko, S.V.; Cherdantsev, A.V.; Cherdantsev, M.V.; Isaenkov, S.V.; Markovich, D.M. Study of formation and development of disturbance waves in annular gas–liquid flow. *Int. J. Multiph. Flow* **2015**, *77*, 65–75. [CrossRef]
11. Schubring, D.; Shedd, T. Wave behavior in horizontal annular air-water flow. *Int. J. Multiph. Flow* **2008**, *34*, 636–646. [CrossRef]
12. Belt, R.J.; Westende, J.M.C.V.; Prasser, H.M.; Portela, L.M. Time and spatially resolved measurements of interfacial waves in vertical annular flow. *Int. J. Multiph. Flow* **2010**, *36*, 570–587. [CrossRef]
13. Asali, J.C.; Hanratty, T.J. Ripples generated on a liquid film at high gas velocities. *Int. J. Multiph. Flow* **1993**, *19*, 229–243. [CrossRef]
14. Holowach, M.J.; Hochreiter, L.E.; Cheung, F.B. A model for droplet entrainment in heated annular flow. *Int. J. Heat Fluid Flow* **2002**, *23*, 807–822. [CrossRef]
15. Wicks, M.; Duckler, A.E. Entrainment and pressure drop in concurrent gas-liquid flow: I. Air-water in horizontal flow. *AIChE J.* **1960**, *6*, 463–468. [CrossRef]
16. Paleev, I.I.; Filippovich, B.S. Phenomena of liquid transfer in two-phase dispersed annular flow. *Int. J. Heat Mass Tran.* **1966**, *9*, 1089–1093. [CrossRef]
17. Xu, G.; Cai, L.; Ullmann, A.; Brauner, N. Trapped water flushed by flowing oil in upward-inclined oil pipelines. In *Volume 1: Upstream Pipelines; Project Management; Design and Construction; Environment; Facilities Integrity Management; Operations and Maintenance; Pipeline Automation and Measurement, Proceedings of the International Pipeline Conference, Calgary, AB, Canada, 24 September 2012*; American Society of Mechanical Engineers: New York, NY, USA, 2012; pp. 637–647.
18. Huang, S.; Zhang, B.; Lu, J.; Wang, D. Study on flow pattern maps in hilly-terrain air-water-oil three-phase flows. *Exp Therm. Fluid Sci.* **2013**, *47*, 158–171. [CrossRef]
19. Parsi, M.; Vieira, R.E.; Torres, C.F.; Kesana, N.R.; McLaury, B.S.; Shirazi, S.A. Experimental investigation of interfacial structures within churn flow using a dual wire-mesh sensor. *Int. J. Multiph. Flow* **2015**, *73*, 155–170. [CrossRef]
20. Magnini, M.; Ullmann, A.; Brauner, N.; Thome, J.R. Numerical study of water displacement from the elbow of an inclined oil pipeline. *J. Petrol. Sci. Eng.* **2018**, *166*, 1000–1017. [CrossRef]
21. Farokhipour, A.; Mansoori, Z.; Saffar-Avval, M.; Ahmadi, G. 3D computational modeling of sand erosion in gas-liquid-particle multiphase annular flows in bends. *Wear* **2020**, *450–451*, 203241. [CrossRef]
22. Pshenin, V.V.; Zakirova, G.S. Improving the efficiency of oil vapor recovery units in the commodity transport operations at oil terminals. *J. Min. Inst.* **2023**, *265*, 121–128. [CrossRef]
23. Korshak, A.A.; Pshenin, V.V. Modeling of water slug removal from oil pipelines by methods of computational fluid dynamics. *Neft. Khozyaystvo Oil Ind.* **2023**, *10*, 117–122. (In Russian) [CrossRef]
24. Zhao, X.Y.; Cao, X.W.; Cao, H.G.; Zhang, J.; Zhang, J.N.; Peng, W.S.; Bian, J. Numerical study of elbow erosion due to sand particles under annular flow considering liquid entrainment. *Particuology* **2023**, *76*, 122–139. [CrossRef]
25. Oliemans, R.V.A.; Pots, B.F.M.; Trompe, N. Modelling of annular dispersed two-phase flow in vertical pipes. *Int. J. Multiph. Flow* **1986**, *12*, 711–732. [CrossRef]
26. Ishii, M.; Mishima, K. Droplet entrainment correlation in annular two-phase flow. *Int. J. Heat Fluid Flow* **1989**, *32*, 1835–1846. [CrossRef]

27. Kolev, N.I. *Multiphase Flow Dynamics. 2: Thermal and Mechanical Interactions*; Springer Science & Business Media: Berlin/Heidelberg, Germany, 2005.
28. Azzopardi, B.J. Drops in annular two-phase flow. *Int. J. Heat Multiph. Flow* **1997**, *23*, 1–53. [CrossRef]
29. Wierzba, A. Deformation and breakup of liquid drops in a gas stream at nearly critical Weber numbers. *Exp. Fluids* **1990**, *9*, 59–64. [CrossRef]
30. Fore, L.B.; Ibrahim, B.B.; Beus, S.G. Visual measurements of droplets size in gas-liquid annular flow. *Int. J. Multiph. Flow* **2002**, *28*, 1895–1910. [CrossRef]
31. Ishii, M.; Grolmes, M.A. Inception criteria foe droplet entrainment in two-phase concurrent film flow. *AIChE J.* **1975**, *21*, 308–318. [CrossRef]
32. Magrini, K.L.; Sarica, C.; Al-Sarkhi, A.; Zhang, H.Q. Liquid entrainment in annular gas/liquid flow in inclined pipes. *SPE J.* **2012**, *17*, 617–630. [CrossRef]
33. *Fluent Theory Guide, Release 19.2*; ANSYS Inc.: Canonsburg, PA, USA, 2018.
34. Zahedi, P.; Jun, Z.; Hadi, A.; McLaury, B.S.; Shirazi, S.A. CFD simulation of multiphase flows and erosion predictions under annular flow and low liquid loading conditions. *Wear* **2017**, *376–377*, 1260–1270. [CrossRef]
35. Kataoka, I.; Ishii, M.; Nakayama, A. Entrainment and deposition rates of droplets in annular two-phase flow. *Int. J. Heat Mass Transf.* **2000**, *43*, 1573–1589. [CrossRef]
36. Berna, C.; Escriva, A.; Munoz-Cobo, J.L.; Herranz, L.E. Review of droplet entrainment in annular flow: Characterization of the entrained droplets. *Prog. Nucl. Energy* **2015**, *79*, 64–86. [CrossRef]
37. Mugele, R.A.; Evans, H.D. Droplet size distribution in sprays. *Ind. Eng. Chem.* **1951**, *43*, 1317–1324. [CrossRef]
38. Chandrasekhar, S. *Hydrodynamic and Hydromagnetic Stability*; Oxford University Press: Oxford, UK, 1981.
39. Henstock, W.H.; Hanratty, T.J. The interfacial drag and the height of the wall layer in annular flows. *AIChE J.* **1976**, *22*, 990–1000. [CrossRef]
40. Vieira, R.E.; Parsi, M.; Zahedi, P.; McLaury, B.S.; Shirazi, S.A. Electrical resistance probe measurements of solid particle erosion in multiphase annular flow. *Wear* **2017**, *382*, 15–28. [CrossRef]

Journal of Marine Science and Engineering

Review

A Review on Submarine Geological Risks and Secondary Disaster Issues during Natural Gas Hydrate Depressurization Production

Xianzhuang Ma [1,2], Yujing Jiang [1,2,*], Peng Yan [2,*], Hengjie Luan [2], Changsheng Wang [2], Qinglin Shan [2] and Xianzhen Cheng [2]

[1] Graduate School of Engineering, Nagasaki University, Nagasaki 852-8521, Japan; mxianzhuang@126.com
[2] State Key Laboratory of Mining Disaster Prevention and Control Co-Founded by Shandong Province and the Ministry of Science and Technology, Shandong University of Science and Technology, Qingdao 266590, China; luanjie0330@126.com (H.L.); cswang0635@163.com (C.W.); shanqinglin2000@163.com (Q.S.); chengxianzhen@sdust.edu.cn (X.C.)
* Correspondence: jiang@nagasaki-u.ac.jp (Y.J.); yanpeng951202@163.com (P.Y.)

Abstract: The safe and efficient production of marine natural gas hydrates faces the challenges of seabed geological risk issues. Geological risk issues can be categorized from weak to strong threats in four aspects: sand production, wellbore instability, seafloor subsidence, and submarine landslides, with the potential risk of natural gas leakage, and the geological risk problems that can cause secondary disasters dominated by gas eruptions and seawater intrusion. If the gas in a reservoir is not discharged in a smooth and timely manner during production, it can build up inside the formation to form super pore pressure leading to a sudden gas eruption when the overburden is damaged. There is a high risk of overburden destabilization around production wells, and reservoirs are prone to forming a connection with the seafloor resulting in seawater intrusion under osmotic pressure. This paper summarizes the application of field observation, experimental research, and numerical simulation methods in evaluating the stability problem of the seafloor surface. The theoretical model of multi-field coupling can be used to describe and evaluate the seafloor geologic risk issues during depressurization production, and the controlling equations accurately describing the characteristics of the reservoir are the key theoretical basis for evaluating the stability of the seafloor geomechanics. It is necessary to seek a balance between submarine formation stability and reservoir production efficiency in order to assess the optimal production and predict the region of plastic damage in the reservoir. Prediction and assessment allow measures to be taken at fixed points to improve reservoir mechanical stability with the numerical simulation method. Hydrate reservoirs need to be filled with gravel to enhance mechanical strength and permeability, and overburden need to be grouted to reinforce stability.

Keywords: natural gas hydrate; depressurization production; submarine geological risks; secondary disaster; review study

Citation: Ma, X.; Jiang, Y.; Yan, P.; Luan, H.; Wang, C.; Shan, Q.; Cheng, X. A Review on Submarine Geological Risks and Secondary Disaster Issues during Natural Gas Hydrate Depressurization Production. *J. Mar. Sci. Eng.* **2024**, *12*, 840. https://doi.org/10.3390/jmse12050840

Academic Editor: Hailong Lu

Received: 23 April 2024
Revised: 8 May 2024
Accepted: 15 May 2024
Published: 17 May 2024

1. Introduction

With rapid economic and social development, the endless global demand for energy has led to the overexploitation of conventional fossil fuel energy, which has become unsustainable and uneconomical. At the same time, natural gas hydrate, as a kind of unconventional fossil fuel energy, is gaining more and more attention from scholars [1–3]. Under high pressure and low temperature, the gas molecules in natural gas are trapped in a cage of water molecules and form ice-like compounds. Natural gas hydrates are mainly found in plateau permafrost areas and deep-sea continental sediments. A total of 1 m^3 of gas hydrate contains about 164 m^3 of natural gas and 0.8 m^3 of water. The total carbon content of natural gas hydrate in nature is about twice the total carbon content of conventional fossil fuel resources, which is considered the most important energy alternative in the world due to its huge reserves and high energy density [4–6]. With such abundant natural gas hydrate resource reserves, a series of feasible extraction methods have been proposed, such as the depressurization production method, the heat injection extraction method, the gas displacement extraction method, and the chemical reagent injection development method [7–9]. The validation of laboratory-scale tests proved the feasibility of these production methods. The depressurization production method has the advantages of simple operation, low cost and high gas production efficiency, and is considered to be one of the most promising extraction methods for commercializing deep-sea gas hydrates at present [10,11]. With the advancement of science and technology, depressurization production methods and their applications have been rapidly developed [12,13]. Breakthroughs have been made in the research of energy use efficiency optimization and hydrate depressurization decomposition based on artificial intelligence. Image recognition by artificial intelligence has been widely applied to hydrate reservoir prediction and experimental data analysis [14]. Many countries have successively formulated research programs on natural gas hydrates and have carried out research on natural gas hydrate production testing. The situation of marine natural gas hydrate depressurization production testing is shown in Table 1 [15–24].

Table 1. Depressurization production test of marine gas hydrate.

Production Test Area	Time	Reservoir Characterization	Production Method	Gas Production
Nankai Trough, Japan	March 2013	Water depth: 1000 m Burial depth: 300–360 m Type: Sand layer Average initial permeability: 20 mD	Depressurization production with vertical well	Cumulative: 11.9×10^4 m^3 Average: 2.0×10^4 m^3/d
Nankai Trough, Japan	May 2017	Water depth: 1000 m Burial depth: 300–360 m Type: Sand layer Average initial permeability: 20 mD	Depressurization production with vertical well	Cumulative: 26.2×10^4 m^3 Average: 0.73×10^4 m^3/d
Shenhu Sea, China	July 2017	Water depth: 1266 m Burial depth: 203–277 m Type: Muddy chalk Average initial permeability: 2.9 mD	Depressurization production with vertical well	Cumulative: 30.9×10^4 m^3 Average: 0.5×10^4 m^3/d
Shenhu Sea, China	April 2020	Water depth: 1225 m Burial depth: 207–253 m Type: Muddy chalk Average initial permeability: 2.38 mD	Depressurization production with horizontal well	Cumulative: 86.14×10^4 m^3 Average: 2.87×10^4 m^3/d

Many studies have shown natural gas hydrate plays a very important role in reservoir mechanical stability during the drilling process, and hydrate decomposition inevitably destroys the reservoir cementation, deteriorates the reservoir mechanical properties, and puts the submarine hydrate formation at risk of destabilization [25]. As pore pressure changes, hydrate decomposition and fluids output cause stress redistribution in the reservoir during depressurization production resulting in reservoir deformation and a wide-scale deformation of the submarine formation, which will lead to seafloor subsidence. Due to poor cementation in the reservoir, accompanied by a large range of plastic deformation during depressurization production, gas and water are prone to carry a large amount of sand particles during production, forming the sand production. Uneven seafloor subsidence during depressurization extraction will deform the production well in the reservoir, and sand production may lead to well clogging and damage. Production well destruction is prone to submarine gas leakage [26]. When a gas hydrate reservoir is located in a submarine slope, hydrate decompression and super pore pressure can significantly reduce the slope stability and may trigger a large-scale submarine landslide. Hydrate depressurization production may cause submarine geological risk issues, such as sand production, wellbore destabilization, seafloor settlement, submarine landslides, and potentially gas leakage risk. Hydrate-bearing sediment reservoirs and overburden are generally characterized by low permeability, so that a large amount of gas generated by hydrate decomposition during depressurization production cannot be removed in time and accumulates in the reservoir resulting in super pore pressure in the reservoir or overburden [27]. When super pore pressure reaches the stress limit that the overburden layer can withstand, the overburden will be damaged and lead to a sudden gas eruption. Hydrate decomposition during depressurization production creates a zone of high permeability, and the hydrate decomposition zone is also a low-pressure center in hydrate formation. An osmotic pressure develops between the low-pressure center and the seawater–seafloor boundary layer, driving seawater towards the reservoir. Submarine geologic risks can destabilize the overburden and reservoir, leading to the creation of seafloor–reservoir channels. This can result in a significant influx of seawater into the reservoir and seawater intrusion. Hence geological risk issues during depressurization production are prone to a secondary disaster problem dominated by gas eruption and seawater intrusion [28–31]. Although natural gas hydrate is an important natural resource with abundant reserves, it is also an important factor affecting the stability of submarine geomechanics. Therefore, a clear understanding of the potential seafloor engineering geohazard problems during depressurization production is of great significance in ensuring environmental and production safety and avoiding secondary disasters dominated by gas eruptions and seawater intrusion, so as to efficiently develop natural gas hydrates.

Natural gas hydrate production tests have verified that hydrates can be exploited through depressurization. However, there are potential submarine geologic risks during production, such as sand production, wellbore destabilization, seafloor subsidence, and landslides. In this paper, we review the research progress of hydrate depressurization production in recent years, summarize the geological risks during production with secondary disasters, and improve the understanding of the geomechanical stability of hydrate reservoirs and the safety of the seabed environment. It is hoped that this study will provide a valuable reference for promoting the safe and efficient production of natural gas hydrates.

2. Submarine Geological Risk Issues

2.1. Sand Production

Sand production during hydrate depressurization is a complex problem involving multiple flows and geomechanical responses. The purpose of depressurization is to decompose the hydrate and collect natural gas, but the weak cementing properties of the hydrate layer cause sand and soil particles, which lose their original stability under pressure release and hydrate decomposition, to be dislodged from the sediment and enter into the production well with the fluid flow. As shown in Figure 1, under the drive of fluid flow, sand particles continue to flow out and form a continuous collapse within the reservoir. In the case of more severe sand production, reservoir subsidence around the well occurs under the action of overburden gravity [32,33]. Discontinuous deformation in the form of faults enhances the connectivity of the sediment skeleton, thereby favoring fluid flow. The increase in fluid flow rate is accompanied by the stripping of particles from the sediment skeleton, thus bringing out more sand particles. Hydrate reservoirs with small and poorly connected pore space typically have low initial permeability. When hydrate is extracted by large-scale depressurization, fluid channels are formed in the sediment skeleton, and sand particles in the reservoir are easily disturbed by fluid flow. During depressurization production, the temperature within hydrate reservoirs changes drastically, and the thermal expansion and contraction effect causes hydrates to undergo volumetric strain, which causes sand particles in formation to move and release the constraints of the sediment skeleton, and then free flow with the fluid occurs. Sand production results from the expansion of the depressurization range and the increase in effective stress in reservoirs, so that the sand grains are stripped from the skeleton due to the destruction of the sediments [34,35]. Low production pressure is commonly used in production to promote hydrate decomposition, but increased production rates can reduce the mechanical stability of reservoir sediment, leading to extensive plastic damage. As the pressure gradient and permeability increase in the reservoir, hydrate decomposition accelerates, leading to reservoir destabilization and an increased fluid flow rate. This makes it easier for the fluid to carry sand out of the reservoir. The sand problem is typically not significant at the start of depressurization production. However, it becomes more severe with production time expansion, particularly after hydrate dissociation around the well. The sand rate is positively correlated with the initial hydrate saturation in the formation [36–39].

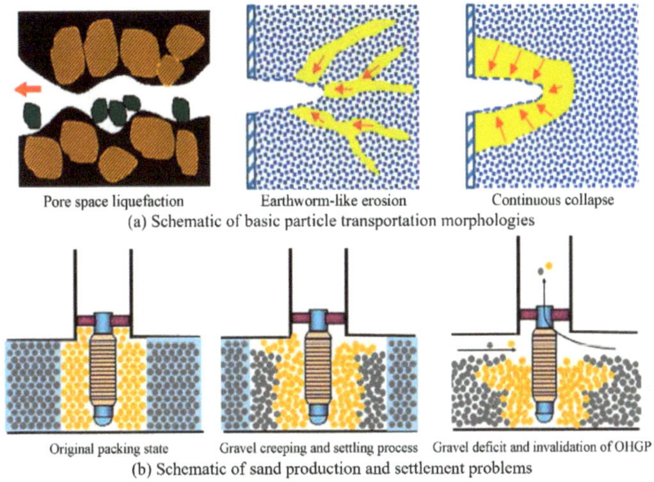

Figure 1. Schematic of the sand production and subsidence problem (cited from Li et al., 2019) [32].

Sand production is one of the key submarine geologic risks limiting the long-term and effective development of gas hydrate resources, and the behavior of sand production has attracted attention since the first marine hydrate development test in Japan was prematurely terminated in 2013 due to wellbore clogging problems caused by sand production. The second production test, at Nankai Trough in Japan, utilized a series of sand control measures to limit sand production. Production tests in 2017 and 2020 in the Shenhu Sea area did not experience wellbore plugging caused by sand outflow due to sand control measures [40–43]. The entry of sand into the production well and pipeline can cause equipment to wear out, reducing its useful life and increasing maintenance and replacement costs. This, in turn, affects overall extraction efficiency and operational stability. If sand collects inside pipelines or equipment, it can form clogs that can interfere with the smooth transportation of natural gas, or the clogs can even trigger blowouts due to pressure buildup [44–46]. The abrasion of sand particles on the equipment may lead to equipment failure and natural gas leakage, which may not only affect the safety of the personnel at the site, but also pollute surrounding environment. When sand production reaches a critical point, it can cause the formation of numerous holes and lead to continuous collapse. This can result in submarine subsidence and trigger engineering geological disasters, such as natural gas leakage or eruptions [41]. Hydrate reservoir collapse caused by sand production will easily destabilize overburden. If the amount of sand production is excessive, the reservoir may connect with the seafloor surface, leading to seawater intrusion due to osmotic pressure. It is important to take timely and effective preventive and management measures for the potential risk of sand production caused by depressurization, which requires theoretical and experimental research on the mechanism.

Simulation and experimental studies of sand production during depressurization production mainly focus on mechanism [47,48]. There are two fundamental reasons for sand production: the stress concentration and strength reduction due to hydrate decomposition. Generally speaking, the sand particle transport behavior belongs to the geomechanical control module in the theoretical model. Establishment of a multi-field coupled model that considers the geomechanical behavior is the key to simulate and study sand production behavior [49,50]. The numerical simulation method and flow chart of sand production during depressurization production are shown in Figure 2, and its reservoir model and study method are shown in Table 2 [42–46]. Multi-field coupled governing equations for sand production typically involve sand detachment, transport, and sediment deformation, and depressurization leads to hydrate decomposition and stress concentration distribution. Depressurization results in hydrate decomposition and stress concentration distribution. Stress concentration causes particles to detach from the sediment skeleton, while fluid flow alters the multiphase fluid pressure and the temperature distribution. The main purpose of experimental studies are usually to determine sand production mechanism and investigate the relationship between sand production and pressure, fluid flow rate, sand properties [51–53]. As shown in Table 3, the model geometries and experimental methods in sand production experiments are listed, and the experimental study on sand production prevention and control is shown in Figure 3 [47–54]. The choice of the depressurization scheme and the use of sand prevention tools changes the gas production behavior. The choice of lower production pressure during depressurization production can promote hydrate decomposition and thus increase the gas production. At the same time, it will increase the risk of sand production. Sand screen tubing impedes sand transportation to the wellbore and increases gas transport resistance, resulting in reduced gas production rates. The contradiction between the stability of sand production control and efficient gas production affects the gas production efficiency of natural gas, and there is a need to explore and research more novel ways of controlling hydrate sand production during depressurization production. In addition to optimizing the depressurization method and using sand control devices, reservoir stability enhancement measures can also be used. For example, the gravel filling method enhances reservoir mechanical stability, and the chemical reagent injection method enhances reservoir cementation. On the basis of reservoir modification and the use of sand control devices, the possibility of sand stripping is reduced by adjusting the depressurization rate to ensure a relatively stable production process, and production well heating needs to be used jointly to prevent clogging of the sand control network. In order to counteract the sand stripping issue, filters or other equipment are required to separate and filter the sand particles to prevent them from entering the extraction pipelines and equipment. Monitoring equipment is required to detect sand problems during the depressurization production to ensure extraction process is conducted safely and efficiently.

Table 2. Research method and content during numerical simulation of sand production.

Reservoir Model Description	Key Control Equations for Sand Production	Main Research Content	Research Objectives	Reference
3D hollow circle	Response of permeability and porosity to sediment denudation and sand production	3D DEM fluid flow model simulation	Effects of boundary stress and fluid flow on sediment denudation and sand production	Cui et al. [42]
Rotating cylinder	Sand migration and clogging of anti-sand devices	Numerical analysis under different working conditions	Balancing sand production control and gas production	Zhu et al. [43]
2D planar symmetry model	Combined equations of control for sand stripping, transport, sediment deformation, and hydrate decomposition	Evaluation of sensitivity of sand production parameters to changes in the volume	Effect of different depressurization rates and production methods on sand production	Uchida et al. [44]
Hollow cylinder	Fluid–solid coupling calculations for sand transport	Fracture energy regularization method was implemented to diminish mesh dependency related to energy dissipation	Improve the accuracy of sand modeling	Shahsavari et al. [45]

Table 2. *Cont.*

Reservoir Model Description	Key Control Equations for Sand Production	Main Research Content	Research Objectives	Reference
Hydrate core model	Multi-field coupled model considering sand mass conservation	Incremental format solution analysis using the IMPES method and cylindrical cores	Investigating the effects of parameters such as wellbore pressure, initial hydrate saturation, and loading stress on fluid flow and sand emergence behavior	Li et al. [46]

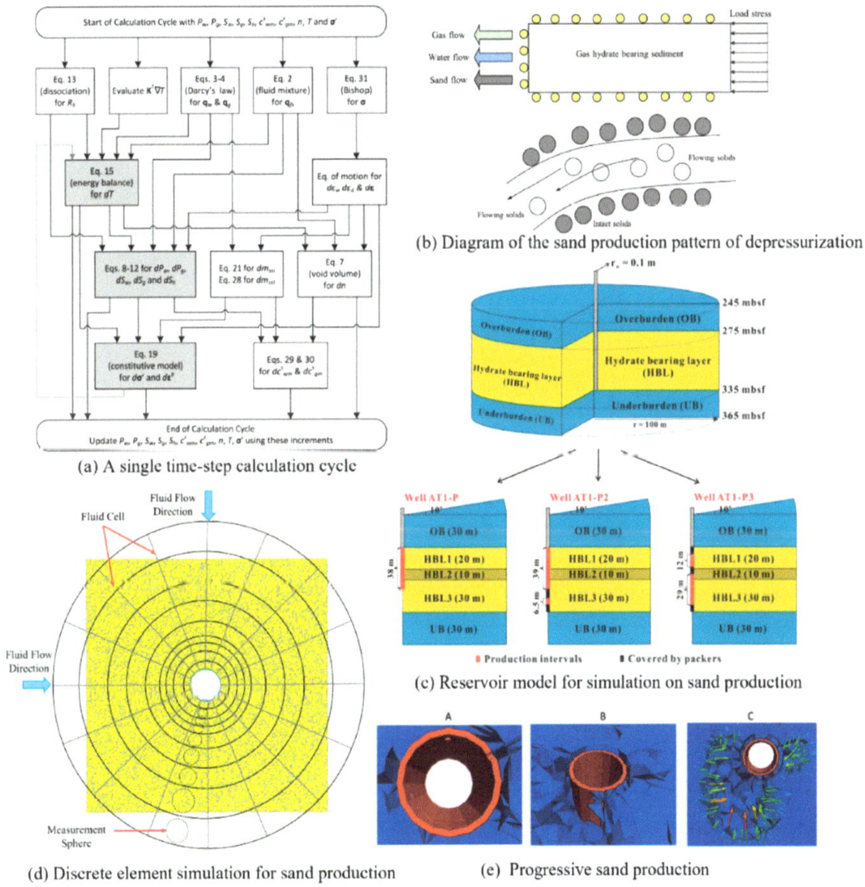

Figure 2. Method and flowchart of numerical simulation study of sand production during depressurization production (cited from Cui et al., 2016; Uchida et al., 2016; Zhu et al., 2020; Shahsavari et al., 2021; Li et al., 2024) [42–46]. A–C in subfigure (**e**) show the weak to violent sand production around well during production.

Table 3. Experimental research method and objective of sand discharge risk.

Experimental Model Description	Average Diameter of Sand Grains	Main Experimental Content	Research Objective	Reference
300 mm (D) 240 mm (L)	0.3 mm	Different overlay stresses and depressurization methods	A correction to the analytical solution for classical steady state flow	Kozhagulova et al. [47]
3.81 cm (D) 5.3 cm (L)	75 μm	Different radial to axial stress ratio conditions	Obtaining a predictive model for the mass-to-stress ratio of the discharged sand	Zivar et al. [48]
12.5 mm (L) 25 mm (D) 2.5 (H)	0.35 mm	Injection of fluid at a given pressure	Determining critical wellbore pressure for reservoir collapse	Song et al. [49]
442.3 mL (V)	27.4 μm 15.99 μm 14.45 μm 3.7 μm 4.37 μm	Different grit screen hole sizes	Analysis of gas production and sand discharge behavior with different anti-sand sieve hole sizes	Li et al. [50]
49 mm (L) 25 mm (D)	4~125 mm	Multi-channel hydration acoustic monitoring	Determining the relationship between sand output and production and designing a sand control network	Ding et al. [51]
390 mm (L) 38 mm (D)	126.4 μm	Experimentation of different anti-sand production methods	Finding effective sand control methods for hydrate reservoir development	Wang et al. [52]
50 mm (D)	15~20 μm	Clogging of sand control screen experiment	Proposed depressurization combined with wellbore heating to prevent plugging of sand control grids	Li et al. [53]

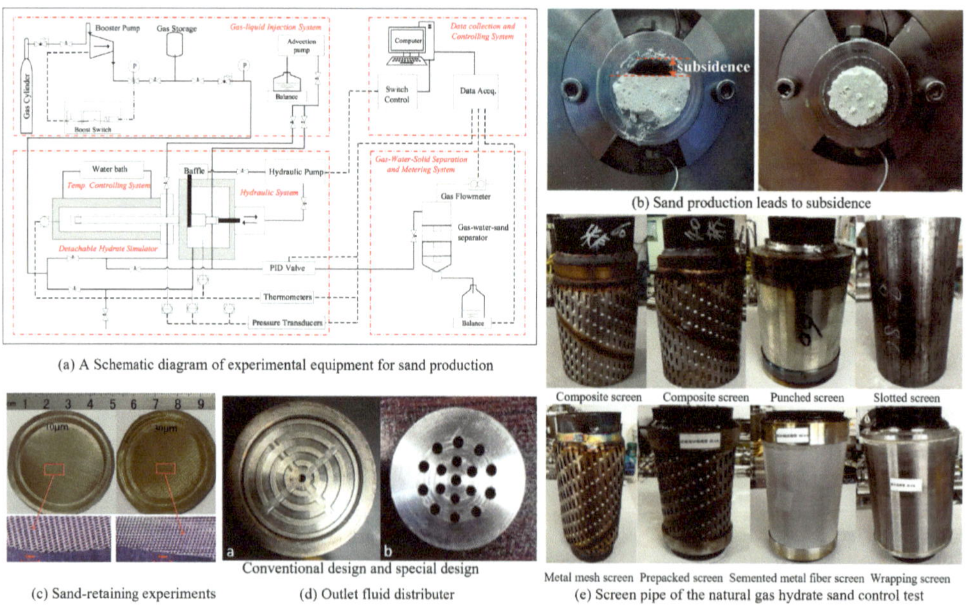

Figure 3. Experimental study on sand production risk (cited from Zivar et al., 2019; Li et al., 2022; Wang et al., 2023; Li et al., 2020; Dong et al., 2018) [48,50,52–54]. (a) and (b) in subfigure (**d**) represent conventional design and special design, respectively.

2.2. Wellbore Instability

Wellbore instability during drilling and depressurization production of marine hydrate reservoir is caused by hydrate formation conditions. If the hydrate formation is weak, prone to collapse, or contains easily dissolvable rock layers, the drilling fluid intrusion process may lead to large-scale hydrate decomposition, thus triggering wellbore instability [55–57]. Fault or fracture development in the formation is likely to accelerate the diffusion of drilling fluid and hydrate decomposition [58–60]. Poor connectivity and mobility of hydrate formation is easy to form super pore pressure gas, and high pore pressure exists in the complex special geological structure area. Drilling fluids play a crucial role in maintaining formation stability, bearing capacity, and drilling efficiency. Improper selection or use fluids can result in the loss of rock cuttings, formation collapse, or dissolution, thereby increasing the risk of wellbore instability or even gas eruption [61,62]. The construction of a well forms a channel that connects the seabed surface with the reservoir. After depressurization, osmotic pressure causes seawater to flow along the channel and into the reservoir, resulting in seawater intrusion. The risk of wellbore destabilization during the drilling process and depressurization production cannot be ignored. Hydrate decomposition can reduce the mechanical strength of the sediment surrounding wellbore, potentially causing it to reach the plastic yield state and become more susceptible to destabilization. During the early stages of depressurization production, the localized area around the horizontal well may not be fast enough to cause wellbore damage, even though it quickly reaches the yield state. From the perspective of reservoir around the well, hydrate decomposition during gas hydrate depressurization production reduces sediment mechanical strength, which may lead to a significant increase in plastic strain. From the perspective of the entire reservoir, the inhomogeneity of hydrate saturation distribution in the formation leads to the variability of reservoir mechanical strength and deformation distribution. Long-term depressurization of the reservoir can cause significant inhomogeneous deformation, leading to shear damage in the wellbore.

Sediments on the seabed that contain hydrates are typically poorly cemented and have a low mechanical strength. Drilling and depressurized production of hydrate reservoirs pose a risk of wellbore instability and damage. Wellbore damage may lead to natural gas leakage from the pipeline into the formation, polluting the surrounding environment and even threatening the safety of offshore platforms by gas eruption accidents [63]. Hydrate formations on the seafloor carry enormous formation and super pore pressures, and wellbore destabilization and damage can occur in the form of subsea gas leaks and eruptions. For example, the 2010 Gulf of Mexico well blowout triggered an oil spill that posed a significant threat to the local marine ecosystem [64]. Therefore, pressure control is a crucial task throughout the production process. Assessing wellbore stability during depressurized production is also essential. Numerical simulation methods have been utilized in many studies to address the challenges of geomechanical response, hydrate decomposition, and the coupled effects of multiple physical fields in wellbore stability analysis during drilling and depressurization production [65–69]. Establishment of theoretical models during numerical simulation needs to consider the coupling between the dynamic heat and mass transfer between the wellbore and the reservoir, the multiphase flow in the sediment, the evolution of mechanical properties and other physical fields. The drilling process accompanied by the intrusion of drilling fluids and the increase in temperature can lead to hydrate decomposition increasing the risk of wellbore destabilization [70–73]. The location of the risk of wellbore instability during depressurized production is affected by the extent of plastic damage to the sediments. The studies on wellbore instability triggered by drilling and depressurized production are shown in Figures 4 and 5, respectively. Reservoir models and simulation method used in the studies are shown in Table 4. The reservoir deformation law for long-term depressurization production of hydrate reservoir shows that the effective stress in the reservoir around the well is centrally distributed, which puts the wellbore at risk of extrusion damage [74–76]. Distribution of hydrate saturation in the formation has obvious inhomogeneity, and the reservoir is susceptible to uneven settlement. This makes the horizontal well vulnerable to shear damage caused by uneven deformation. The consolidation degree of deep reservoir is generally higher than in shallow reservoirs, and there is a tendency for the mechanical strengths, such as Young's modulus, of hydrate-bearing sediment to increase gradually with depth. This can have a positive impact on reservoir stability when exploiting deep hydrate reservoirs, but the high stress effects of deep formations can pose a challenge to wellbore stability.

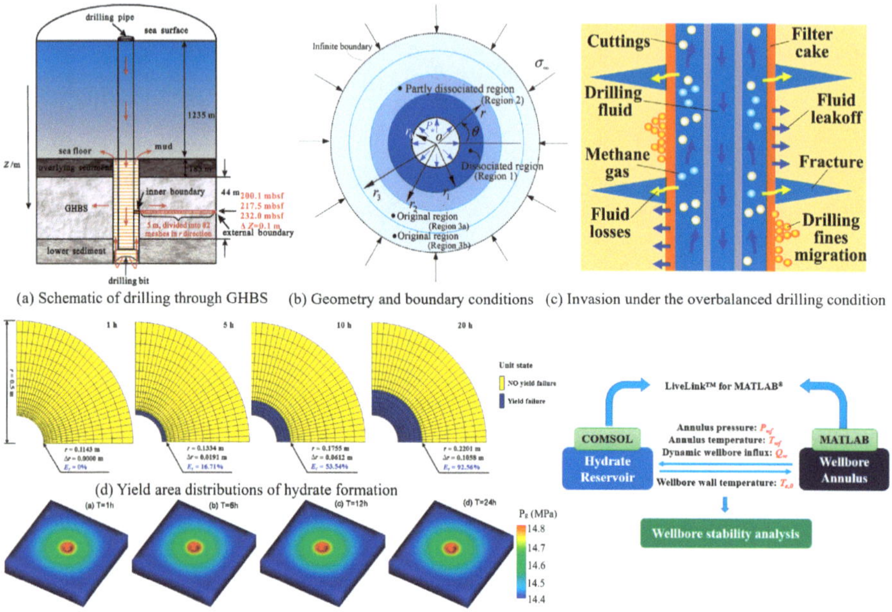

(a) Schematic of drilling through GHBS (b) Geometry and boundary conditions (c) Invasion under the overbalanced drilling condition

(d) Yield area distributions of hydrate formation

(e) Distribution of pore pressure after drilling (f) Numerical solution of the wellbore-reservoir

Figure 4. Numerical simulation of wellbore instability problem during drilling of natural gas hydrate reservoir (cited from Wang et al., 2019; Liao et al., 2021; Sun et al., 2018; Dong et al., 2021) [66,67,70,71].

Table 4. Reservoir models and simulation methods used in the study of wellbore instability risk problems.

Hydrate Reservoir Model	Key Theoretical Models	Main Research Contents	Research Objectives	Reference
Simplified 2D axisymmetric wellbore modeling	A drilling model considering the degradation of reservoir mechanical properties due to hydrate decomposition	Elastic-plastic intervals of reservoirs obtained by closed-form solutions of computational mechanical fields	Exploring the mechanism of the reduction of drilling fluid temperature, pressure, and elastic modulus in the decomposition zone on wellbore stability	Wang et al. [66]
Plane strain model	Consider dynamic heat and mass transfer between the wellbore and the reservoir	Elasto-plastic analysis of wellbore during drilling process	Analyze the heat and mass transfer law between wellbore and reservoir and the mechanism of wellbore yield damage behavior	Liao et al. [67]
Simplified 2D axisymmetric wellbore modeling	Considering the reduction of formation stiffness and strength after hydrate dissociation	Elasto-plastic analysis of wellbore during drilling process	Analysis of the mechanical response of wellbore in elastic-plastic formations and summarization of the destabilization mechanism	Guo et al. [68]

Table 4. Cont.

Hydrate Reservoir Model	Key Theoretical Models	Main Research Contents	Research Objectives	Reference
Symmetric plane strain model	A theoretical model to characterize the deformation field using rock mechanics theory	Elasto-plastic analysis of wellbore during drilling process	Summarize the major influences on wellbore and hydrate formation instability	Zhang et al. [69]
Simplified 2D axisymmetric wellbore modeling	Considering the effects of drilling mud intrusion and pore water salinity	Elasto-plastic analysis of wellbore during drilling process	Guidance for the design of drilling fluids based on elasto-plastic distribution laws	Sun et al. [70]
3D reservoir model	Development of a three-dimensional multi-field coupled model describing drilling fluid intrusion	Dynamic response of drilling fluid intrusion processes and reservoirs	Characterizing the reservoir response to drilling fluid intrusion in conjunction with hydrate decomposition	Dong et al. [71]
2D planar reservoir model	Improvement of elasto-plasticity intrinsic model based on the saturation of hydrate and ice in the pore space	Analyzing the yielding of the wellbore and reservoir around the well during depressurized production in horizontal and vertical wells	Predicting stress concentrations and yield zones in wellbores and reservoirs during mining	Rutqvist et al. [72]
Simplified 2D axisymmetric wellbore modeling	A model considering the evolution of the plastic zone due to changes in temperature and pressure fields	Analyzing the effect of gas hydrate decomposition on wellbore stress and plastic zone distribution	Providing a theoretical basis for wellbore design from a mechanical point of view	Li et al. [73]
2D planar reservoir model	Consideration of multi-field coupled models based on consolidation theory	Effective principal stress concentration distribution and its stress path analysis	Prediction of stress field evolution and wellbore stability during the mining process	Yuan et al. [74]
2D planar reservoir model	Theoretical modeling of the Mohr–Coulomb criterion combined with multi-field coupled models	Mechanical behavior of hydrate reservoirs and wellbores during one year of production	Predicting deformation characteristics of producing wells due to hydrate decomposition	Sun et al. [75]
Multilayer hydrate reservoir modeling	Theoretical models considering coupled fluid and geomechanics	Effective principal stress concentration distribution and its stress path analysis	Geomechanical response and reservoir stability analysis around wells under coupled effects	Dong et al. [76]

With the help of numerical simulation method, the risk areas of geomechanical response of gas hydrate reservoirs can be assessed and predicted, so that corresponding preventive and control measures can be taken when designing hydrate depressurization production methodology. There is a transient heat transfer model between the wellbore and the sediment during drilling, which requires numerical simulation based on its thermodynamic stability and gives an optimized drilling method. Numerical simulation can be used to predict the extent of plastic damage in the reservoir and the location of the wellbore instability risk when designing and exploiting, corresponding to which the method of enhancing the mechanical properties of the reservoir and the wellbore protection measures can be taken. At the initial stage of well construction, a solid plugging agent or cement slurry is injected around the wellhead to reinforce the well wall and fill the possible leakage gaps, so as to improve the sealing and stability of the wellbore. Pressure balancing measures are used during production to maintain the balance of pressure inside and outside the wellbore, reduce the impact of changes in formation pressure on the wellbore. It is also necessary to implement a regular wellbore inspection and maintenance program to monitor the condition of the wellbore and repair or replace any possible damage in a timely manner to maintain the stability of the wellbore. Measures to plug and seal the wellbore after production is completed are taken to avoid possible leakage or pressure imbalance and to prevent environmental contamination or damage to the wellbore.

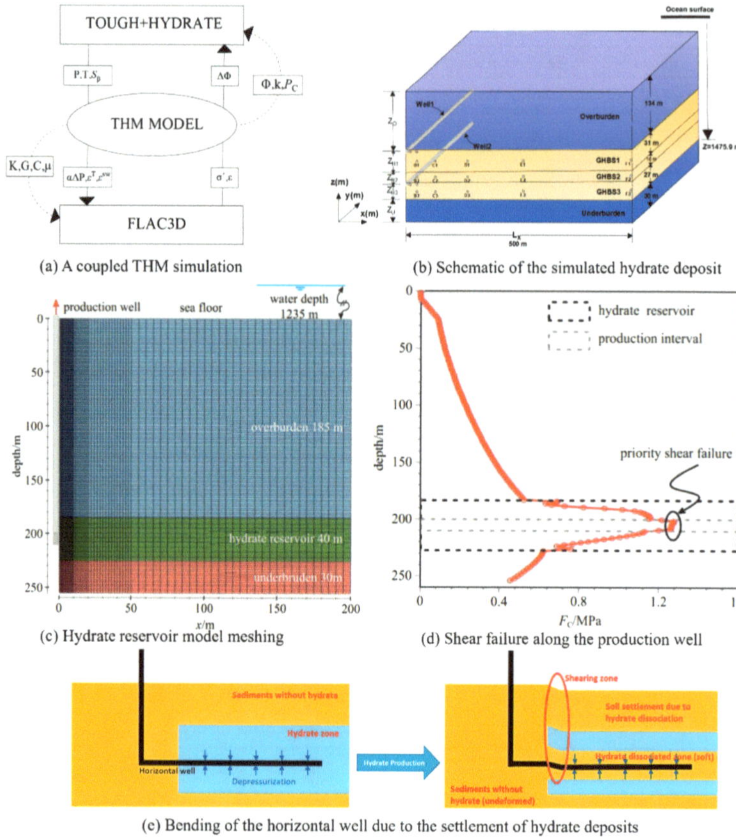

Figure 5. Numerical simulation study of wellbore instability problem during depressurization production (cited from Yuan et al., 2020; Sun et al., 2019; Dong et al., 2022) [74–76].

2.3. Seafloor Subsidence

Marine natural gas hydrate is shallowly buried in the formation, and overburden is dominated by low permeability clays, which do not have a dense and hard rock structure. Hydrate-bearing sediment is poorly cemented, and hydrate decomposition leads to a significant reduction in the mechanical strength of the reservoir, and the range of the reduction will continue to expand with hydrate decomposition [77–79]. Depressurization causes the hydrate in the sediment pore space to decompose into natural gas and water. The compressibility of fluid is much larger than that of solid hydrate, so the pore space of the sediment is compressed under the action of overburden gravity and seawater pressure, which deforms the entire hydrate reservoir. A low pressure region may form in the reservoir, causing a redistribution of internal pressure combined with ongoing gas and liquid recovery, which increases the effective stress in the formation. The range of influence of the effective stress expands over time as hydrate extraction continues. Continuous discharge of fluids in the hydrate reservoir causes the overburden to lose its support, and produces deformation and subsidence in the seafloor surface under the action of seawater pressure. From the mechanical stability analysis of the hydrate formation, the main cause of subsidence is the compression of sediment pore space during the depressurization process, so the main manifestation of subsidence is the compression and deformation of the seafloor surface [80,81]. The process of sediment compression and seafloor subsidence causes internal compaction of the hydrate reservoir, which alters the physical properties of the formation and can impact the sustainability of gas production [82,83]. Overburden stability is disturbed by the uneven settlement of the seafloor causing sediment damage, and seawater pushes through the overburden into the reservoir in large quantities under the action of osmotic pressure. Decompression causes stress to concentrate around the wellbore, resulting in significant sediment compression and seafloor subsidence in the area surrounding the vertical well. Additionally, it causes extensive seafloor subsidence in the upper area of the horizontal well [84–86]. Seafloor subsidence can affect the stability of facilities such as submarine pipelines and cables, and increase the cost of extraction. The inhomogeneity of hydrate distribution leads to uneven reservoir deformation and seafloor subsidence, resulting in wellbore shear damage that triggers gas leakage and thus contamination of the seafloor environment.

As shown in Figure 6, research methodology for settlement risk problem is dominated by numerical simulation. From the studies of long-term depressurization production, the seafloor subsidence can reach several meters, as shown in Table 5 [72,87–91]. Seafloor settlement is different in different case studies, indicating that the seafloor surface settlement behavior is related to the mechanical properties of hydrate formation and production scheme. When hydrate formation is subjected to mechanical constraints, the vertical strain is much larger than the horizontal strain. Although the reservoir pressure drops very rapidly, the subsidence is gradual over time [88,89]. Overburden and underburden are compressed by seawater gravity and geostatic stresses, respectively, and move in the direction of the production wells, resulting in seafloor surface subsidence and underburden uplift. Seafloor surface subsidence with pore pressure reduction and the low mechanical strength of weakly cemented sediments make the seafloor surface subsidence behavior highly sensitive to production pressure. Seafloor subsidence involves the complex mechanical behavior of hydrate-bearing sediments during depressurization production [90,91]. Therefore, in addition to numerical simulation assessment there is a need to study the relationship between hydrate decomposition and reservoir mechanical properties through triaxial tests to improve the understanding of seafloor subsidence problems [92–95].

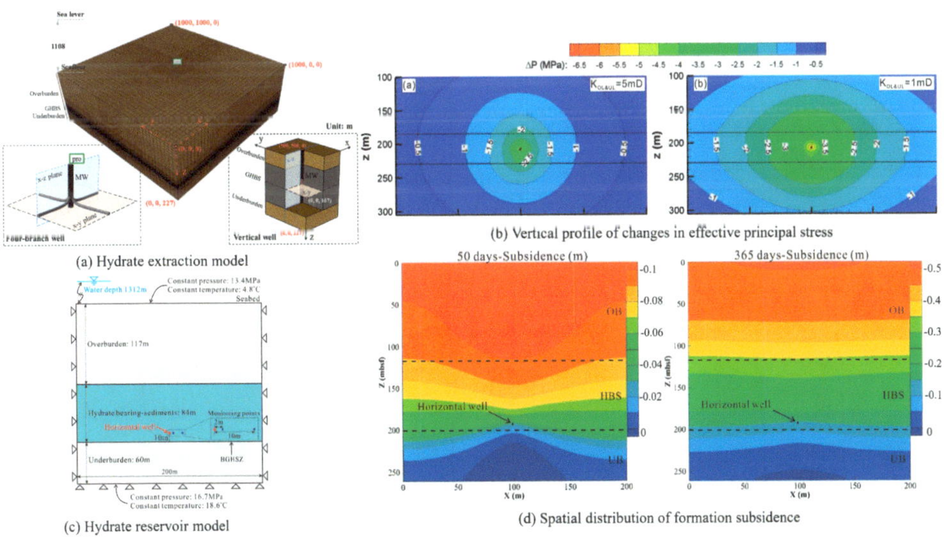

Figure 6. Numerical simulation of seafloor subsidence risk issues (cited from Jin et al., 2018; Zhang et al., 2023; Yuan et al., 2021) [89–91].

At present, the world has not carried out long-term large-scale depressurization production of natural gas hydrate. In 2020, in the Shenhu area of the South China Sea, a 30-day hydrate production test was carried out; the test used included the "four-in-one" on-site real-time monitoring research method, as shown in Figure 7 [17]. Using a variety of technical means such as hydrate reservoir temperature monitoring, seafloor sediment pore pressure monitoring, seafloor natural gas leakage, seafloor stratigraphic stability monitoring (subsidence and landslide), full-profile seawater environmental monitoring, and water–gas–methane exchange flux monitoring of the production test platform and the surrounding area, we carried out all-around environmental monitoring from the seafloor to the middle seawater up to the sea level, and objectively evaluated the environmental impacts of the depressurization production of natural gas hydrates. From the mechanical nature of hydrate-bearing sediments, the seafloor subsidence issue in long-term production of gas hydrate is difficult to avoid. It is necessary to consider the balance between hydrate formation stability and gas production in order to design the optimal production program. Appropriate measures to improve the mechanical properties of the seafloor surface, such as overburden injection and reinforcement methods, can be taken prior to hydrate extraction. The numerical simulation method can be used to assess the stability of hydrate formation and predict potential hazard problems that may occur.

Table 5. Behavior of seafloor settlement during depressurization production.

Site	Production Time	Wellbore Selection	Seafloor Subsidence	Reference
Gulf of Mexico, Mexico	2 years	Depressurization through a horizontal and vertical well	4.5 m (horizontal well) 2.3 m (vertical well)	Rutqvist et al. [72]
Eastern Nankai Trough, Japan	6 days	Depressurization through a vertical well	15 cm	Uchida et al. [41]
Eastern offshore India, India	90 days	Depressurization through a vertical well	1.7 cm	Lin et al. [88]
SH2 drill hole, South China Sea, China	2 years	Depressurization through a horizontal well	2.5 m	Jin et al. [89]
South China Sea, China	2 years	Depressurization through a vertical well and a four-branch multilateral well	1.2 m	Zhang et al. [90]
South China Sea, China	1 years	Depressurization through a horizontal well	0.5 m	Yuan et al. [91]

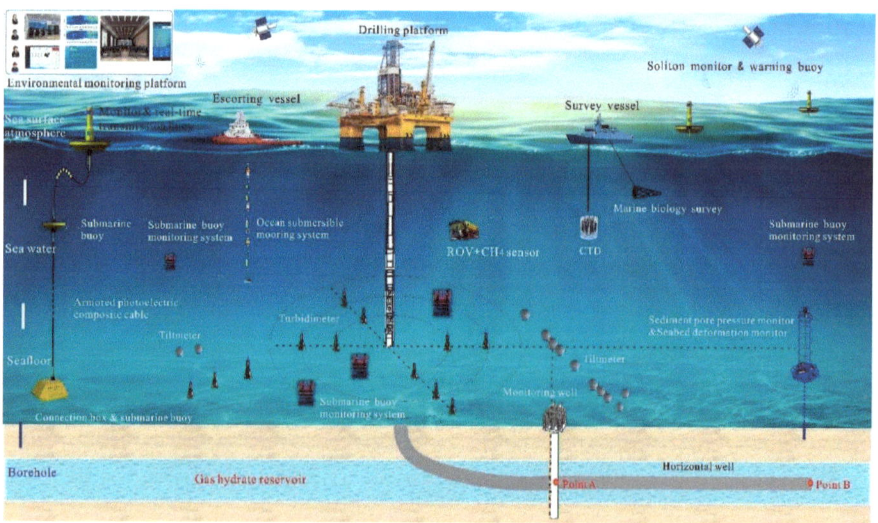

Figure 7. "Four-in-one" real-time on-site monitoring system (cited from Ye et al., 2020) [17].

2.4. Submarine Landslide

Natural gas hydrate is abundantly stored in the submarine slope, and the factors that cause submarine landslides in hydrate formations can be categorized into natural environmental factors and anthropogenic factors [96,97]. Natural environmental factors mainly include the natural hydrate decomposition caused by global warming, and anthropogenic factors are mainly the temperature and pressure perturbation of hydrate reservoir caused by hydrate extraction. Continuous deposition of sediment decreases overburden permeability, increases hydrate saturation and weakens reservoir permeability, and overburden permeability is much lower than reservoir permeability. Capacity enhancement method such as depressurization and heat injection can promote hydrate decomposition to generate large quantities of gas, but in low-permeability reservoir gas are difficult to produce in a timely manner from production well. If a large amount of gas and liquid generated by hydrate decomposition cannot be discharged in a timely manner, super pore pressures will be generated between the reservoir and the overburden. Hydrate decomposition decreases the mechanical strength of the reservoir, and the super porous pressure gas collects in the slope, which makes the submarine slope more unstable [98,99]. Seawater warming due to global warming induces widespread decomposition of hydrate reservoir, releasing large quantities of methane gas. Strong load and high temperature fluid from earthquakes and volcanic eruptions can cause shear or tensile damage to seafloor slope and hydrate decomposition, resulting in large impact damage to the seafloor environment. During hydrate extraction, different extraction method directly and significantly affects seafloor settlement and seafloor slope stability [100,101]. When a horizontal well is used to depressurize the extraction method to improve the decomposition efficiency of gas hydrate and obtain higher production, it will cause a wide range of deformation in the hydrate reservoir, especially when the hydrate reservoir is in the seafloor slope, it is easy to cause the mechanical instability of the slopes. Long-term and large-scale gas hydrate extraction poses a potential threat to the stability of the seafloor surface, especially to the hydrate slope stratigraphy [102,103]. Hydrate production causes overburden and reservoir deformation to slide toward the production well, which can trigger a submarine landslide. Large-scale submarine landslides can separate the hydrate reservoir from the overburden, resulting in direct exposure of the reservoir to seawater, which can lead to widespread hydrate decomposition and natural gas leakage [104,105]. They not only cause natural gas leaks, but also damage deep-sea oil and gas wells and pipelines, threatening the safety of the undersea environment.

Hydrate decomposition is one of the key factors for inducing a submarine landslide, and natural factors and anthropogenic perturbations induced seafloor landslides, and the on-site monitoring systems are shown in Figure 8 [106,107]. Geological records show that the increase in seafloor temperature leads to hydrate decomposition and destroys the cementation between gas hydrate and seafloor slopes, destabilizing the slopes. The historically famous Storegga submarine landslide and the Amazon Fan failure, which occurred on the Norwegian continental margin in the northeastern Atlantic Ocean, have been attributed to hydrate decomposition [108,109]. When gas hydrate is endowed in inclined reservoirs, the decomposition of hydrates will cause a weakening of the mechanical strength of the reservoir, and the inability of gas to be discharged in a timely manner may generate pore pressure buildup, inducing submarine slope instability [110–112]. Current research on natural gas hydrate formation landslides is dominated by numerical simulations and field observations, which often increase reservoir instability when production enhancement measures are planned to increase natural gas production, which requires mechanical characterization of submarine slopes during depressurization production to fully understand the potential geomechanical response issues that come with increased production [113,114]. A numerical simulation study of the seafloor landslide risk in the hydrate formation is shown in Figure 9, and a study on the hydrate extraction scheme and seafloor slope destabilization mechanism is shown in Table 6. The average inclination angle of hydrate reservoir in the South China Sea is approximately 3.3 to 3.6°, and the maximum angle reaches 25° [114].

The destabilization of seafloor slopes induced by artificially disturbed depressurization production is likely to cause the collapse of inclined reservoirs and trigger large-scale hydrate decomposition. Hydrate decomposition leads to a decrease in sediment cohesion, which puts the slope at risk of destabilization, and the most dangerous area covers the hydrate decomposition zone. The placement of production well at the bottom of subsea slopes to increase production efficiency may affect reservoir stability. From the numerical simulation study of seafloor geological risk issues, the theoretical model of multi-field coupling can be used to describe and evaluate the seafloor geological risk issues during depressurization production. The theoretical model needs to be established from the perspective of multi-field coupling, and the controlling equations that accurately describe the characteristics of the reservoir are the key theoretical basis for evaluating the stability of seafloor geomechanics.

In order to reveal the landslide mechanism of hydrate formation, geophysical field observation techniques are indispensable. Acoustic waves can propagate within the hydrate formation and be reflected, refracted, or absorbed by different strata, and acoustic wave monitoring technology is an important means to assess the production of marine gas hydrate reservoirs. Gas production in natural gas hydrate reservoirs involves an in situ phase transition of solid hydrate decomposition, and the use of non-contact observation means of acoustic wave propagation speed, reflection characteristics, and other information that can be inferred via in situ hydrate decomposition and stratum deformation. Resistivity sensors are buried in the reservoir to obtain the temperature and pressure characteristics of the reservoir and the in situ hydrate decomposition information in real time, so as to strengthen the ability to recognize the hydrate generation and decomposition process in the formation. Moreover, resistivity imaging technology is utilized to measure the spatial distribution of underground resistivity in order to obtain three-dimensional imaging of the formation structure and the hydrate distribution in real time, thus inferring the stability of formation slopes. In addition, natural gas hydrate formation landslides are often accompanied by changes in formation vibration signals, and monitoring the vibration of the formation can provide landslide precursor information to predict the landslide risk that may be triggered by formation deformation. The comprehensive use of acoustic, electrical, vibration and other geophysical data, as well as the development of multi-parameter cross-analysis technical means, can more comprehensively and finely characterize the deformation and slippage of natural gas hydrate formations.

Reservoir deformation and super porous pressure usually occur inside the hydrate formation, and the limitations of the submarine environment restrict the traditional monitoring means, and there is an urgent need to develop monitoring equipment and techniques applicable to the submarine environment and the complex hydrate formation structure. The research and development of the submarine landslide field monitoring system faces many technical difficulties. Hydrate reservoir deformation is usually a slow and gradual process, and the monitoring accuracy is required to be high. The displacement generated by submarine landslides is large, and although the existing monitoring technology can realize accurate monitoring of small centimeter-level deformation, it is difficult to improve its monitoring range. Therefore, it is necessary to develop high-precision, large-range, and long-term stable monitoring means to sense the deformation signals of the hydrate reservoir and its overburden in real time. For on-site inspection, real-time monitoring devices such as turbidity meters, pore pressure sensors, resistivity probes, methane leak detectors and acoustic measurements need to be deployed at points on the surface of the overburden, and displacement monitoring devices need to be deployed in the overburden and the interior of the hydrate reservoir, to form a full-profile, refined real-time inspection system.

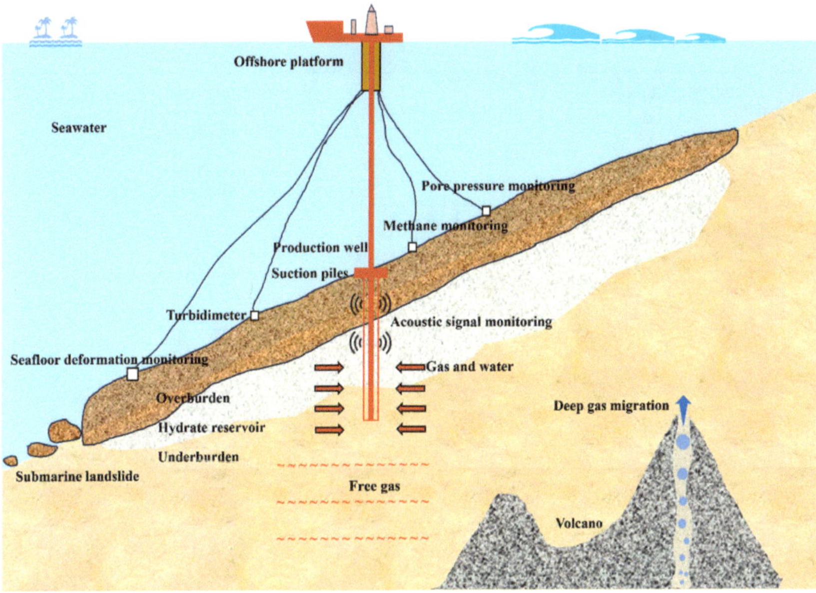

Figure 8. Natural factors and anthropogenic disturbances inducing a submarine landslide (cited from Yan et al., 2020; Vanneste et al., 2014) [2,106].

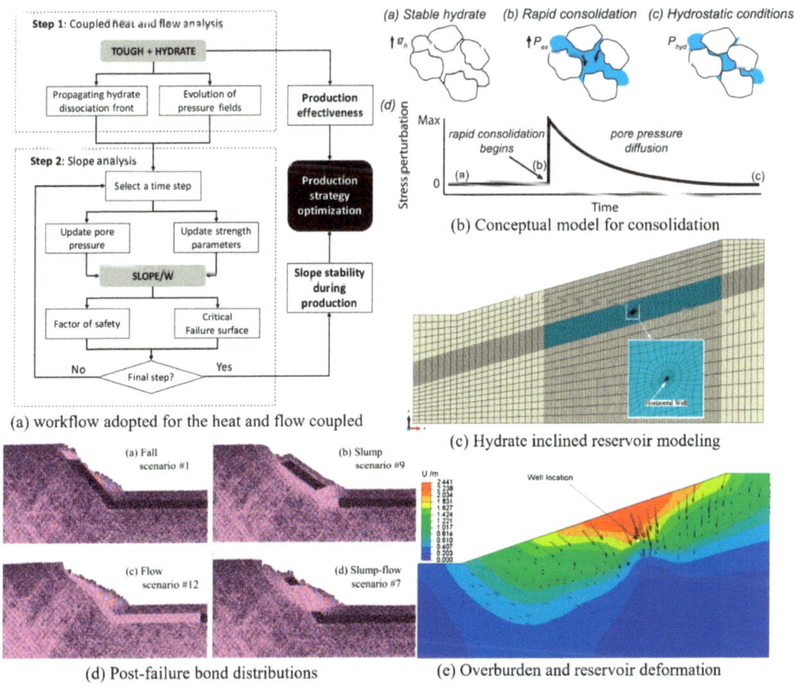

Figure 9. Numerical simulation study of the submarine landslide risk issues in hydrate formations (cited from Tan et al., 2021; Jiang et al., 2015; Handwerger et al., 2017; Song et al., 2019) [111–114].

Table 6. Hydrate depressurization production scheme and seafloor slope stability evaluation methodology.

Main Theoretical Model	Slope Stability Analysis Methods	Production Method	Mechanism of Slope Destabilization in Hydrate Formations	Inclination Angle	Reference
Multi-field coupled modeling of transient pore pressure due to hydrate decomposition	Limit equilibrium slope analysis method	Warming waters at the bottom of the slope trigger hydrate decomposition	Gas migration to overburden to form super pore pressure	3°, 20°	Liu et al. [110]
Modeling the dynamics of pore pressure and slope strength parameters during the production process	Limit equilibrium slope analysis method	Huff–puff method through a horizontal well	Strength loss due to super pore pressure formation and hydrate decomposition	10°	Tan et al. [111]
Coupling computational fluid dynamics and CFD-DEM model	Characterization of microfracture evolution	Instantaneous thermal dissociation of a methane hydrate	Dissipation of elastic strain energy in a short period of time driven by super pore pressure after hydrate decomposition	45°	Jiang et al. [112]
Ontological relationships for changes in sediment strength before and after hydrate decomposition	Rate and state friction model	Anomalous destabilization of hydrate	The presence of hydrates can significantly affect sediment strength increasing slope stability	5°~10°	Handwerger et al. [113]
Multi-field coupled model considering nonlinear theory and Moore–Cullen criterion	Orthogonal experimental design and the strength reduction method	Depressurization through a horizontal well	Both the extent of hydrate decomposition and the thickness of the overburden affect slope stability	15°	Song et al. [114]

3. Submarine Geological Risk Trigger Secondary Disaster

3.1. Gas Eruption

The factors inducing the generation of gas eruptions can be classified as natural geological activities and hydrate extraction disturbances. According to the scale of gas eruption, it can be classified into slow fluid migration caused by stratigraphic fractures and violent gas release caused by superporous pressure. Geological activity at depth causes natural gas to be transported upward and accumulate in the pores of this confined space, where gas hydrates are formed at low temperatures as the deposition process continues and pore pressure increases. However, geological activities such as earthquakes can create fissures in the strata that break up the confined space and create channels to the seafloor surface, leading to a sudden release of pore pressure and hydrate decomposition. Observational studies at the seafloor site in the Black Sea have shown that hydrate can decompose in situ into natural gas and rise to sea level in the form of gas bubbles [115], as shown in Figure 10. It is hypothesized that the behavior of gas hydrate decomposition in the Black Sea is due to natural factors, as geologic activity on the seafloor has created fractures in the reservoir buried beneath the overburden that are connected to the seafloor surface. Under the action of the natural environment, high-temperature fluid from geological activities such as volcanic eruptions increase hydrate reservoir temperature, leading to hydrate decomposition. Deep warm fluid has an obvious promoting effect on hydrate decomposition, changing the temperature and pressure conditions of the formation and affecting the internal structure of the formation, leading to the instability of the hydrate formation. Gas and water released from hydrate decomposition will gather and cannot be discharged from the formation in time, resulting in super pore pressure, and the pressure will gather to a certain extent, causing the formation to have fissure channels and resulting in gas eruption. Submarine earthquakes and volcanic activity can damage hydrate reservoirs and create fissures that connect to the seafloor surface. Natural gas is always less dense than seawater, and when there is a dense concentration of bubbles on the seafloor, the bubbles rise to form a plume. As the bubble plume rises, the bubble volume expands due to the decrease in pressure. Under the effect of buoyancy, the natural gas is eventually released into the atmosphere [116–119]. Natural gas hydrates endowed on the seafloor are dominated by methane hydrate, a greenhouse effect gas that has 21–25 times the ability to influence global temperatures than carbon dioxide of the same mass [120,121]. If the methane gas stored in the seabed strata is released, it will cause incalculable harm to the global environment.

Figure 10. Natural gas leakage due to in situ decomposition of submarine hydrate (cited from Pape et al., 2011) [115]. (**a**) Free gas bubble escape from seafloor (**b**) Gas bubbles escape from diverse orifices.

The disturbance of depressurization production causes gas hydrate decomposition in the reservoir to produce gas, due to the low permeability of hydrate-bearing reservoir, and the permeability of the overburden layer is lower than the permeability of the reservoir, so that the gas cannot be discharged in a timely manner and is easy to produce super pore pressure in the formation. When a combined depressurization and reservoir heating method is used, the high-temperature environment promotes large scale decomposition in the hydrate reservoir. Heating causes heat to move faster through the reservoir, while low permeability causes low pore pressure to move slowly through the reservoir. Low permeability hydrate-bearing sediment is present between the hydrate decomposition region and the low pore pressure region. This prevents the gas from flowing to the production well in a timely manner and tends to cause the gas to accumulate inside the reservoir, which in turn leads to the formation of super pore pressure. Sand production risk clogging wellbores or transport pipelines, and wellbore instability risk during depressurization production damaging production well, undoubtedly impede the normal extraction of natural gas from the reservoir and increase the risk of gas build-up in the reservoir. Sand production and wellbore instability during depressurization production affects gas production and prevents gas from escaping in time, leading to the formation of excess pore pressures. When the pore pressure reaches a limit that the overburden cannot withstand, fracture channels will be generated, causing sudden gas release and violent gas eruption [122,123].

Studying the slope instability of hydrate formation triggered by super porous pressure associated with gas hydrate development is of great significance in exploring the mechanism of formation instability triggered by gas eruptions. In an experimental study, the gas eruption phenomenon is triggered by hydrate formation destabilization caused by super porous pressure, as shown in Figure 11. Liu et al. [124] established a visual observation device to simulate the slope damage process, which can apply high pressure gas to the low-permeability chalk layer sediment to simulate the super pore pressure generated by hydrate decomposition. At the same time, a data acquisition system was used to monitor the physical and morphological damage processes of overburden. The data acquisition system was also used to monitor the physical and morphological processes of overlying seafloor damage. Under the action of super pore pressure caused by hydrate decomposition, the typical phenomena of overlying seafloor damage are pockmark deformation and shear damage. Zhang et al. [125] established a centrifugal simulation experiment system for seafloor landslides, and a high-pressure fluid can be passed into the slope device to simulate the super pore pressure phenomenon caused by gas hydrate decomposition. The test equipment helps to study the tensile damage behavior of hydrate formation slopes induced by super pore pressure and the formation shear instability caused by the dissipation of pore pressure due to crack formation during the downward movement of steep slopes. Climate warming and seafloor temperature rise have a relatively slow effect on hydrate decomposition, while the temperature at the bottom of the reservoir rises rapidly when the thermal stimulation method is used to extract hydrates. Song et al. [126] established a test system to simulate the effect of deep fluid migration on the stability of seafloor slopes. With the accumulation of gas in the device, the overburden of the hydrate reservoir deforms continuously, forming a dome on the seafloor, and the increase of super pore pressure will trigger the formation of hydraulic fractures at the edge of the dome, so that the gas is ejected from the seafloor to cause gas eruptions at the seafloor hydrate system landslides. In addition to indoor experimental studies, Sun et al. [127] established complex deep-sea engineering geologic in situ monitoring equipment for hydrate reservoir landslide monitoring and early warning technology, which utilizes seafloor three-dimensional electrical and acoustic measurements for in situ monitoring in order to obtain the parameters of sediment engineering properties. Sediment index parameters such as grain size, bulk weight, water content, porosity, and other sediment index parameters can be obtained through in situ long-term observation of spatial and temporal variations in seafloor sediment resistivity, acoustic velocity, and acoustic attenuation, and geophysical intrusion analysis [128]. Field monitoring data combined with indoor physical and mechanical characterization tests to establish the relationship between seafloor resistivity, acoustic parameters, and soil deformation strength indicators can be used to quantitatively describe the dynamic process of the hydrate decomposition-induced seafloor stratigraphic gas eruption disaster.

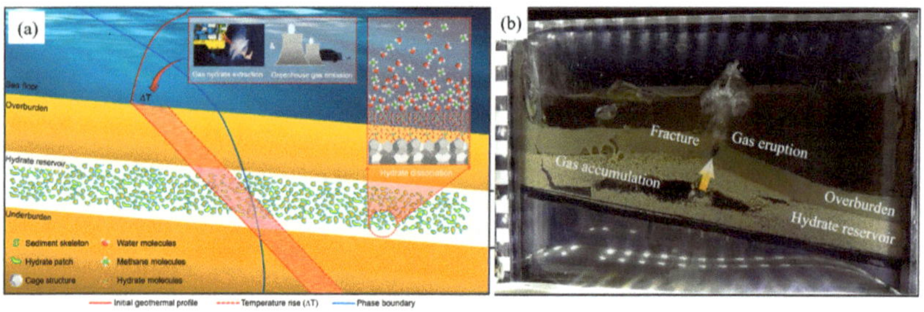

Figure 11. Gas eruption triggered by super pore pressure (cited from Song et al., 2023) [126]. (**a**) Effects of warming on hydrate reservoir (**b**) Gas eruptions cause slope damage.

3.2. Seawater Intrusion

Depressurization production releases natural gas from the reservoir and creates low pressure zones in the formation centered on the wellbore, with low pressure propagation expanding as extraction time increases. Under the influence of depressurization, osmotic pressure is generated between low pressure zones inside the reservoir and the seafloor–seawater boundary layer, which encourages seawater to flow through the low permeability overburden to the reservoir. The depressurization induces hydrate decomposition around the wellbore, and unlike overburden and hydrate-bearing sediments, the wellbore generally has higher permeability around the wellbore after hydrate decomposition. The pore pressure around the wellbore is the center of low pressure on the seafloor, and the strength of the sediments has been severely weakened after hydrate decomposition. In particular, the seafloor around the wellbore has a higher risk of overburden instability. Damage from overburden destabilization can easily form a connection between the reservoir and the seafloor, enhancing the permeability of the overburden and triggering an influx of seawater into the reservoir under osmotic pressure. If the high pore pressure area is connected to the seafloor, gas eruption will be formed; if low pressure area is connected to the seafloor, seawater intrusion will be formed, and the seawater intrusion schematic diagram is shown in Figure 12.

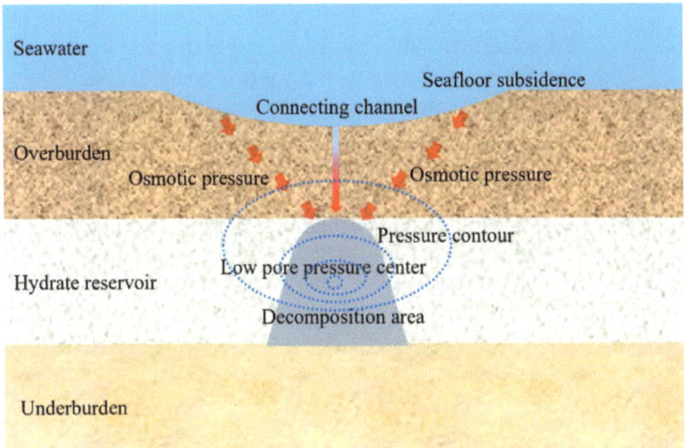

Figure 12. Schematic of seafloor subsidence caused by horizontal well depressurization production leading to seawater intrusion.

The entry of seawater into the reservoir changes the physical properties of the reservoir, increasing the water content of the reservoir and increasing the water production, which affects the extraction effect. From previous numerical simulation studies, it can be seen that the reservoir around the production wells may collapse due to stress concentration as mining continues. The hydrate decomposition range gradually increases, leading to large deformation in the overlying sediment layer. Uneven seafloor subsidence leads to destabilization or cracking in the overburden, and if the fracture connects the seafloor to the reservoir, then seawater intrusion will be triggered by osmotic pressure [102,104,129,130]. The destabilization or damage of the wellbore will also easily cause seawater to flow into the reservoir through the wellbore or the surrounding sediments under osmotic pressure, which will lead to seawater backflow. In order to effectively prevent wellbore instability, seafloor subsidence, and the risk of seawater backup, it is necessary to adopt mechanical property enhancement measures for the overburden layer above the wellbore. For example, grouting can be used to reinforce the overburden and reduce permeability at the same time.

3.3. Prospects

During depressurization production, there are potential submarine geological risks of sand production, wellbore instability, seafloor subsidence, and submarine landslides, and risk issues can cause secondary disasters dominated by gas eruptions and seawater intrusion. The current research on the risk of sand production includes field observation, numerical simulation and experimental research methods. Research on wellbore instability mainly adopts numerical simulation methods, and the research on seafloor subsidence and submarine landslides mainly adopts numerical simulation methods with experimental and a small amount of field observation research. There are fewer studies on secondary hazards, among which the experimental study of gas eruption lacks consideration of the disturbance of hydrate reservoir by depressurization production, and the study of seawater intrusion is basically in the blank. The field observation study of marine hydrate extraction is limited due to site and cost reasons. Experimental studies on the mechanical properties of hydrate-bearing sediment can help us to understand the evolution of the mechanical properties of the reservoir during depressurization production and reveal the mechanism of seafloor subsidence and submarine landslides. At present, numerical simulations and experimental studies are the mainstays of research on submarine geological risks and secondary disasters during depressurization production. Methods for studying the geological risks arising from submarine hydrate extraction cover a wide range of aspects, yet there is a gap between field and indoor studies, mainly due to the fact that indoor tests fail to adequately reproduce the in situ geological and engineering environments in the field. By comprehensively utilizing a variety of research tools on the geological risk of gas hydrate, it is possible to gain a more comprehensive and in-depth understanding of the mechanism and characteristics of the submarine geological risk triggered by the perturbation of gas hydrate extraction, so as to provide a scientific basis for preventing and responding to the related geological hazards. A large-scale three-dimensional natural gas hydrate reservoir submarine geological risk test system should be developed to realistically reproduce the process of stratum instability and secondary geological hazards triggered by hydrate production disturbance in a high pressure and low temperature environment. Main influencing factors inducing the disasters should be sensed with the help of a full-profile and refined multi-physical quantity characterization system of the on-site geophysical observation, and the differences between indoor tests and the actual submarine environment can be overcome by combining these with numerical simulation to obtain an accurate and precise understanding. Combined with numerical simulation to overcome the differences between indoor experiments and the actual seabed environment, accurate and reliable experimental results can be obtained to improve the understanding of the mechanism of submarine geologic risk triggered by the disturbance of natural gas hydrate exploitation. The first and second production test projects of natural gas hydrates were carried out in the South China Sea in 2017 and 2020, and significant technological breakthroughs were achieved. Currently, the Ministry of Natural Resources is making every effort to push forward the preparatory work for the third hydrate production test project in the South China Sea. In future studies, the laboratory needs to build 3D, large-size equipment to simulate the real seafloor environment. Due to the complexity of hydrate decomposition and its multi-field coupling behavior, refined AI algorithms are needed for numerical calculations in future simulation studies.

4. Conclusions

Depressurization production of marine gas hydrates is prone to geological risks such as sand production, wellbore instability, seafloor subsidence, and seafloor landslides, as well as potential secondary hazards such as gas eruption and seawater intrusion. The research methods for submarine geological risk issues cover many aspects, including geophysical observation, indoor experimental research, and numerical simulation analysis. This paper reviews the progress of theoretical modeling and the potential geological risk problems during depressurization production on the basis of previous research on submarine engineering geological issues, and mainly obtains the following conclusions and outlooks:

(1) The threat level of potential geologic risk issues in hydrate formation destabilization can be classified into four aspects from shallow to deep: sand production, wellbore instability, seafloor subsidence, and submarine landslides. Geologic risk issues, in turn, cause secondary disasters dominated by natural gas eruptions and seawater intrusion. When assessing the geologic risks, theoretical modeling needs to be based on the perspective of multi-field coupling, in which the accurate description of reservoir permeability, temperature, geomechanics, and other control models are the key basis.

(2) Sand production is sensitive to production pressure and sediment mechanical properties, and excessive sand production seriously affects gas production. A large amount of sand production leads to the creation of holes in the reservoir and even affects overburden stability, which is not conducive to the maintenance of stratigraphic stability. If the development of holes is serious, it will damage the seafloor surface and lead to seawater intrusion into the reservoir.

(3) Hydrate decomposition reduces the mechanical strength of the sediment around the wellbore, causing it to reach a plastic yield state, and the plastic state of the sediment makes the wellbore susceptible to instability. Wellbore destabilization leads to gas leakage and environmental contamination and can also lead to seawater intrusion due to the connection between the seabed surface and the reservoir.

(4) The gas and water released from the hydrate decomposition are aggregated and cannot be discharged in time to generate super pore pressure, and the pressure is aggregated to a certain level to cause gas eruption through the fracture channels in the formation under the effects of overburden failure. The pore pressure around the wellbore is a low-pressure center on the seabed that triggers seawater influx into the reservoir under the promotion of osmotic pressure.

(5) Hydrate decomposition weakens the mechanical strength of the reservoir and may generate super porous pressure, inducing a submarine landslide or a gas eruption. Hydrate reservoirs need to be filled with gravel to enhance their mechanical strength and permeability, and overburden needs to be grouted to reinforce stability. The study of submarine geological risk and its secondary disasters requires the cross-application of various research tools.

Author Contributions: Writing—original draft preparation, X.M., Y.J. and P.Y.; conceptualization, P.Y. and H.L.; methodology, X.M. and Y.J.; data curation, C.W., Q.S. and X.C.; investigation, X.C. and P.Y.; resources, Y.J. and H.L.; writing—review and editing, P.Y. and X.M.; visualization, H.L. and X.M. All authors have read and agreed to the published version of the manuscript.

Funding: This research was funded by the Shandong Provincial Natural Science Foundation (Grant No. ZR2019ZD14) and the Youth Innovation Team of the Shandong Higher Education Institutions (2022KJ214).

Institutional Review Board Statement: Not applicable.

Informed Consent Statement: Not applicable.

Data Availability Statement: Data will be made available on request.

Conflicts of Interest: The authors declare that they have no known competing financial interests or personal relationships that could have appeared to influence the work reported in this paper.

References

1. Sloan, E.D. Fundamental principles and applications of natural gas hydrates. *Nature* **2003**, *426*, 353–359. [CrossRef]
2. Yan, C.; Ren, X.; Cheng, Y.; Song, B.; Li, Y.; Tian, W. Geomechanical issues in the exploitation of natural gas hydrate. *Gondwana Res.* **2020**, *81*, 403–422. [CrossRef]
3. Dong, L.; Wu, N.; Zhang, Y.; Liao, H.; Hu, G.; Li, Y. Improved Duncan-Chang model for reconstituted hydrate-bearing clayey silt from the South China Sea. *Adv. Geo-Energy Res.* **2023**, *8*, 136–140. [CrossRef]
4. Li, Y.; Dong, L.; Wu, N.; Nouri, A.; Liao, H.; Chen, Q.; Sun, J.; Liu, C. Influences of hydrate layered distribution patterns on triaxial shearing characteristics of hydrate-bearing sediments. *Eng. Geol.* **2021**, *294*, 106375. [CrossRef]
5. Wan, Y.; Yuan, Y.; Zhou, C.; Liu, L. Multiphysics coupling in exploitation and utilization of geo-energy: State-of-the-art and future perspectives. *Adv. Geo-Energy Res.* **2023**, *10*, 7–13. [CrossRef]
6. Dong, L.; Li, Y.; Wu, N.; Wan, Y.; Liao, H.; Wang, H.; Zhang, Y.; Ji, Y.; Hu, G.; Leonenko, Y. Numerical simulation of gas extraction performance from hydrate reservoirs using double-well systems. *Energy* **2023**, *265*, 126382. [CrossRef]
7. Wu, N.; Zhang, H.; Yang, S.; Zhang, G.; Liang, J.; Lu, J.; Su, X.; Schultheiss, P.; Holland, M.; Zhu, Y. Gas hydrate system of Shenhu area, Northern South China Sea: Geochemical Results. *J. Geol. Res.* **2011**, *2011*, 370298. [CrossRef]
8. Moridis, G.J.; Reagan, M.T.; Kim, S.J.; Seol, Y.; Zhang, K. Evaluation of the gas production potential of marine hydrate deposits in the Ulleung basin of the Korean East Aea. *SPE J.* **2007**, *14*, 759–781. [CrossRef]
9. Li, Y.; Wu, N.; He, C.Q.; Sun, Z.; Zhang, Z.; Hao, X.; Chen, Q.; Bu, Q.; Liu, C.; Sun, J. Nucleation probability and memory effect of methane-propane mixed gas hydrate. *Fuel* **2021**, *291*, 120103. [CrossRef]
10. Zhao, J.; Zhu, Z.; Song, Y.; Liu, W.; Zhang, Y.; Wang, D. Analyzing the process of gas production for natural gas hydrate using depressurization. *Appl. Energy* **2015**, *142*, 125–134. [CrossRef]
11. Yan, P.; Luan, H.; Jiang, Y.; Liang, W.; Liu, M.; Chen, H. Influence of depressurization mode on natural gas hydrate production characteristics: One-dimensional experimental study. *J. Petrol. Sci. Eng.* **2024**, *234*, 212671. [CrossRef]
12. Yang, Z.; Si, H.; Zhong, D. AI-based composition model for energy utilization efficiency optimization of gas hydrate recovery by combined method of depressurization and thermal stimulation. *J. Nat. Gas Sci. Eng.* **2021**, *92*, 104001. [CrossRef]
13. Song, Y.; Zhou, H.; Wang, P.; Yang, M. Prediction of clathrate hydrate phase equilibria using gradient boosted regression trees and deep neural networks. *J. Chem. Thermodyn.* **2019**, *135*, 86–96. [CrossRef]
14. Sadi, M.; Fakharian, H.; Ganji, H.; Kakavand, M. Evolving artificial intelligence techniques to model the hydrate-based desalination process of produced water. *J. Water Reuse. Desal.* **2019**, *9*, 372–384. [CrossRef]

15. Li, J.; Ye, J.; Qin, X.; Qiu, H.; Wu, N.; Lu, H.; Xie, W.; Lu, J.; Peng, F.; Xu, Z.; et al. The first offshore natural gas hydrate production test in South China Sea. *China Geol.* **2018**, *1*, 5–16. [CrossRef]
16. Xue, K.; Liu, Y.; Yu, T.; Yang, L.; Zhao, J.; Song, Y. Numerical simulation of gas hydrate production in shenhu area using depressurization: The effect of reservoir permeability heterogeneity. *Energy* **2023**, *271*, 126948. [CrossRef]
17. Ye, J.; Qin, X.; Xie, W.; Lu, H.; Ma, B.; Qiu, H.; Liang, J.; Lu, J.; Kuang, Z.; Lu, C.; et al. The second natural gas hydrate production test in the South China Sea. *China Geolo.* **2020**, *3*, 197–209. [CrossRef]
18. Yu, T.; Guan, G.; Wang, D.; Song, Y.; Abudula, A. Numerical evaluation on the effect of horizontal-well systems on the long-term gas hydrate production behavior at the second Shenhu test site. *J. Nat. Gas Sci. Eng.* **2021**, *95*, 104200. [CrossRef]
19. Xiao, C.W.; Li, X.S.; Li, G.; Yu, Y.; Yu, J.; Lv, Q. Numerical analysis of production behaviors and permeability characteristics on the second gas hydrate production test in the South China Sea. *Energy Fuels* **2022**, *36*, 10960–10974. [CrossRef]
20. Zhu, Y.; Wang, P.; Pang, S.; Zhang, S.; Xiao, R. A review of the resource and test production of natural gas hydrates in China. *Energy Fuels* **2021**, *35*, 9137–9150. [CrossRef]
21. Sun, J.; Ning, F.; Zhang, L.; Liu, T.; Peng, L.; Liu, Z.; Li, C.; Jiang, G. Numerical simulation on gas production from hydrate reservoir at the 1st offshore test site in the eastern Nankai Trough. *J. Nat. Gas Sci. Eng.* **2016**, *30*, 64–76. [CrossRef]
22. Chen, L.; Feng, Y.; Kogawa, T.; Okajima, J.; Komiya, A.; Maruyama, S. Construction and simulation of reservoir scale layered model for production and utilization of methane hydrate: The case of Nankai Trough Japan. *Energy* **2018**, *143*, 128–140. [CrossRef]
23. Yu, T.; Guan, G.; Wang, D.; Song, Y.; Abudula, A. Numerical investigation on the long-term gas production behavior at the 2017 Shenhu methane hydrate production site. *Appl. Energy* **2021**, *285*, 116466. [CrossRef]
24. Yin, F.; Gao, Y.; Chen, Y.; Sun, B.; Li, S.; Zhao, D. Numerical investigation on the long-term production behavior of horizontal well at the gas hydrate production site in South China Sea. *Appl. Energy* **2022**, *311*, 118603. [CrossRef]
25. Jiang, Y.; Zhang, R.; Ye, R.; Zhou, K.; Gong, B.; Golsanami, N. Mechanical properties of nodular natural gas hydrate-bearing sediment. *Adv. Geo-Energy* **2024**, *11*, 41–53. [CrossRef]
26. Yang, Y.; He, Y.; Zheng, Q. An analysis of the key safety technologies for natural gas hydrate exploitation. *Adv. Geo-Energy* **2017**, *1*, 100–104. [CrossRef]
27. Yang, L.; Wang, J.; Jiang, Y. Experimental study and numerical simulation of overlying layer soil failure caused by hydrate decomposition. *ACS Omega* **2020**, *5*, 31244–31253. [CrossRef] [PubMed]
28. Yan, C.; Cheng, Y.; Li, M.; Han, Z.; Zhang, H.; Li, Q.; Teng, F.; Ding, J. Mechanical experiments and constitutive model of natural gas hydrate reservoirs. *Int. J. Hydrogen Energy* **2017**, *42*, 19810–19818. [CrossRef]
29. Chong, Z.R.; Yang, S.H.B.; Babu, P.; Linga, P.; Li, X. Review of natural gas hydrates as an energy resource: Prospects and challenges. *Appl. Energy* **2016**, *162*, 1633–1652. [CrossRef]
30. Wegner, S.A.; Campbell, K.J. Drilling hazard assessment for hydrate bearing sediments including drilling through the bottom-simulating reflectors. *Mar. Petrol. Geol.* **2014**, *58*, 382–405. [CrossRef]
31. Best, A.I.; Richardson, M.D.; Boudreau, B.P.; Alan, G.B.; Ira, L.J.; Lyons, A.P.; Martens, C.S.; Orange, D.L.; Wheeler, S.J. Shallow seabed methane gas could pose coastal hazard. *Trans. Am. Geophys.* **2006**, *87*, 213–217. [CrossRef]
32. Li, Y.; Wu, N.; Ning, F.; Hu, G.; Liu, C.; Dong, C.; Lu, J. A sand-production control system for gas production from clayey silt hydrate reservoirs. *China Geol.* **2019**, *2*, 121–132. [CrossRef]
33. Wu, N.; Li, Y.; Chen, Q.; Liu, C.; Jin, Y.; Tan, M.; Dong, L.; Hu, G. Sand production management during marine natural gas hydrate exploitation: Review and an innovative solution. *Energy Fuels* **2021**, *35*, 4617–4632. [CrossRef]
34. Deng, F.; Huang, B.; Li, X.; Liu, J.; Li, G.; Xu, Y.; Yin, B. Review of Sand Control and Sand Production in a Gas Hydrate Reservoir. *Energy Fuels* **2022**, *36*, 11712–11723. [CrossRef]
35. Li, Y.; Wu, N.; Gao, D.; Chen, Q.; Liu, C.; Yang, D.; Jin, Y.; Ning, F.; Tan, M.; Hu, G. Optimization and analysis of gravel packing parameters in horizontal wells for natural gas hydrate production. *Energy* **2021**, *219*, 119585. [CrossRef]
36. Fang, X.; Ning, F.; Wang, L.; Liu, Z.; Lu, H.; Yu, Y.; Li, Y.; Sun, J.; Shi, H.; Zhao, Y.; et al. Dynamic coupling responses and sand production behavior of gas hydrate-bearing sediments during depressurization: An experimental study. *J. Petrol. Sci. Eng.* **2021**, *201*, 108506. [CrossRef]
37. Li, X.S.; Xu, C.G.; Zhang, Y.; Ruan, X.K.; Li, G.; Wang, Y. Investigation into gas production from natural gas hydrate: A review. *Appl. Energy* **2016**, *172*, 286–322. [CrossRef]
38. Yan, C.; Li, Y.; Cheng, Y.; Wang, W.; Song, B.; Deng, F.; Feng, Y. Sand production evaluation during gas production from natural gas hydrates. *J. Nat. Gas Sci. Eng.* **2018**, *57*, 77–88. [CrossRef]
39. Terao, Y.; Duncan, M.; Hay, B.; Dang, L. Deepwater methane hydrate gravel packing completion results and challenges. In Proceedings of the Offshore Technology Conference, OTC, D031S034R007, Houston, TX, USA, 5–8 May 2014. [CrossRef]
40. Li, Y.L.; Liu, C.L.; Liu, L.; Sun, J.; Liu, H.; Meng, Q. Experimental study on evolution behaviors of triaxial-shearing parameters for hydrate-bearing intermediate fine sediment. *Adv. Geo-Energy Res.* **2018**, *2*, 43–52. [CrossRef]
41. Liu, P.; Zhang, W.; Mao, S.; Su, P.; Chen, H.; Hu, L. Study on the Mechanism of Natural Gas Hydrate Decomposition and Seabed Seepage Triggered by Mass Transport Deposits. *J. Mar. Sci. Eng.* **2024**, *12*, 646. [CrossRef]
42. Cui, Y.; Nouri, A.; Chan, D.; Rahmati, E. A new approach to DEM simulation of sand production. *J. Petrol. Sci. Eng.* **2016**, *147*, 56–67. [CrossRef]
43. Uchida, S.; Klar, A.; Yamamoto, K. Sand production model in gas hydrate-bearing sediments. *Int. J. Rock Mech. Min.* **2016**, *86*, 303–316. [CrossRef]

44. Zhu, H.; Xu, T.; Yuan, Y.; Xia, Y.; Xin, X. Numerical investigation of the natural gas hydrate production tests in the Nankai Trough by incorporating sand migration. *Appl. Energy* **2020**, *275*, 115384. [CrossRef]
45. Shahsavari, M.H.; Khamehchi, E.; Fattahpour, V.; Molladavoodi, H. Investigation of sand production prediction shortcomings in terms of numerical uncertainties and experimental simplifications. *J. Petrol. Sci. Eng.* **2021**, *207*, 109147. [CrossRef]
46. Li, X.; Wan, Y.; Lei, G.; Sun, J.; Cheng, W.; Dou, X.; Zhao, Y.; Ning, F. Numerical investigation of gas and sand production from hydrate-bearing sediments by incorporating sand migration based on IMPES method. *Energy* **2024**, *288*, 129556. [CrossRef]
47. Kozhagulova, A.; Minh, N.H.; Zhao, Y.; Fok, S.C. Experimental and analytical investigation of sand production in weak formations for multiple well shut-ins. *J. Petrol. Sci. Eng.* **2020**, *195*, 107628. [CrossRef]
48. Zivar, D.; Shad, S.; Foroozesh, J.; Salmanpour, S. Experimental study of sand production and permeability enhancement of unconsolidated rocks under different stress conditions. *J. Petrol. Sci. Eng.* **2019**, *181*, 106238. [CrossRef]
49. Song, Y.; Ranjith, P.G.; Wu, B. Development and experimental validation of a computational fluid dynamics-discrete element method sand production model. *J. Nat. Gas Sci. Eng.* **2020**, *73*, 103052. [CrossRef]
50. Li, X.; Hu, H.; Wang, Y.; Li, X. Experimental study of gas-liquid-sand production behaviors during gas hydrates dissociation with sand control screen. *Energy* **2022**, *254*, 124414. [CrossRef]
51. Ding, J.; Cheng, Y.; Yan, C.; Song, B.; Sun, H.; Teng, F. Experimental study of sand control in a natural gas hydrate reservoir in the South China sea. *Int. J. Hydrogen Energy* **2019**, *44*, 23639–23648. [CrossRef]
52. Wang, Y.; Long, F.; Li, X.; Xu, Y.; Kou, X. Experimental Study of Sand Migration under Distinct Sand Control Methods during Gas Hydrate Decomposition by Depressurization. *Energy Fuels* **2023**, *37*, 12966–12979. [CrossRef]
53. Li, Y.; Wu, N.; Ning, F.; Gao, D.; Hao, X.; Chen, Q.; Liu, C.; Sun, J. Hydrate-induced clogging of sand-control screen and its implication on hydrate production operation. *Energy* **2020**, *206*, 118030. [CrossRef]
54. Dong, C.Y.; Zhong, Y.X.; Wu, Y.X.; Zhou, Y.; Zeng, S.; Yan, Q. Experimental study on sand retention mechanisms and feasibility evaluation of sand control for gas hydrate reservoirs with highly clayey fine sands. *J. China Univ. Pet.* **2018**, *42*, 79–87. [CrossRef]
55. Fereidounpour, A.; Vatani, A. Designing a polyacrylate drilling fluid system to improve wellbore stability in hydrate bearing sediments. *J. Nat. Gas Sci. Eng.* **2015**, *26*, 921–926. [CrossRef]
56. Aregbe, A.G. Wellbore stability problems in deepwater gas wells. *World J. Eng. Technol.* **2017**, *5*, 626. [CrossRef]
57. Golmohammadi, S.M.; Nakhaee, A. A cylindrical model for hydrate dissociation near wellbore during drilling operations. *J. Nat. Gas Sci. Eng.* **2015**, *27*, 1641–1648. [CrossRef]
58. Li, Q.; Liu, J.; Wang, S.; Guo, Y.; Han, X.; Li, Q.; Cheng, Y.; Dong, Z.; Li, X.; Zhang, X. Numerical insights into factors affecting collapse behavior of horizontal wellbore in clayey silt hydrate-bearing sediments and the accompanying control strategy. *Ocean Eng.* **2024**, *297*, 117029. [CrossRef]
59. Li, B.; Li, H.; Guo, B.; Cai, X. A new numerical solution to predict the temperature profile of gas-hydrate-well drilling. *SPE J.* **2017**, *22*, 1201–1212. [CrossRef]
60. Motghare, P.D.; Musale, A. Unconventional hydrocarbons: Gas hydrates-drilling challenges and suitable technology. In Proceedings of the SPE Oil and Gas India Conference and Exhibition, Mumbai, India, 4–6 April 2017; p. D021S009R002. [CrossRef]
61. Guo, Y.; Sun, B.; Zhao, K.; Zhang, H. A prediction method of natural gas hydrate formation in deepwater gas well and its application. *Petroleum* **2016**, *2*, 296–300. [CrossRef]
62. Liu, M.; Jin, Y.; Lu, Y.; Chen, M.; Hou, B.; Chen, W.; Wen, X.; Yu, X. A wellbore stability model for a deviated well in a transversely isotropic formation considering poroelastic effects. *Rock Mech. Rock Eng.* **2016**, *49*, 3671–3686. [CrossRef]
63. Qiu, K.; Yamamoto, K.; Birchwood, R.; Chen, Y. Well-integrity evaluation for methane-hydrate production in the deepwater Nankai Trough. *SPE Drill Complet.* **2015**, *30*, 52–67. [CrossRef]
64. Joye, S.B.; Bracco, A.; Ozgokmen, T.M.; Chanton, J.P.; Grosell, M.; MacDonald, I.R.; Cordes, E.E.; Montoya, J.P.; Passow, U. The Gulf of Mexico ecosystem, six years after the Macondo oil well blowout. *Deep-sea Res. Pt II* **2016**, *129*, 4–19. [CrossRef]
65. Wei, R.; Xia, Y.; Wang, Z.; Li, Q.; Lv, X.; Leng, S.; Zhang, L.; Zhang, Y.; Xiao, B.; Yang, S.; et al. Long-term numerical simulation of a joint production of gas hydrate and underlying shallow gas through dual horizontal wells in the South China Sea. *Appl. Energy* **2022**, *320*, 119235. [CrossRef]
66. Wang, H.N.; Chen, X.P.; Jiang, M.J.; Guo, Z.Y. Analytical investigation of wellbore stability during drilling in marine methane hydrate-bearing sediments. *J. Nat. Gas Sci. Eng.* **2019**, *68*, 102885. [CrossRef]
67. Liao, Y.; Wang, Z.; Chao, M.; Sun, X.; Wang, J.; Zhou, B.; Sun, B. Coupled wellbore–reservoir heat and mass transfer model for horizontal drilling through hydrate reservoir and application in wellbore stability analysis. *J. Nat. Gas Sci. Eng.* **2021**, *95*, 104216. [CrossRef]
68. Guo, Z.Y.; Wang, H.N.; Jiang, M.J. Elastoplastic analytical investigation of wellbore stability for drilling in methane hydrate-bearing sediments. *J. Nat. Gas Sci. Eng.* **2020**, *79*, 103344. [CrossRef]
69. Zhang, H.; Cheng, Y.; Li, Q.; Yan, C.; Han, X. Numerical analysis of wellbore instability in gas hydrate formation during deep-water drilling. *J. Ocean. Univ. China* **2018**, *17*, 8–16. [CrossRef]
70. Sun, J.; Ning, F.; Lei, H.; Gai, X.; Sanchez, M.; Lu, J.; Li, Y.; Liu, L.; Liu, C.; Wu, N.; et al. Wellbore stability analysis during drilling through marine gas hydrate-bearing sediments in Shenhu area: A case study. *J. Petrol. Sci. Eng.* **2018**, *170*, 345–367. [CrossRef]
71. Dong, L.; Wan, Y.; Li, Y.; Liao, H.; Liu, C.; Wu, N.; Leonenko, Y. 3D numerical simulation on drilling fluid invasion into natural gas hydrate reservoirs. *Energy* **2022**, *241*, 122932. [CrossRef]

72. Rutqvist, J.; Moridis, G.J.; Grover, T.; Silpngarmlert, S.; Collett, T.S.; Holdich, S.A. Coupled multiphase fluid flow and wellbore stability analysis associated with gas production from oceanic hydrate-bearing sediments. *J. Petrol. Sci. Eng.* **2012**, *92*, 65–81. [CrossRef]
73. Li, Y.; Cheng, Y.; Yan, C.; Song, L.; Liu, H.; Tian, W.; Ren, X. Mechanical study on the wellbore stability of horizontal wells in natural gas hydrate reservoirs. *J. Nat. Gas Sci. Eng.* **2020**, *79*, 103359. [CrossRef]
74. Yuan, Y.L.; Xu, T.F.; Xin, X.; Xia, Y.; Li, B. Mechanical stability analysis of strata and wellbore associated with gas production from oceanic hydrate-bearing sediments by depressurization. *Chin. J. Theor. Appl. Mech.* **2020**, *52*, 544–555. [CrossRef]
75. Sun, X.; Luo, T.; Wang, L.; Wang, H.; Song, Y.; Li, Y. Numerical simulation of gas recovery from a low-permeability hydrate reservoir by depressurization. *Appl. Energy* **2019**, *250*, 7–18. [CrossRef]
76. Dong, B.C.; Xiao, P.; Sun, Y.F.; Kan, J.Y.; Yang, M.K.; Peng, X.W.; Sun, C.Y.; Chen, G.J. Coupled flow and geomechanical analysis for gas production from marine heterogeneous hydrate-bearing sediments. *Energy* **2022**, *255*, 124501. [CrossRef]
77. Li, Y.; Liu, L.; Jin, Y.; Wu, N. Characterization and development of natural gas hydrate in marine clayey-silt reservoirs: A review and discussion. *Adv. Geo-Energy* **2021**, *5*, 75–86. [CrossRef]
78. Ding, Y.; Qian, A.; Lu, H.; Li, Y.; Zhang, Y. DEM investigation of the effect of hydrate morphology on the mechanical properties of hydrate-bearing sands. *Comput. Geotech.* **2022**, *143*, 104603. [CrossRef]
79. Chen, H.; Du, H.; Shi, B.; Shan, W.; Hou, J. Mechanical properties and strength criterion of clayey sand reservoirs during natural gas hydrate extraction. *Energy* **2022**, *242*, 122526. [CrossRef]
80. Yan, C.; Chen, Y.; Tian, W.; Cheng, Y.; Li, Y. Effects of methane-carbon dioxide replacement on the mechanical properties of natural gas hydrate reservoirs. *J. Clean Prod.* **2022**, *354*, 131703. [CrossRef]
81. Lee, Y.; Deusner, C.; Kossel, E.; Choi, W.; Seo, Y.; Haeckel, M. Influence of CH4 hydrate exploitation using depressurization and replacement methods on mechanical strength of hydrate-bearing sediment. *Appl. Energy* **2020**, *277*, 115569. [CrossRef]
82. Wang, L.; Li, Y.; Wu, P.; Shen, S.; Liu, T.; Leng, S.; Chang, Y.; Zhao, J. Physical and mechanical properties of the overburden layer on gas hydrate-bearing sediments of the South China sea. *J. Petrol. Sci. Eng.* **2020**, *189*, 107020. [CrossRef]
83. Wang, X.; Sun, Y.; Li, B.; Zhang, G.; Guo, W.; Li, S.; Jiang, S.; Peng, S.; Chen, H. Reservoir stimulation of marine natural gas hydrate-a review. *Energy* **2023**, *263*, 126120. [CrossRef]
84. Yin, F.; Gao, Y.; Zhang, H.; Sun, B.; Chen, Y.; Gao, D.; Zhao, X. Comprehensive evaluation of gas production efficiency and reservoir stability of horizontal well with different depressurization methods in low permeability hydrate reservoir. *Energy* **2022**, *239*, 122422. [CrossRef]
85. Jiang, Y.; Ma, X.; Luan, H.; Liang, W.; Yan, P.; Song, W.; Shan, Q. Numerical simulation on the evolution of physical and mechanical characteristics of natural gas hydrate reservoir during depressurization production. *J. Nat. Gas Sci. Eng.* **2022**, *108*, 104803. [CrossRef]
86. Fang, H.L. A fully coupled thermo-hydro-mechanical model for methane hydrate reservoir simulations. In *Advances in Environmental Geotechnics: Proceedings of the International Symposium on Geoenvironmental Engineering in Hangzhou, China, 8–10 September 2009*; Springer: Berlin/Heidelberg, Germany, 2010; pp. 455–461. [CrossRef]
87. Uchida, S.; Klar, A.; Yamamoto, K. Sand production modeling of the 2013 Nankai offshore gas production test. In Proceedings of the International Conference on Energy Geotechnics, Kiel, Germany, 29–31 August 2016; pp. 451–458. [CrossRef]
88. Lin, J.S.; Uchida, S.; Myshakin, E.M.; Seol, Y.; Rutqvist, J.; Boswell, R. Assessing the geomechanical stability of interbedded hydrate-bearing sediments under gas production by depressurization at NGHP-02 Site 16. *Mar. Petrol. Geol.* **2019**, *108*, 648–659. [CrossRef]
89. Jin, G.; Lei, H.; Xu, T.; Xin, X.; Yuan, Y.; Xia, Y.; Juo, J. Simulated geomechanical responses to marine methane hydrate recovery using horizontal wells in the Shenhu area, South China Sea. *Mar. Petrol. Geol.* **2018**, *92*, 424–436. [CrossRef]
90. Zhang, Y.; Zhang, P.; Hui, C.; Tian, S.; Zhang, B. Numerical analysis of the geomechanical responses during natural gas hydrate production by multilateral wells. *Energy* **2023**, *269*, 126810. [CrossRef]
91. Yuan, Y.; Xu, T.; Jin, C.; Zhu, H.; Gong, Y.; Wang, F. Multiphase flow and mechanical behaviors induced by gas production from clayey-silt hydrate reservoirs using horizontal well. *J. Clean Prod.* **2021**, *328*, 129578. [CrossRef]
92. Liu, Z.; Kim, J.; Hu, G.; Hu, W.; Ning, F. Geomechanical property evolution of hydrate-bearing sediments under dynamic loads: Nonlinear behaviors of modulus and damping ratio. *Eng. Geol.* **2021**, *295*, 106427. [CrossRef]
93. You, Z.; Hao, Y.; Hu, W.; Shen, S.; Wu, P.; Li, Y. Strength analysis of hydrate-bearing sandy sediments in excess gas and excess water based on drained triaxial compression tests. *Eng. Geol.* **2023**, *325*, 107292. [CrossRef]
94. Liu, T.; Tang, H.; Wu, P.; Wang, H.; Song, Y.; Li, Y. Acoustic characteristics on clayey-silty sediments of the South China Sea during methane hydrate formation and dissociation. *Energy* **2023**, *282*, 128978. [CrossRef]
95. Yoneda, J.; Jin, Y.; Muraoka, M.; Oshima, M.; Suzuki, K.; Walker, M.; Otsuki, S.; Kumagai, K.; Collett, T.S.; Boswell, R.; et al. Multiple physical properties of gas hydrate-bearing sediments recovered from Alaska North Slope 2018 Hydrate-01 Stratigraphic Test Well. *Mar. Petrol. Geol.* **2021**, *123*, 104748. [CrossRef]
96. McAdoo, B.G.; Watts, P. Tsunami hazard from submarine landslides on the Oregon continental slope. *Mar. Geol.* **2004**, *203*, 235–245. [CrossRef]
97. Masson, D.G.; Harbitz, C.B.; Wynn, R.B.; Pedersen, G.; Lovholt, F. Submarine landslides: Processes, triggers and hazard prediction. Philosophical Transactions of the Royal Society A: Mathematical. *Phys. Eng. Sci.* **2006**, *364*, 2009–2039. [CrossRef] [PubMed]
98. Fell, R. Landslide risk assessment and acceptable risk. *Can. Geotech. J.* **1994**, *31*, 261–272. [CrossRef]

99. Fell, R.; Corominas, J.; Bonnard, C.; Cascini, L.; Leroie, E.; Savage, W.Z. Guidelines for landslide susceptibility, hazard and risk zoning for land use planning. *Eng. Geol.* **2008**, *102*, 85–98. [CrossRef]
100. Dai, F.C.; Lee, C.F.; Ngai, Y.Y. Landslide risk assessment and management: An overview. *Eng. Geol.* **2002**, *64*, 65–87. [CrossRef]
101. Kleinberg, R.L.; Flaum, C.; Griffin, D.D.; Brewer, P.G.; Malby, G.E.; Peltzer, E.T.; Yesinowski, J.P. Deep sea NMR: Methane hydrate growth habit in porous media and its relationship to hydraulic permeability, deposit accumulation, and submarine slope stability. *J. Geophys. Res. Sol. Earth* **2003**, *108*, 2508. [CrossRef]
102. Nixon, M.F.; Grozic, J.L.H. Submarine slope failure due to gas hydrate dissociation: A preliminary quantification. *Can. Geotech. J.* **2007**, *44*, 314–325. [CrossRef]
103. Zhang, X.H.; Lu, X.B.; Shi, Y.H.; Xia, Z.; Liu, W.T. Centrifuge experimental study on instability of seabed stratum caused by gas hydrate dissociation. *Ocean Eng.* **2015**, *105*, 1–9. [CrossRef]
104. Zhang, X.H.; Lu, X.B.; Chen, X.D.; Zhang, L.M.; Shi, Y.H. Mechanism of soil stratum instability induced by hydrate dissociation. *Ocean Eng.* **2016**, *122*, 74–83. [CrossRef]
105. Zhang, H.; Luo, X.; Bi, J.; He, G.; Guo, Z. Submarine slope stability analysis during natural gas hydrate dissociation. *Mar. Georesour. Geotec.* **2019**, *37*, 467–476. [CrossRef]
106. Vanneste, M.; Sultan, N.; Garziglia, S.; Forsberg, C.F.; L'Heureux, J.S. Seafloor instabilities and sediment deformation processes: The need for integrated, multi-disciplinary investigations. *Mar. Geol.* **2014**, *352*, 183–214. [CrossRef]
107. Mienert, J.; Vanneste, M.; Bunz, S.; Forsberg, C.F.; L'Heureux, J.S. Ocean warming and gas hydrate stability on the mid-Norwegian margin at the Storegga Slide. *Mar. Petrol. Geol.* **2005**, *22*, 233–244. [CrossRef]
108. Micallef, A.; Masson, D.G.; Berndt, C.; Stow, D.A.V. Morphology and mechanics of submarine spreading: A case study from the Storegga Slide. *J. Geophy. Res. Earth* **2007**, *112*. [CrossRef]
109. Maslin, M.; Mikkelsen, N.; Vilela, C.; Haq, B. Sea-level–and gas-hydrate–controlled catastrophic sediment failures of the Amazon Fan. *Geology* **1998**, *26*, 1107–1110. [CrossRef]
110. Liu, F.; Tan, L.; Crosta, G.; Huang, Y. Spatiotemporal destabilization modes of upper continental slopes undergoing hydrate dissociation. *Eng. Geol.* **2020**, *264*, 105286. [CrossRef]
111. Tan, L.; Liu, F.; Huang, Y.; Crosta, G.; Frattini, P.; Cen, X. Production-induced instability of a gentle submarine slope: Potential impact of gas hydrate exploitation with the huff-puff method. *Eng. Geol.* **2021**, *289*, 106174. [CrossRef]
112. Jiang, M.; Sun, C.; Crosta, G.B.; Zhang, W. A study of submarine steep slope failures triggered by thermal dissociation of methane hydrates using a coupled CFD-DEM approach. *Eng. Geol.* **2015**, *190*, 1–16. [CrossRef]
113. Handwerger, A.L.; Rempel, A.W.; Skarbek, R.M. Submarine landslides triggered by destabilization of high-saturation hydrate anomalies. *Geochem. Geophy. Geosy.* **2017**, *18*, 2429–2445. [CrossRef]
114. Song, B.; Cheng, Y.; Yan, C.; Han, Z.; Ding, J.; Li, Y.; Wei, J. Influences of hydrate decomposition on submarine landslide. *Landslides* **2019**, *16*, 2127–2150. [CrossRef]
115. Pape, T.; Bahr, A.; Klapp, S.A.; Abegg, F.; Bohrmann, G. High-intensity gas seepage causes rafting of shallow gas hydrates in the southeastern Black Sea. *Earth Planet Sci. Lett.* **2011**, *307*, 35–46. [CrossRef]
116. McGinnis, D.F.; Greinert, J.; Artemov, Y.; Beaubien, S.E.; Wuest, A. Fate of rising methane bubbles in stratified waters: How much methane reaches the atmosphere? *J. Geophys. Res. Oceans* **2006**, *111*. [CrossRef]
117. Naudts, L., Greinert, J.; Artemov, Y.; Staelens, P.; Poort, J.; Rensbergen, P.V.; Batist, M.D. Geological and morphological setting of 2778 methane seeps in the Dnepr paleo-delta, northwestern Black Sea. *Mar. Geol.* **2006**, *227*, 177–199. [CrossRef]
118. Zhang, Y. Methane escape from gas hydrate systems in marine environment, and methane-driven oceanic eruptions. *Geophys. Res. Lett.* **2003**, *30*. [CrossRef]
119. Paull, C.K.; Brewer, P.G.; Ussler, W.; Peltzer, E.T.; Rehder, G.; Clague, D. An experiment demonstrating that marine slumping is a mechanism to transfer methane from seafloor gas-hydrate deposits into the upper ocean and atmosphere. *Geo. Mar. Lett.* **2002**, *22*, 198–203. [CrossRef]
120. Milkov, A.V. Global estimates of hydrate-bound gas in marine sediments: How much is really out there? *Earth Sci. Rev.* **2004**, *66*, 183–197. [CrossRef]
121. Buffett, B.A. Clathrate hydrates. *Annu. Rev. Earth Pla. Sci.* **2000**, *28*, 477–507. [CrossRef]
122. Holtzman, R.; Juanes, R. Thermodynamic and hydrodynamic constraints on overpressure caused by hydrate dissociation: A pore-scale model. *Geophys. Res. Lett.* **2011**, *38*. [CrossRef]
123. Zhang, M.; Niu, M.; Shen, S.; Dai, S.; Xu, Y. Review of natural gas hydrate dissociation effects on seabed stability. *Nat. Hazards* **2021**, *107*, 1035–1045. [CrossRef]
124. Liu, T.; Lu, Y.; Zhou, L. Experiment and analysis of submarine landslide model caused by elevated pore pressure. *J. Mar. Sci. Eng.* **2019**, *7*, 146. [CrossRef]
125. Zhang, J.H.; Lin, H.L.; Wang, K.Z. Centrifuge modeling and analysis of submarine landslides triggered by elevated pore pressure. *Ocean Eng.* **2015**, *109*, 419–429. [CrossRef]
126. Song, X.; Nian, T.; Mestdagh, T.; Batist, M.D. Long-and short-term dynamic stability of submarine slopes undergoing hydrate dissociation. *Gas Sci. Eng.* **2023**, *111*, 204934. [CrossRef]
127. Sun, Z.; Jia, Y.; Shan, H.; Fan, Z.; Song, X.; Xue, L.; Li, K. Monitoring and early warning technology of hydrate-induced submarine disasters. *Earth Environ. Sci.* **2020**, *570*, 062030. [CrossRef]

128. Yasuhiro, Y.; Yoshihiko, Y.; Yuzuru, Y. Submarine landslides at subduction margins: Insights from physical models. *Tectonophysics* **2010**, *484*, 156–167. [CrossRef]
129. Zhang, X.H.; Lu, X.B.; Shi, Y.H.; Xia, Z. Study on the mechanical properties of hydrate-bearing silty clay. *Mar. Petrol. Geol.* **2015**, *67*, 72–80. [CrossRef]
130. Yoneda, J.; Takiguchi, A.; Ishibashi, T.; Yasui, A.; Mori, J.; Kakumoto, M.; Aoki, K.; Tenma, N. Mechanical response of reservoir and well completion of the first offshore methane-hydrate production test at the eastern Nankai Trough: A coupled thermo-hydromechanical analysis. *SPE J.* **2019**, *24*, 531–546. [CrossRef]

energies

MDPI

Article

Multiphysics Measurements for Detection of Gas Hydrate Formation in Undersaturated Oil Coreflooding Experiments with Seawater Injection

Bianca L. S. Geranutti, Mathias Pohl, Daniel Rathmaier, Somayeh Karimi and Manika Prasad and Luis E. Zerpa *

Center for Rock and Fluid Multiphysics, Colorado School of Mines, Golden, CO 80401, USA;
drathmaier@mines.edu (D.R.); mprasad@mines.edu (M.P.)
* Correspondence: lzerpa@mines.edu; Tel.: +1-303-384-2627

Abstract: A solid phase of natural gas hydrates can form from dissolved gas in oil during cold water injection into shallow undersaturated oil reservoirs. This creates significant risks to oil production due to potential permeability reduction and flow assurance issues. Understanding the conditions under which gas hydrates form and their impact on reservoir properties is important for optimizing oil recovery processes and ensuring the safe and efficient operation of oil reservoirs subject to waterflooding. In this work, we present two fluid displacement experiments under temperature control using Bentheimer sandstone core samples. A large diameter core sample of 3 inches in diameter and 10 inches in length was instrumented with multiphysics sensors (i.e., ultrasonic, electrical conductivity, strain, and temperature) to detect the onset of hydrate formation during cooling/injection steps. A small diameter core sample of 1.5 inches in diameter and 12 inches in length was used in a coreflooding apparatus with high-precision pressure transducers to determine the effect of hydrate formation on rock permeability. The fluid phase transition to solid hydrate phase was detected during the displacement of live-oil with injected water. The experimental procedure consisted of cooling and injection steps. Gas hydrate formation was detected from ultrasonic measurements at 7 °C, while strain measurements registered changes at 4 °C after gas hydrate concentration increased further. Ultrasonic velocities indicated the pore-filling morphology of gas hydrates, resulting in a high hydrate saturation of theoretically up to 38% and a substantial risk of intrinsic permeability reduction in the reservoir rock due to pore blockage by hydrates.

Keywords: gas hydrate; multiphysics measurements; acoustics; monitoring; rock physics; saturation estimation

Citation: Geranutti, B.L.S.; Pohl, M.; Rathmaier, D.; Karimi, S.; Prasad, M.; Zerpa, L.E. Multiphysics Measurements for Detection of Gas Hydrate Formation in Undersaturated Oil Coreflooding Experiments with Seawater Injection. *Energies* **2024**, *17*, 3280. https://doi.org/10.3390/en17133280

Academic Editors: Lunxiang Zhang and Hongsheng Dong

Received: 5 June 2024
Revised: 18 June 2024
Accepted: 27 June 2024
Published: 4 July 2024

1. Introduction

Gas hydrates are crystalline solids formed by gas and water molecules at high pressures and low temperatures (typically above the ice point), where small gas molecules (between 0.35 and 0.9 nm) are trapped within the cavities of hydrogen-bonded water molecules. The trapped gas is in the size range between nitrogen and normal pentane [1,2]. Gas hydrates occur naturally onshore below the permafrost and at or below the seafloor within sediments [1]. This paper presents an experimental study of gas hydrate formation in undersaturated oil reservoirs during cold water injection, which could result in formation damage (i.e., permeability decrease by the formation of solid gas hydrate within the rock pore network). Normally, the temperature and pressure profiles of petroleum reservoirs are such that hydrates are not stable. However, in colder regions, conditions can be favorable for hydrate formation at the water-hydrocarbon interface [3,4].

In the literature, there are reported experimental studies connecting the reduction in rock permeability to hydrate concentration [5–10]. The permeability reduction is directly related to the gas hydrates' deposition morphology and rock properties. Previous work, however, considered water and natural gas systems but most conventional commercial hydrocarbon reservoirs worldwide target the oil phase which requires at a certain stage of production some sort of secondary recovery mechanism, such as waterflooding. To the best of the authors' knowledge, looking at the multiphase problem of gas hydrate formation from the gaseous compounds entirely dissolved in the oil phase, has not yet been researched. Therefore, this work contributes to the understanding of gas hydrate formation in oil reservoirs during cold seawater injection. The physical properties of gas hydrate-bearing porous media depend on the volume fraction and spatial distribution of the hydrate phase. The grain size and the effective stress determine the hydrates' distribution in sediments [11].

There are two main morphologies for hydrate deposition in the pore space. The first one is when hydrates form along the grain surfaces. This type can only be seen in the contact of the grains (i.e., cementing) and enveloping the grain (i.e., grain/pore coating). In the second morphology, hydrates are initially deposited in the middle of the pores resulting in a pore-filling morphology [12,13]. When hydrates form as grain coating, changes in permeability are expected to be small. When hydrates form as pore filling, the permeability loss can be more severe with the potential of completely plugging the pore throats [5,9,10].

The hydrate morphology in sediments was connected to acoustic velocities in previous studies [12,14–17]. In both cases where gas hydrates are formed at the pore walls (i.e., grain/pore coating and cementing), P- and S-wave velocities increased with hydrate saturation. For pore-filling hydrates, the P-wave velocity increased while the S-wave velocity is practically independent of hydrate concentration [12].

Gas hydrates are also considered electrical insulators. Their formation replaces the conductive pore fluid, restricting the flow of electric current [18–21]. However, there is a competing effect due to the salinity increase in the remaining water in the pore space [22].

In this paper, we demonstrate experimental hydrate formation from an undersaturated oil reservoir using coreflooding experiments. Multiphysics measurements (ultrasonic, strain, and temperature) were implemented for the detection of fluid phase transitions into a solid phase (i.e., gas hydrates) and to investigate the distribution of gas hydrates inside the core.

2. Materials and Methods

The materials used in this research were two Bentheimer sandstone core samples, one of 3 inches in diameter and 10 inches in length (large diameter core, required for the installation sensors on the core), and the other of 1.5 inches in diameter and 12 inches in length (small diameter core, required for accurate pressure drop measurements). The fluid system consisted of synthetic formation water, seawater, and a recombined crude oil with synthetic gas to mimic a live-oil composition. The experimental procedure to assess hydrate formation consisted of cooling, injection, and multiphysics measurement steps.

2.1. Core Sample Preparation

The selected rock for the experiments was the Bentheimer sandstone provided by Kocurek Industries from the Valaginian formation. This sandstone is considered to be ideal for standard laboratory experiments because of its limited amount of minerals (with 91.7 wt% quartz, 4.9 wt% feldspars, 2.7 wt% clay minerals, 0.4 wt% carbonates, and 0.2 wt% pyrite and iron hydroxides), a constant grain size distribution, porosity, permeability, and dielectric values [23].

Kocurek Industries reported the average porosity of the Bentheimer to be between 23 and 26%, and the permeability between 1500 and 3000 mD. To check the porosity and permeability of Bentheimer samples for the pressures used in this study (8270 kPa to 13,790 kPa), a core plug sample (1.5 inches in diameter and 2 inches in length) was placed in the CMS-300 apparatus. The CMS-300 performs computer-controlled measurements of reservoir rock samples to determine pore volume and Klinkenberg permeability. The core sample was pressurized with helium and the measurements at experimental pressure conditions yielded a porosity of 25.5% and permeability of 2750 mD.

The first coreflooding test used a core sample of 3 inches in diameter and 10 inches in length, with an estimated pore volume of 290 mL. Four types of multiphysics sensors were installed on the core sample: 1 MHz piezo-electrical crystals (P- and S-wave for acoustic velocity measurements), conductivity electrode rings, 1000 Ohm strain gauges for deformation, and K-Type thermocouple sensors for temperature recording along the core. The core preparation consisted of the lateral isolation, made with K20 epoxy, to avoid leaks from the confining hydraulic oil into the core. Previous to the installation of the sensors, grooves were cut to create a contact area for the conductivity electrode rings. Figure 1 shows the core after the lateral isolation process and the core after the grooves were made.

Twelve strain gauges were installed, followed by the installation of nine electrode rings inside the grooves. The elastic wave crystals were prepared and tested before the installation on the core. In total, 40 crystals (20 P-waves and 20 S-waves) were installed along the core. The last sensors installed were 12 temperature sensors. Figure 2 shows the sensor installation steps of the core sample.

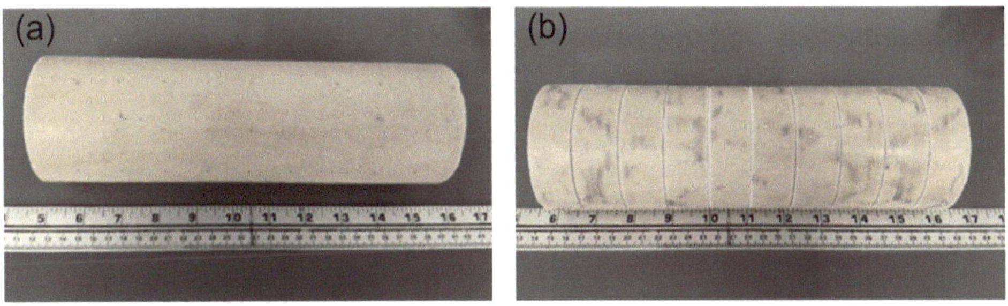

Figure 1. Core sample preparation. Lateral isolation with K20 epoxy in (**a**) and grooves creation for conductivity rings in (**b**).

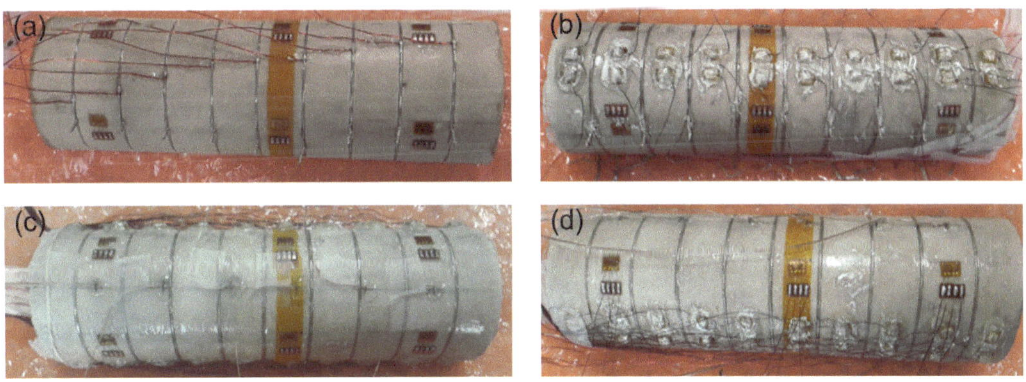

Figure 2. Sensor installation in the core sample: strain gauges and electrode rings installed in (**a**), wave crystals installed in (**b**), temperature sensors installed in (**c**), and all sensors finalized in (**d**).

The entire core was sandwiched between two end-caps that were glued onto the core. A PVC ring and a foil were installed on the bottom cap to create a void space to be filled with soft epoxy. The soft epoxy has the purpose of protecting the sensors from the confining hydraulic oil. Figure 3 shows the soft epoxy process.

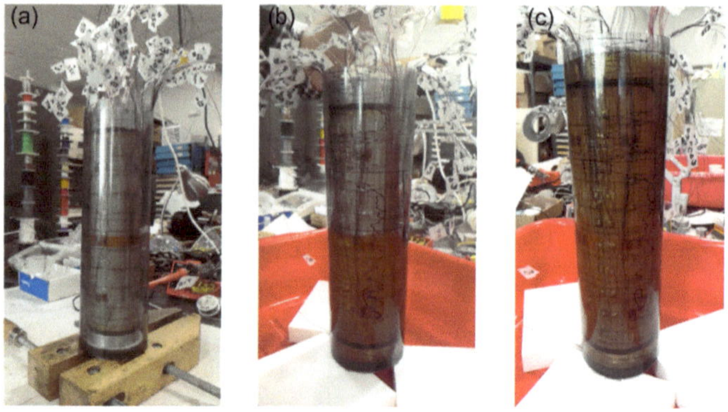

Figure 3. Soft epoxy process: PVC and foil was placed on the core in (**a**), followed by the soft epoxy deposition in (**b**), and finalized by the hardening of soft epoxy in (**c**).

2.2. Synthetic Fluids Preparation

Synthetic formation brine and seawater were prepared using the compositions presented in Table 1. The total salinity was 74.6 g/L for the formation water and 39.3 g/L for the seawater.

Table 1. Composition of the synthetic formation water and seawater used in the coreflood experiment. The core sample is initially fully saturated with formation water before displacement by live-oil (drainage), and seawater is used for the subsequent coreflood to represent waterflooding displacing the oil (imbibition).

Salt Types	Formation Water Composition (g/L)	Seawater Composition (g/L)
NaCl	56.930	25.690
$MgCl_2 \cdot 6H_2O$	8.450	11.040
$CaCl_2 \cdot 2H_2O$	7.250	1.560
$SrCl_2 \cdot 6H_2O$	0.736	0.024
$BaCl_2 \cdot 2H_2O$	0.697	-
KCl	0.243	0.784
$NaBO_2 \cdot 4H_2O$	0.043	0.201
Na_2SO_4	0.030	-

A dead oil sample with a gravity of 38.3 °API and a viscosity of 2.45 cp was recombined with a synthetic gas mixture to mimic an undersaturated oil reservoir. The composition of the natural gas is presented in Table 2. The live-oil preparation occurred in the PVT 400/1000 Analysis System from Sanchez technologies by mixing the dead oil with the synthetic gas. The first step was to inject the dead oil at atmospheric pressure and initial experimental temperature (17 °C) into the cell. The second step was the pressurization of the cell to the experimental pressure (8270 kPa), followed by the injection of synthetic gas. The oil–gas mixture was pressurized to 13,790 kPa for 24 h to allow for the gas to completely dissolve into the oil phase. The measured bubble point of the live-oil was 7580 kPa.

Table 2. Composition of synthetic gas mixture used for live-oil recombination.

Component	Composition (mol%)
CO_2	7.0
CH_4	80.0
C_2H_6	7.0
C_3H_8	4.0
C_4H_{10}	2.0

2.3. Experimental Procedure for the Large Core

Figure 4 shows the coreflooding setup designed for the large diameter core with the physical dimensions of 3 inches in diameter and 10 inches in length. The pressure vessel was vertically placed in a barrel for temperature isolation. The vessel was surrounded by a copper coil that was connected to the recirculating chiller for temperature control. Three Teledyne 500D ISCO pumps (Teledyne ISCO Inc., Lincoln, NE, USA) were incorporated in this setup providing the net confining stress, the pore pressure backup, and for fluid injection. Four types of data acquisition were planned for this setup: (1) elastic waves were generated using a pulser and measured with Tektronix TDS 3014C oscilloscope (Tektronix Inc., Beaverton, OR, USA), (2) complex conductivities were measured with the SIP-Lab from Radic Research (Radic Research, Berlin, Germany), (3) static deformation, and (4) temperature measured continuously. Before saturating the core with formation water, a vacuum was applied to the entire core (using a Fisher Scientific PU1309-N840.0-9.01 vacuum pump from Fisher Scientific International, Inc., Hampton, NH, USA).

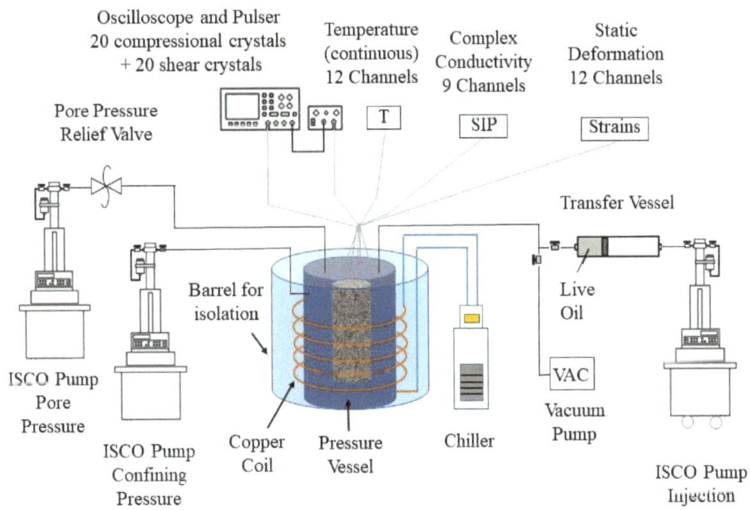

Figure 4. Coreflood setup constituted of a pressure vessel, vacuum pump, three ISCO pumps, chiller, and data acquisition equipment.

The coreflooding experiment was performed at constant pore pressure of 8270 kPa and a net confining stress of 15,270 kPa. Initially, the core was fully saturated with formation water. Then, three pore volumes of live-oil were injected to displace the formation water to irreducible water saturation. The hydrate detection procedure consisted of a series of cooling steps with an injection period, followed by multiphysics measurements. For the cooling steps, the temperature was set to 15, 12, 10, 8, 7, 6, 4, 3, and 2 °C, respectively. Each cooling step was represented by a cooling period and temperature stabilization for 24 h, followed by seawater injection from the bottom of the sample at a volumetric rate of 0.2 cm^3/min. The formation of hydrates within the core is expected to increase P-wave as well as S-wave velocity measurements depending on the hydrate deposition morphology, i.e., whether hydrates will become load bearing or remain in the center of the pores. Furthermore, and because of hydrates exhibit a volume expansion upon formation, it is expected that the in situ formation of hydrates causes a deformation of the core.

For the ultrasonic measurements, each P-wave crystal was paired up with a S-wave crystal. Each pair was separated by 1 inch with the following pair along the core length. The signal was transmitted from one crystal (P or S) to another one of the respective crystals across the diameter of the sample (180°). The standard deviation for the velocities among the crystals was around 30 m/s. The measurement error was determined at 1% for the P-waves and 3% for the S-waves.

Omega SGT-3/1000-XY11 strain gauges (Omega Engineering, Inc., Norwalk, CT, USA) were used for deformation measurements across the core length. Each strain gauge had two strain directions: horizontal and vertical. The measurements of the effective strain were calculated by Wheatstone bridge circuit voltages recorded by a computer with MCC DAQ software version 4.2.1 (Measurement Computing, Norton, MA, USA). The main purpose of strain measurements was to evaluate the possible expansion of the rock due to gas hydrate formation. Similarly to deformation, the temperature sensors were placed along the whole length of the core. The temperature sensors were connected to KTA-259K thermocouple shields that passed the information to the Mega 2560 R3 Arduino boards (Arduino, Ivrea, Italy) controlled by a computer with the Arduino IDE software version 2.3.2. The temperature was recorded by the CoolTerm application from Roger Meier. The temperature sensors measured the temperature across the core and provided supporting information during gas hydrate formation as it is an exothermic reaction [2].

2.4. Experimental Procedure for the Small Core

Figure 5 shows the coreflooding setup designed for the small diameter core with the physical dimensions of 1.5 inches in diameter and 12 inches in length. The pressure vessel setup including the chiller for temperature control and pumps for pore and confining pressure, respectively, was similar to the experimental procedure for the large core. In addition, pressure gauges of the type Omega DPG210-5K (Omega Engineering, Inc., Norwalk, CT, USA) with an accuracy of 0.1% and a pressure range from 0 to 5000 psig, and a differential pressure transducer of the type Omega PX3005-25 (Omega Engineering, Inc., Norwalk, CT, USA) with an accuracy of 0.075% and a pressure range from −1.0 to 1.0 psi, measure the differential pressure between the top and the bottom of the core. This measured differential pressure gives indirect information about the formation of hydrates within the porous media of the sample as hydrates clog the pore channels which increases the resistance to fluid flow and manifests in a higher differential pressure.

The coreflooding experiment followed the same procedure as for the large core with the adjustment of temperature cooling steps to 15, 11, 8, 7, 6, 5, 4, 3, and 2 °C, and a volumetric injection rate to 0.05 cm^3/min.

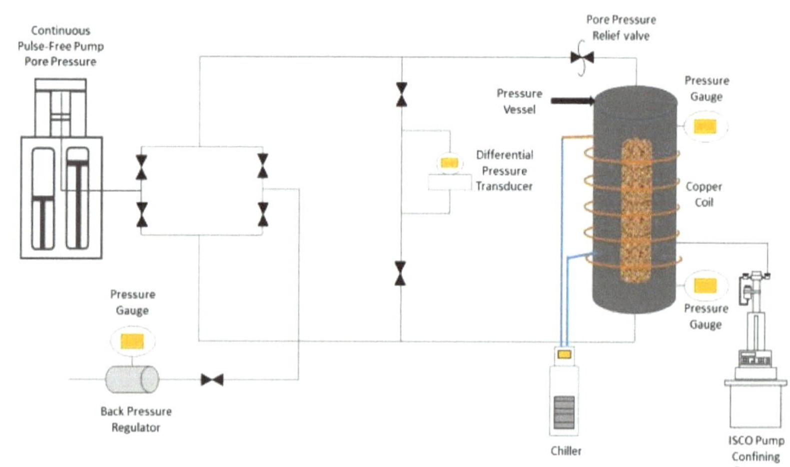

Figure 5. Coreflood setup constituted of a pressure vessel, differential pressure transducer, pressure gauges, ISCO pump, continuous pulse-free pump, chiller, and back pressure regulator.

3. Results

Before performing the coreflooding experiment to detect hydrate formation with multiphysics measurements, we calculated the hydrate equilibrium conditions in terms of pressure and temperature for different water salinities (i.e., fresh water, seawater, and formation water). We performed these calculations with the software PVTSim Nova version 5.1 (Calsep A/S, Copenhagen, Denmark) using the synthetic live-oil and water compositions with varying salinities as input parameters and the Peng–Robinson (PR) Peneloux Equation of State (EOS). Figure 6 shows the calculated hydrate equilibrium curves. The region to the right of the curves, at higher temperatures and lower pressures, represent conditions were gas hydrates are not stable while the region to the left of the curves represents the hydrate stable conditions. The increase in water salinity shifts the hydrate equilibrium conditions to lower temperatures.

During a cooling process that takes a hydrocarbon water system below the hydrate equilibrium temperature, there will be a phase transition into a solid hydrate phase which begins with the nucleation of a hydrate crystal. Hydrate nucleation requires several degrees of subcooling below the equilibrium temperature. It is expected that hydrates will start to form when the system is several degrees colder than the calculated hydrate equilibrium temperature. For the system considered in this work, the hydrate equilibrium temperature is 11 °C for the salinity of formation water.

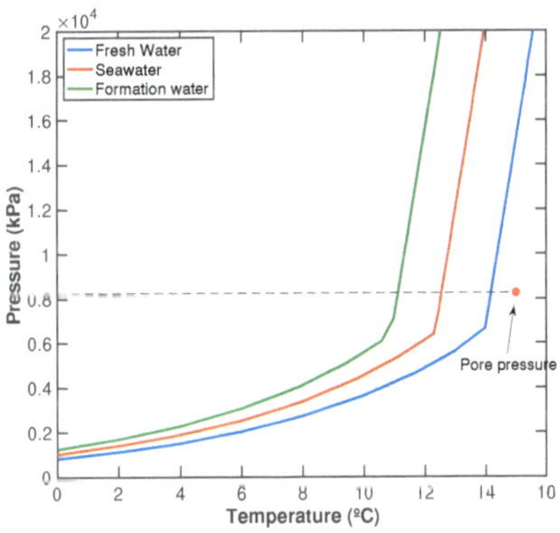

Figure 6. Hydrate equilibrium curves for fresh water, seawater, and formation water, indicating the initial experiment pressure and temperature conditions

3.1. *Large Diameter Core Sample Results*

The formation of a solid gas hydrate phase was detected by changes in the arrival time of elastic waves. Initially, no significant changes in elastic waves were observed during cooling steps from 15 to 8 °C. A small decrease in the P-wave arrival time was noticed at the bottom of the sample due to an increase in water saturation because of oil displacement by seawater. Changes in amplitude were visible in the raw data, however, the arrival time was very similar. The waterfront was visible at the P4 crystal (fourth acoustic crystal location from the top to the bottom of the sample) for the 8 °C seawater injection step. There was no significant change in the S-waves. The period of injection was reduced from 60 min to 20 min at the same injection rate (0.2 cm^3/min) for the following cooling steps to maintain a sufficient amount of live-oil in the sample.

Significant changes in the arrival time of elastic waves due to gas hydrate formation were observed between 8 and 7 °C. Figure 7 shows the raw P- and S-waves measured for the 8 and 7 °C seawater injection steps for representative crystals at the top (P1), middle (P5), and bottom (P10) of the core sample. It is possible to see a decrease in the P-wave arrival time for the 7 °C curves at the three representative locations (P1, P5, and P10). The S-waves were slightly affected at 7 °C, but the difference remained inside the error margin of the measurements. The S-wave in the middle of the core was the least affected, probably because of a lower oil saturation due to a water imbibition effect. The following cooling steps had the purpose of evaluating hydrate growth. For the following steps, the P-wave arrival time kept decreasing.

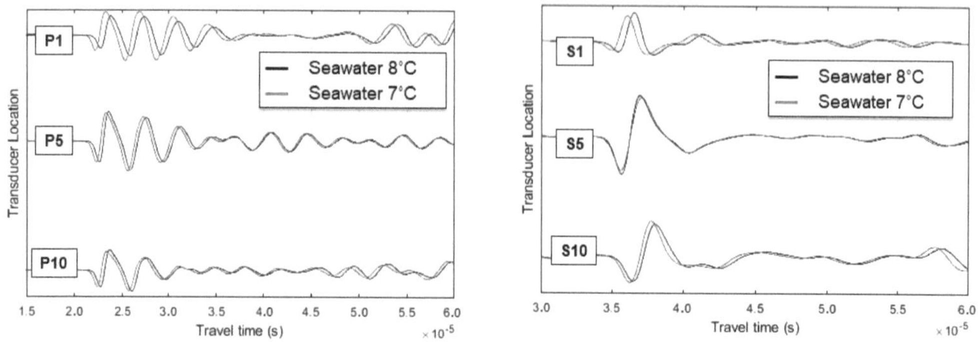

Figure 7. Raw P-waves (**left**) and S-waves (**right**) comparison for seawater injection at 8 °C and seawater injection at 7 °C indicating gas hydrate formation.

The acoustic velocities were calculated based on the arrival time and sample dimensions. No delay time was considered as the crystals were glued directly onto the sample surface. A linear temperature correction to remove temperature effects was performed based on previous results from the evaluation of temperature effects with a water-saturated core [24]. Figure 8 shows the P- and S-wave velocities measured at transducers along the length of the core for conditions before injection and for cooling/injection steps at 8 °C, 7 °C, and 4 °C. From before seawater injection to 8 °C, an increase in P-wave velocity in the bottom part (P7 to P10) of the sample due to an increase in water saturation was observed. The velocity increase due to water injection was in the order of 10 to 20 m/s, which are smaller than the error margin.

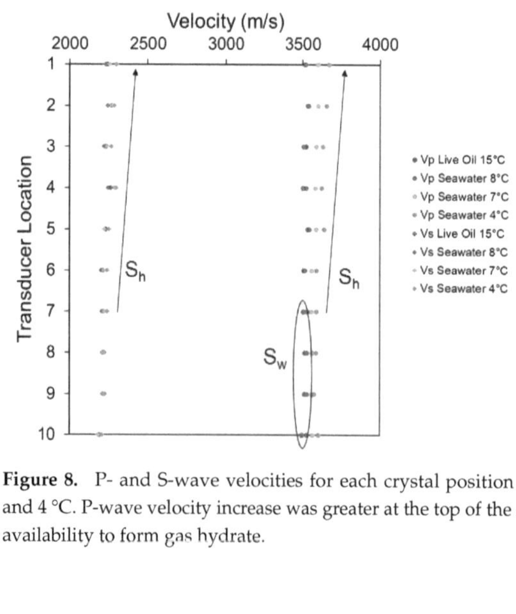

Figure 8. P- and S-wave velocities for each crystal position of the large core sample at 15, 8, 7, and 4 °C. P-wave velocity increase was greater at the top of the sample due to a higher dissolved gas availability to form gas hydrate.

From 8 °C to 7 °C, the P-wave velocity increased all over the sample indicating hydrate formation. The top crystal showed a higher increase, approximately 80 m/s. The bottom crystal had a smaller increase, approximately 40 m/s, associated also with a greater water saturation. The S-wave velocity increase, even in the top portion, was smaller than the error margin, approximately 20 m/s. As a solid phase, gas hydrates increase the compressional wave velocity propagation, as observed in P-wave velocities. The same does not happen for S-waves because a water film exists between the rock surface and hydrates (i.e., in a pore-filling hydrate morphology), which does not allow the S-wave to be transmitted. The velocity increase is higher in the top portion (P1 to P7) because of the availability of more dissolved gas (i.e., higher oil saturation towards the top of the core sample) leads to more hydrate formation. From 7 °C to 4 °C, the P-wave velocity increased all over the sample indicating the formation of additional gas hydrates. The top acoustic crystal showed a higher increase, approximately 70 m/s. The bottom crystal was around 40 m/s. The S-wave velocity started to have a small impact on the top of the sample, approximately 40 m/s (close to the error margin).

Figure 9 shows the average P-wave velocity for every experimental procedure step. The average velocity increase due to water saturation increase was very small, around 5 m/s. From 8 °C to 7 °C, the average velocity increased 50 m/s, indicating the formation of gas hydrate. For lower temperatures, the velocity kept increasing due to additional hydrate formation with values between 10 and 30 m/s for each subsequent cooling step. Figure 10 shows the average S-wave velocity for every experimental procedure step. No significant change was observed due to an increase in water saturation or due to gas hydrate formation.

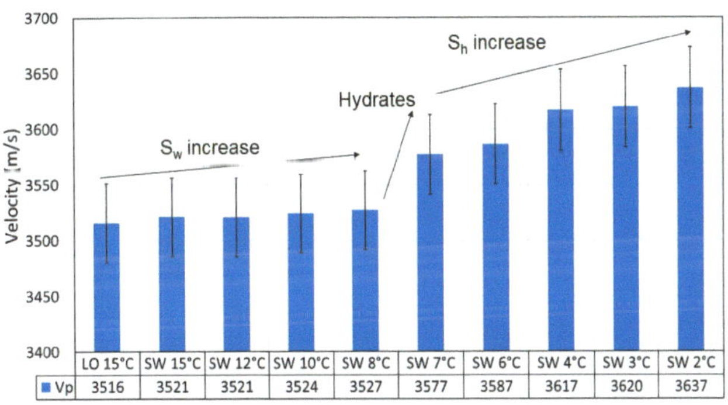

Figure 9. P-wave average velocities for hydrate detection experimental procedure. From 8 to 7 °C an increase in velocity indicated gas hydrate formation. Gas hydrates kept growing for lower temperatures.

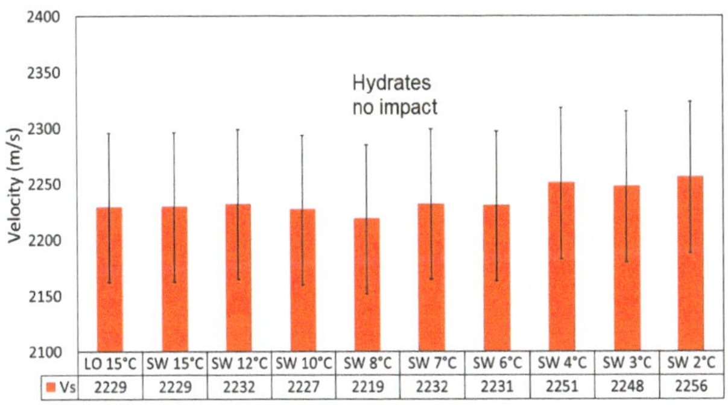

Figure 10. S-wave average velocities for hydrate detection experimental procedure. No significant change was observed.

Figure 11 shows the strain values for the gas hydrate experimental procedure. The readings were performed in two directions (vertical and horizontal) at three locations along the core (top, middle, and bottom). Only half of the strain gauges worked properly until the end of the procedure because of cable rupture problems. Bottom A-C vertical at 15 °C and Top B-D vertical at 3 °C presented values higher than 2% strain and were inconsistent with the other values, so these are not displayed. The reduction in strain was observed from 4 °C to 2 °C, indicating that the core sample expanded. Similar to ice, the transition of fluid phases into solid gas hydrates expand in volume [2]. The ultrasonic measurements show that 4 °C was the temperature at which hydrate saturation increased significantly in the core, so the strain measurements are in agreement with ultrasonic data.

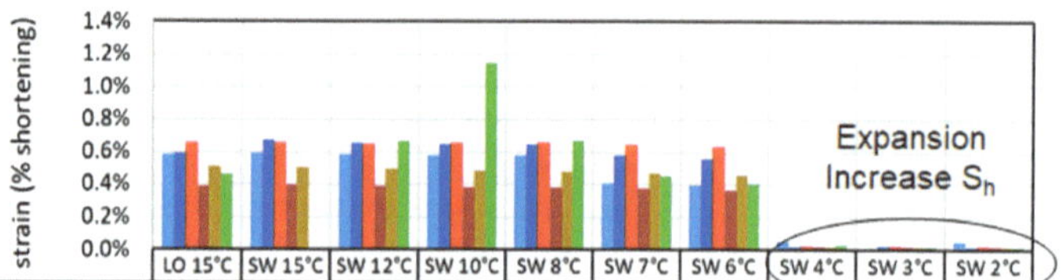

	LO 15°C	SW 15°C	SW 12°C	SW 10°C	SW 8°C	SW 7°C	SW 6°C	SW 4°C	SW 3°C	SW 2°C
Top B-D vertical	0.6%	0.6%	0.6%	0.6%	0.6%	0.4%	0.4%	0.1%		0.1%
Top B-D horizontal	0.6%	0.7%	0.7%	0.6%	0.6%	0.6%	0.6%	0.0%	0.0%	0.0%
Middle B-D vertical	0.7%	0.7%	0.6%	0.7%	0.7%	0.7%	0.6%	0.0%	0.0%	0.0%
Middle B-D horizontal	0.4%	0.4%	0.4%	0.4%	0.4%	0.4%	0.4%	0.0%	0.0%	0.0%
Middle A-C vertical	0.5%	0.5%	0.5%	0.5%	0.5%	0.5%	0.5%	0.0%	0.0%	0.0%
Bottom A-C vertical	0.5%		0.7%	1.1%	0.7%	0.5%	0.4%	0.0%	0.0%	0.0%

Figure 11. Strain values for gas hydrate detection experimental procedure. Expansion of the core sample was notice at 4 ° C.

Figure 12 shows the temperature readings from 8 °C to 7 °C from the gas hydrate detection experimental procedure. The temperature was measured in opposite diameters (A and B) along the length of the core (1 at the top until 6 at the bottom). The increase in temperature due to the exothermic hydrate formation reaction was visible in the top to the middle portion of the sample (T1, T2, T3, and T4). The sensor T3A presented a reading issue, but a variation was noticed between 17 and 18 h. To maintain the temperature of the core sample the closest possible to the respective seawater injection step temperature, the cooler sometimes had the set temperature adjusted 4 h prior to the data acquisition. This happened and is observed for the 7 °C seawater injection step approximately at 20 h of the recorded data. The bottom part of the sample only had temperature fluctuations after this adjustment. In the bottom part, gas hydrate formation could not have been enough to increase the temperature or it happened simultaneously with the cooler adjustment. The room temperature was stable during the cooling. The Arduinos' temperatures represent the inside board temperature reading and were a few degrees higher than the room temperature.

Elastic waves and temperature measurements supported that gas hydrates formed at 7 °C. The strain and pressure differential showed significant hydrates impact at 4 °C. The P-waves and temperature were more sensitive to initial gas hydrates change, while the other measurements only detected or were affected by gas hydrates when the saturation increased.

The pressure differential between the injection ISCO pump (set to maintain a constant injection rate) and the backpressure ISCO pump (set to maintain a constant pressure) was recorded during the coreflood experiment. From 6 °C to 4 °C, an increase of 20 kPa pressure differential was noticed. The pressure differential increase could indicate resistance to flow due to gas hydrate saturation increase. ISCO pumps do not have a good precision to measure pressure differentials in temperature oscillating environments; therefore, another experiment with high-accuracy pressure gauges was performed to corroborate this hypothesis on the smaller core sample of the Bentheimer sandstone outcrop.

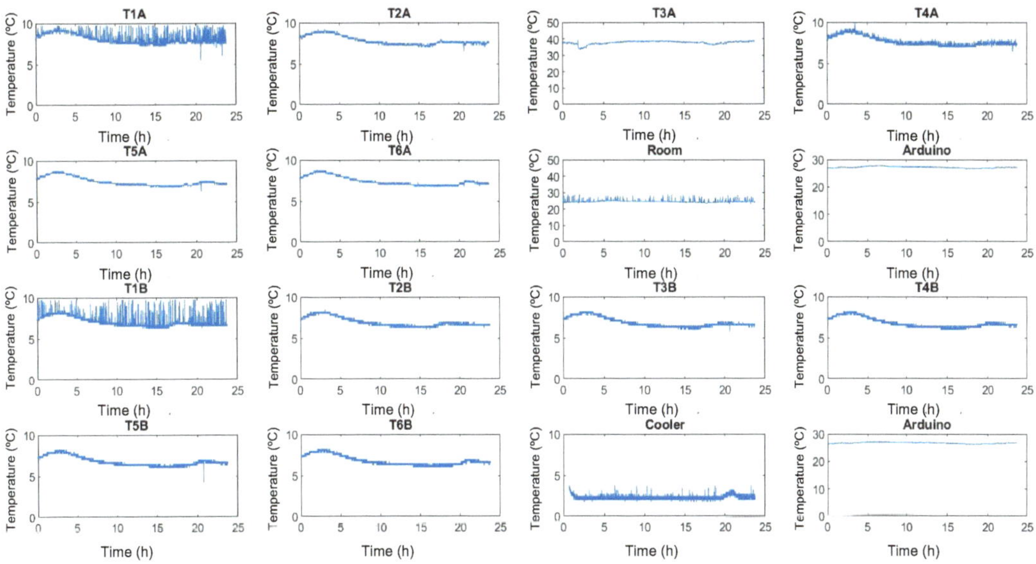

Figure 12. Temperature readings from 8 °C to 7 °C from the gas hydrate detection experimental procedure. Gas hydrate exothermic reaction was noticed after 17 h of cooling.

3.2. Small Diameter Core Sample Results

The coreflooding experiment with the smaller core sample equipped with highly accurate pressure gauges and a differential pressure transducer gives information about the saturation of gas hydrates and resulting permeability effects created under similar conditions as for the larger core sample described above.

To obtain the water permeability, pressure transducer differential measurements at water flow rates of 2, 1, and 0.5 mL/min were performed. The linear relation of differential pressure with flow rate shown in Figure 13 gives a positive slope of 0.4189 with a coefficient of determination equal to 1. With this relationship, the water permeability at flow rates below or above the possible measurements can be calculated through extrapolation. Because of the physical size of the smaller core sample, we reduced the injection rate to 0.05 mL/min to not reach water breakthrough too soon. At this flow rate, the pressure differential according to the relationship gives a value of 0.02080 psi resulting in a water permeability calculated according to Darcy's law (Equation (1)) with the physical dimensions of the smaller core sample of 1572.92 mD.

$$kw = \frac{q_{inj}\mu_w L}{A\Delta p} \tag{1}$$

At each of the temperatures following the cooling temperature ramp, pressure differential measurements were conducted and average values over the time of injection were calculated. The absolute permeability was calculated using Darcy's law at each of the temperature steps, and by relating these values to the water permeability, a relative hydrate permeability was calculated. Figure 14 plots the measured pressure differential and calculated relative hydrate permeability versus the respective temperature. The experiment started at a temperature of 15 °C. There was a slight decrease in the measured pressure differential from 15 to 8 °C, with a corresponding decrease in permeability. From 8 to 7 °C, there is a significant increase in the pressure differential, with a corresponding decrease in permeability, which coincides with the formation of solid hydrates within the pore space. Further decrease in temperature promoted the formation of more hydrate phase manifested by the increase in the measured pressure differential and decrease in permeability.

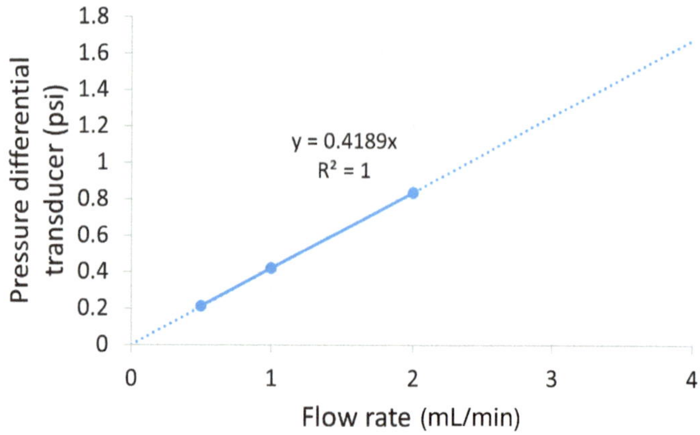

Figure 13. Pressure transducer differential measurements at water flow rates of 2, 1, and 0.5 mL/min including a linear trend line with a coefficient of determination of 1.

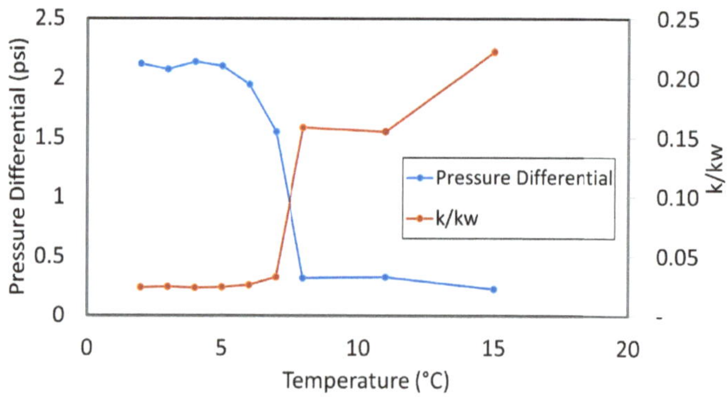

Figure 14. Measured pressure differential and relative hydrate permeability plotted against the temperature step in the cooling temperature ramp. A sudden increase in differential pressure manifesting in a steep drop of relative hydrate permeability at the hydrate forming temperature of 7 °C can be seen.

Different theoretical permeability models plotted in Figure 15 based on different hydrate deposition morphology show the relation of relative permeability with hydrate saturation. The models are either based on the idealistic mathematical assumption of the capillary bundle model or on the empirically derived relationship based on Kozeny-type equations. The derivation of both models for each of the two different hydrate formation deposition morphologies is summarized in Appendix A. Theoretical Permeability Models Derivation. Following our hypothesis of hydrate deposition in the center of the pores supported by the experiments on the large diameter core sample as described above, the hydrate saturation can be back-calculated with the theoretical models. Due to the non-linear relationship, this back-calculation was conducted using numerical methods. The results of this calculation are summarized in Table 3. The relative permeability was normalized to the 15 °C temperature step, where we have no hydrates and represents the baseline pressure differential. A normalized or corrected hydrate saturation was obtained using the theoretical models presented in Figure 15. A hydrate saturation as high as 38% is estimated for the temperature step of 2 ° C.

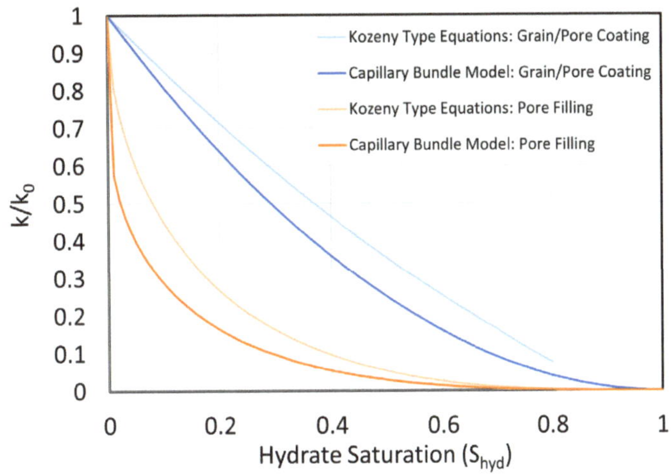

Figure 15. Theoretical permeability models based on the capillary bundle assumption and the Kozeny-type equations depending on the hydrate deposition morphology of pore filling and grain/pore coating. The equations yielding these models are derived in Appendix A. Theoretical Permeability Models Derivation.

Table 3. Calculated relative hydrate permeability based on Darcy's law with pressure differential measurements and resulting hydrate saturation based on the pore-filling deposition morphology according to the Kozeny-type equations.

Temperature (°C)	k/kw	k/kw Normalized	S_{hyd}	S_{hyd} Normalized
15	0.2222	1.0000	0.2417	0.0000
11	0.1553	0.6986	0.3135	0.0255
8	0.1586	0.7136	0.3093	0.0229
7	0.0328	0.1477	0.5670	0.3231
6	0.0262	0.1178	0.5959	0.3658
5	0.0243	0.1092	0.6052	0.3797
4	0.0239	0.1074	0.6072	0.3826
3	0.0246	0.1107	0.6036	0.3772
2	0.0240	0.1082	0.6063	0.3813

4. Future Work

The presented experimental procedures and results of gas hydrate formation in an undersaturated sandstone oil sample using multiphysics measurements delivers data for holistic numerical analyses and the development of a workflow to upscale the findings to the field-scale in the future. This will involve a numerical replication of the experimental procedures to match the observations from the laboratory coreflooding scale before the upscaling of virtual models can deliver predictions of gas hydrate formation in an oil reservoir that is subject to waterflooding. Furthermore, future work will include pore-scale modeling efforts to create a realistic distribution of gas hydrates in the porous media of a reservoir rock. This will be able to test the hydrate formation deposition morphology hypothesis and related permeability effects. Thereby, we would be able to validate idealistic theoretical permeability models with hydrate saturation using the actual pore structure in the framework of Digital Rock Physics modeling.

5. Conclusions

Gas hydrate formation at undersaturated oil reservoir conditions was detected through the coreflood experiments executed in this work. Gas hydrates formed during the 7 °C cooling/injection step of the experimental procedure for the large diameter core sample. This initial gas hydrate formation was captured by P-waves and temperature measurements.

Strain measurements demonstrated that at 4 °C, the hydrate saturation was high enough to promote more significant changes in the interactions between rock and fluid. Additional hydrate formation at 4 °C promoted the expansion of the rock observed by the response of strain gauges.

Complex conductivity did not detect the initial gas hydrate formation because of the competition between the increase in electrical resistivity by solid hydrates and a decrease in electrical resistivity due to the increase in water salinity in remaining unconverted water.

From elastic wave analysis in the experiments, S-waves were not initially affected by the formation of gas hydrates, indicating a pore-filling morphology. This morphology has a greater permeability decrease than a pore-coating morphology, promoting the fluid flow resistance observed via pressure differential measurements between the inlet and outlet of the cores.

A highly accurate pressure differential transducer and the application of Darcy's law resulted in the calculation of the relative hydrate permeability which gave a theoretical quantification of the hydrate saturation by the assumption of the pore-filling deposition morphology based on Kozeny-type equations.

The findings of this research show a risk of performing waterflooding with cold seawater in shallow (colder) undersaturated oil reservoirs and give a method to estimate hydrate saturation and potential permeability decrease based on pressure measurements and theory.

Author Contributions: Conceptualization, L.E.Z. and M.P. (Manika Prasad); methodology, B.L.S.G., M.P. (Mathias Pohl), S.K., M.P. (Manika Prasad) and L.E.Z.; analysis, B.L.S.G., M.P. (Mathias Pohl), D.R., S.K. M.P. (Manika Prasad) and L.E.Z.; investigation, B.L.S.G., M.P. (Mathias Pohl), D.R., S.K., M.P. (Manika Prasad) and L.E.Z.; resources, L.E.Z.; writing—original draft preparation, B.L.S.G., D.R. and L.E.Z.; writing—review and editing, B.L.S.G., M.P. (Mathias Pohl), D.R., M.P. (Manika Prasad) and L.E.Z.; supervision, M.P. (Mathias Pohl), M.P. (Manika Prasad) and L.E.Z.; project administration, L.E.Z.; funding acquisition, M.P. (Manika Prasad) and L.E.Z. All authors have read and agreed to the published version of the manuscript.

Funding: This research was funded by OMV Norge AS, grant number 470121.

Data Availability Statement: The data that support the findings of this study are available on request from the corresponding author. The data are not publicly available due to privacy or ethical restrictions.

Acknowledgments: The authors would like to acknowledge Carolyn Koh, Ahmad Majid, Jyoti Behura, and Weiping Wang for the support through this research.

Conflicts of Interest: The authors declare no conflicts of interest.

Appendix A. Theoretical Permeability Model Derivation

In this paper, different models to relate relative gas hydrate permeability with hydrate saturation are presented. These models can be classified according to the hydrate deposition morphology hypothesis. Here, we show the derivation of the relative hydrate permeability as a function of hydrate saturation based on two main deposition morphologies: the grain/pore-coating morphology and the pore-filling morphology.

Appendix A.1. Capillary Bundle Model

The flux in a simple capillary bundle model consisting of n parallel and straight cylindrical capillaries is given by the Hagen-Poiseuille Equation (Equation (A1)) derived from the Navier-Stokes equations.

$$q = \frac{n\pi R^4}{8\mu}\frac{\Delta p}{L} \tag{A1}$$

where q is the flux in m^3/s, n is the number of capillaries, R is the radius of the capillaries in m, Δp is the pressure differential in Pa, μ is the viscosity in Pa.s, and L is the length of the capillaries in m.

Combining this equation with the flux calculated based on Darcy's law (Equation (1)) gives the permeability as

$$k = \frac{n\pi R^4}{8} \tag{A2}$$

Relating that to the porosity definition in the capillary bundle model of $\phi = n\pi R^2$, the permeability without hydrates becomes

$$k_0 = \frac{\phi R^2}{8} \tag{A3}$$

Appendix A.1.1. Grain/Pore-Coating Morphology

A uniform coating of the walls around the capillaries reduces the radius of the capillaries to R_1 and the permeability becomes

$$k_1 = \frac{n\pi R_1^4}{8} \tag{A4}$$

According to the definition of the porosity, the number of capillaries remain the same with

$$n = \frac{\phi}{\pi R^2} \tag{A5}$$

Inserting (Equation (A5)) in (Equation (A4)) yields the reduction in permeability due to the coating of hydrates, as follows:

$$k_1 = \frac{\frac{\phi}{R^2} R_1^4}{8} = \frac{\phi R_1^4}{8R^2} \tag{A6}$$

The reduced radius can be written as a function of the original radius in terms of the hydrate saturation as reducing fraction such as $R_1^2 = R^2 * (1 - S_h)$, resulting in

$$k_1 = \frac{\phi R^4 (1 - S_h)^2}{8R^2} = \frac{\phi R^2 (1 - S_h)^2}{8} \tag{A7}$$

The relative hydrate permeability is defined as the ratio of the reduced permeability due to hydrates to the original one without hydrates and can be expressed by combining (Equation (A7)) with (Equation (A3)), as follows:

$$\frac{k_1}{k_0} = \frac{\frac{\phi R^2 (1-S_h)^2}{8}}{\frac{\phi R^2}{8}} = (1 - S_h)^2 \tag{A8}$$

Appendix A.1.2. Pore-Filling Morphology

Assuming that hydrates form in the center of the capillaries creating a cylinder within a cylinder, the annular flux of a single capillary is [25]

$$q = \frac{\pi}{8\mu}\frac{\Delta p}{L}(R^4 - R_1^4 - \frac{(R^2 - R_1^2)^2}{log(\frac{R}{R_1})}) \tag{A9}$$

where R is the radius of the capillary cylinder and R_1 is the radius of the hydrate cylinder.

Combining this expression with Darcy's flux (Equation (1)) and multiplying it with n for a bundle of capillaries gives the absolute permeability as

$$k = \frac{n\pi}{8}(R^4 - R_1^4 - \frac{(R^2 - R_1^2)^2}{log(\frac{R}{R_1})}) \tag{A10}$$

Dividing both sides by R^4 results in a better mathematical expression for later on, as follows:

$$k = \frac{n\pi R^4}{8}(1 - \frac{R_1^4}{R^4} - \frac{(1-\frac{R_1^2}{R^2})^2}{log(\frac{R}{R_1})}) \tag{A11}$$

The hydrate saturation as percentage of the total pore volume assuming cylindrical shapes can be written as

$$S_h = \frac{2\pi L R_1^2}{2\pi L R^2} = \frac{R_1^2}{R^2} \tag{A12}$$

This gives the permeability as a function of the hydrate saturation by combining (Equation (A12)) with (Equation (A11)), as follows:

$$k_1 = \frac{n\pi R^4}{8}(1 - S_h^2 - \frac{(1-S_h)^2}{log(\sqrt{\frac{1}{S_h}})}) \tag{A13}$$

To include the porosity with the definition of n (Equation (A5)), the final form for the hydrate permeability becomes

$$k_1 = \frac{\phi R^2}{8}(1 - S_h^2 - \frac{(1-S_h)^2}{log(\sqrt{\frac{1}{S_h}})}) \tag{A14}$$

Following the approach to relate (Equation (A14)) with (Equation (A3)) to each other, the relative hydrate permeability in the pore-filling morphology becomes

$$\frac{k_1}{k_0} = \frac{\frac{\phi R^2}{8}(1 - S_h^2 - \frac{(1-S_h)^2}{log(\sqrt{\frac{1}{S_h}})})}{\frac{\phi R^2}{8}} = 1 - S_h^2 - \frac{(1-S_h)^2}{log(\sqrt{\frac{1}{S_h}})} \tag{A15}$$

Appendix A.2. Kozeny-Type Equation

Due to the irregular shape of pores and flow paths that results usually in a greater length than a straight path, the Kozeny family of hydraulic permeability introduces shape factors and tortuosity [26], as follows:

$$k = \frac{\phi}{\nu\tau(\frac{A}{V})^2_{pore}} \tag{A16}$$

where ν is the shape factor and τ is the tortuosity defined as

$$\tau = (\frac{L_1}{L_0})^2 \tag{A17}$$

The tortuosity is also related to the electrical formation factor F and the porosity with $\tau = F\phi$ [27].

Assuming that only the area, pore volume, and electrical formation factor change, and the shape factor ν stays constant with the formation of hydrates, the relative permeability becomes then by relating

$$\frac{k_1}{k_0} = \frac{\frac{\phi}{\nu F_1 \phi(\frac{A_1}{V_1})^2}}{\frac{\phi}{\nu F_0 \phi(\frac{A_0}{V_0})^2}} = \frac{F_0}{F_1}(\frac{A_0 V_1}{A_1 V0})^2 \tag{A18}$$

The formation factor change between hydrate saturation F_1 and water saturation F_0 can be described with the hydrate saturation and Archie's exponent n [28] such that

$$\frac{F_1}{F_0} = (1 - S_h)^{-n} \tag{A19}$$

Now, the relative permeability depends only on the surface area ratio depending on the morphological model assumption, as follows:

$$\frac{k_1}{k_0} = (1 - S_h)^{n+2}(\frac{A_0}{A_1})^2 \tag{A20}$$

Appendix A.2.1. Grain/Pore-Coating Morphology

In the morphological assumption of hydrates coating the grains, the surface area decreases with increasing hydrate saturation, as follows:

$$\frac{A_0}{A_1} = \frac{2\pi LR}{2\pi LR_1} = \frac{R}{R_1} \tag{A21}$$

As the volume of a cylinder is $V = 2\pi LR^2$, the surface area relation as a function of the hydrate saturation becomes

$$S_h = \frac{V_1}{V_0} = \frac{2\pi L(R^2 - R_1^2)}{2\pi LR^2} = 1 - \frac{R_1^2}{R^2} \tag{A22}$$

Rearranging the equation then gives

$$(1 - S_h) = \frac{R_1^2}{R^2} \tag{A23}$$

Furthermore, combining (Equation (A23)) with (Equation (A21)) yields

$$\frac{A_0}{A_1} = \frac{1}{\sqrt{1 - S_h}} \tag{A24}$$

This gives the relative hydrate permeability as a function of the hydrate saturation resulting in

$$\frac{k_1}{k_0} = (1 - S_h)^{n+2}\frac{1}{1 - S_h} = (1 - S_h)^{n+1} \tag{A25}$$

According to [28], Archie's saturation exponent n equals 1.5 for $0 < S_h < 0.8$ with a diverging value at a hydrate saturation greater than 80%.

Appendix A.2.2. Pore-Filling Morphology

As hydrates grow in the center of the capillary, the surface area grows as well according to (Equation (A21)).

The hydrate saturation is related to the volume of the capillaries with

$$S_h = \frac{V_1}{V_0} = \frac{2\pi LR_1^2}{2\pi L(R - R_1)^2} = \frac{R_1^2}{R^2 - R_1^2} = \frac{R_1^2}{R^2} - 1 \tag{A26}$$

$$\frac{R_1}{R} = \sqrt{S_h} + 1 \tag{A27}$$

Combining (Equation (A21)) with (Equation (A27)) gives

$$\frac{A_0}{A_1} = \frac{1}{1 + \sqrt{S_h}} \tag{A28}$$

Then, the relative hydrate permeability is expressed as

$$\frac{k_1}{k_0} = (1 - S_h)^{n+2}(\frac{1}{1 + \sqrt{S_h}})^2 = \frac{(1 - S_h)^{n+2}}{(1 + \sqrt{S_h})^2} \tag{A29}$$

According to [28], the n exponent increases approximately linearly from $n = 0.4$ at $S_h = 0.1$ to $n = 1$ at $S_h = 1$ yielding the linear dependency of $n(S_h) = \frac{2}{3} * S_h + \frac{1}{3}$.

References

1. Sloan, E.D. Introductory Overview: Hydrate Knowledge Development. *Am. Mineral.* **2004**, *89*, 1155–1161. [CrossRef]
2. Sloan, E.D.; Koh, C.A. *Clathrate Hydrates of Natural Gases*; CRC Press: Boca Raton, FL, USA, 2007.
3. Holder, G.D.; Katz, D.L.; Hand, J.H. Hydrate Formation in Subsurface Environments. *AAPG Bull.* **1976**, *60*, 981–988.
4. Verma, V.K.; Hand, J.H.; Katz, D.L. Gas Hydrates from Liquid Hydrocarbons (Methane-Propane-Water System). In Proceedings of the GVC/AIChE Joint Meeting, Munich, Germany, 17–20 September 1974.
5. Kumar, A.; Maini, B.B.; Bishnoi, P.R.; Clarke, M.; Zatsepina, O.; Srinivasan, S. Experimental Determination of Permeability in the Presence of Hydrates and its Effect on the Dissociation Characteristics of Gas Hydrates in Porous Media. *J. Pet. Sci. Eng.* **2010**, *70*, 114–122. [CrossRef]
6. Delli, M.L.; Grozic, J.L.H. Experimental Determination of Permeability of Porous Media in the Presence of Gas Hydrates. *J. Pet. Sci. Eng.* **2014**, *120*, 1–9. [CrossRef]
7. Johnson, A.; Patil, S.; Dandekar, A. Experimental Investigation of Gas-Water Relative Permeability for Gas-Hydrate-Bearing Sediments from the Mount Elbert Gas Hydrate Stratigraphic Test Well, Alaska North Slope. *Mar. Pet. Geol.* **2011**, *28*, 419–426. [CrossRef]
8. Nimblett, J.; Ruppel, C.D. Permeability Evolution During the Formation of Gas Hydrates in Marine Sediments. *J. Geophys. Res. Solid Earth* **2003**, *108*, 2420–2467. [CrossRef]
9. Liu, X.; Flemings, P.B. Dynamic Multiphase Flow Model of Hydrate Formation in Marine Sediments. *J. Geophys. Res. Solid Earth* **2007**, *112*, 3101–3132. [CrossRef]
10. Waite, W.F.; Santamarina, J.C.; Cortes, D.D.; Dugan, B.; Espinoza, D.N.; Germaine, J.; Jang, J.; Jung, J.W.; Kneafsey, T.J.; Shin, H.; et al. Physical Properties of Hydrate-Bearing Sediments. *Rev. Geophys.* **2009**, *47*, 4003–4041. [CrossRef]
11. Dai, S.; Santamarina, J.C.; Waite, W.F.; Kneafsey, T.J. Hydrate Morphology: Physical Properties of Sands with Patchy Hydrate Saturation. *J. Geophys. Res. Solid Earth* **2012**, *117*, B11205. [CrossRef]
12. Ecker, C.; Dvorkin, J.; Nur, A. Sediments with gas hydrates: Internal structure from seismic AVO. *Geophysics* **1998**, *63*, 1659–1669. [CrossRef]
13. Priegnitz, M.; Thaler, J.; Spangenberg, E.; Schicks, J.M.S.; Schrötter, J.; Abendroth, S. Characterizing Electrical Properties and Permeability Changes of Hydrate Bearing Sediments Using ERT Data. *Geophys. J. Int.* **2015**, *202*, 1599–1612. [CrossRef]
14. Hyndman, R.D.; Spence, G.D. A Seismic Study of Methane Hydrate Marine Bottom Simulating Reflectors. *J. Geophys. Res. Solid Earth* **1992**, *97*, 6683–6698. [CrossRef]
15. Katzman, R.; Holbrook, W.S.; Paull, C.K. Combined Vertical-Incidence and Wide-Angle Seismic Study of a Gas Hydrate Zone, Blake Ridge. *J. Geophys. Res. Solid Earth* **1994**, *99*, 17975–17995. [CrossRef]
16. Sahoo, S.K.; Madhusudhan, B.N.; Marín-Moreno, H.; North, L.J.; Ahmed, S.; Falcon-Suarez, I.H.; Minshull, T.A.; Best, A.I. Laboratory Insights Into the Effect of Sediment-Hosted Methane Hydrate Morphology on Elastic Wave Velocity From Time-Lapse 4-D Synchrotron X-Ray Computed Tomography. *Geochem. Geophys. Geosyst.* **2018**, *19*, 4502–4521. [CrossRef]
17. Singh, S.C.; Minshull, T.A.; Spence, G.D. Velocity Structure of a Gas Hydrate Reflector. *Science* **1993**, *260*, 204–207. [CrossRef]
18. Edwards, R.N. On the Resource Evaluation of Marine Gas Hydrate Deposits Using Sea-floor Transient Electric Dipole-dipole Methods. *Geophysics* **1997**, *62*, 63–74. [CrossRef]
19. Du Frane, W.L.; Stern, L.A.; Weitemeyer, K.A.; Constable, S.; Pinkston, J.C.; Roberts, J.J. Electrical Properties of Polycrystalline Methane Hydrate. *Geophys. Res. Lett.* **2011**, *38*, L09313. [CrossRef]
20. Du Frane, W.L.; Stern, L.A.; Constable, S.; Weitemeyer, K.A.; Smith, M.M.; Roberts, J.J. Electrical Properties of Methane Hydrate+ Sediment Mixtures. *J. Geophys. Res. Solid Earth* **2015**, *120*, 4773–4783. [CrossRef]
21. Yuan, J.; Edwards, R. The Assessment of Marine Gas Hydrates through Electrical Remote Sounding: Hydrate Without a BSR? *Geophys. Res. Lett.* **2000**, *27*, 2397–2400. [CrossRef]
22. Pohl, M. Ultrasonic and Electrical Properties of Hydrate-Bearing Sediments. Ph.D. Thesis, Colorado School of Mines, Golden, CO, USA, 2018.
23. Peksa, A.E.; Wolf, K.H.A.A.; Zitha, P.L.J. Bentheimer Sandstone Revisited for Experimental Purposes. *Mar. Pet. Geol.* **2015**, *67*, 701–719. [CrossRef]
24. Geranutti, B.L.S. Investigation of Gas Hydrate Formation during Sea Water Injection Based on Undersaturated Oil Core Flooding Experiments. Master's Thesis, Colorado School of Mines, Golden, CO, USA, 2020.
25. Lamb, H. *Hydrodynamics*; Dover: Mineola, NY, USA, 1945; pp. 331–332.
26. Scheidegger, A.E. *The Physics of Flow through Porous Media*; Macmillan: Old Tappan, NJ, USA, 1960.
27. Hearst, J.R.; Nelson, P.H.; Paillet, F.L. *Well Logging for Physical Properties*; McGraw-Hill: New York, NY, USA, 2000.
28. Spangenberg, E. Modeling of the Influence of Gas Hydrate Content on the Electrical Properties of Porous Sediments. *J. Geophys. Res.* **2001**, *106*, 6535–6548. [CrossRef]

Journal of
Marine Science and Engineering

MDPI

Article

Numerical Simulation of Gas Production Behavior Using Radial Lateral Well and Horizontal Snake Well Depressurization Mining of Hydrate Reservoir in the Shenhu Sea Area of the South China Sea

Tinghui Wan [1,2], Mingming Wen [1,2,*], Hongfeng Lu [1,2], Zhanzhao Li [1,2], Zongheng Chen [1,2], Lieyu Tian [1,2], Qi Li [1,2], Jia Qu [1,2] and Jingli Wang [1,2,*]

1 Guangzhou Marine Geology Survey, China Geological Survey, Ministry of Natural Resources, Guangzhou 511458, China; atomion@126.com (T.W.); gmgslhf@126.com (H.L.); 13650780173@163.com (Z.L.); czhgs@126.com (Z.C.); tianlieyu23@163.com (L.T.); liqi3412@163.com (Q.L.); qujia2261520@163.com (J.Q.)

2 National Engineering Research Center for Gas Hydrate Exploration and Development, Guangzhou 511458, China

* Correspondence: wmingming@mail.cgs.gov.cn (M.W.); wjl06012527@126.com (J.W.)

Abstract: Improving the production capacity of natural gas hydrates (NGHs) is crucial for their commercial development. Based on the data of the first on-site testing production of NGHs in the Shenhu Sea area, numerical methods were used to analyze the production behavior of radial lateral well (RLW) and horizontal snake well (HSW) with different completion lengths when they deployed at different layers of the Class-1 type hydrate reservoir (with a fixed pressure difference of 6 MPa and continuous production for 360 days). The results indicate that compared with the single vertical well production, RLW and HSW can effectively increase production capacity by enlarging drainage area and the productivity is directly proportional to the total completion length. The RLW and HSW deployed at the three-phase layer (TPL) have optimal mining performance within a 360-day production period. Different to the previous research findings, during a short-term production period of 360 days, regardless of the deployment layer, the overall production capacity of HSW is better than RLW's. The total gas production of HSW-2 circles well type is about four times that of a single vertical well, reaching 1.554×10^7 ST m^3. Moreover, the HSW-1 lateral well type stands out with an average Q_g of 3.63×10^4 ST m^3/d and a specific production index J of 16.93; it has the highest J-index among all well types, which means the best mining efficiency. It is recommended to choose the HSW-1 circle well type, if the coiled tubing drilling technique is used for on-site testing production of NGHs in the future. The research results provide insights into the potential applications of RLW and HSW in this sea area.

Keywords: natural gas hydrate; continuous tubing drilling; radial lateral well; horizontal snake well; numerical simulation

Citation: Wan, T.; Wen, M.; Lu, H.; Li, Z.; Chen, Z.; Tian, L.; Li, Q.; Qu, J.; Wang, J. Numerical Simulation of Gas Production Behavior Using Radial Lateral Well and Horizontal Snake Well Depressurization Mining of Hydrate Reservoir in the Shenhu Sea Area of the South China Sea. *J. Mar. Sci. Eng.* **2024**, *12*, 1204. https://doi.org/10.3390/jmse12071204

Academic Editors: Lunxiang Zhang and Hongsheng Dong

Received: 22 June 2024
Revised: 14 July 2024
Accepted: 16 July 2024
Published: 17 July 2024

1. Introduction

Natural gas hydrates (NGHs) as an unconventional clean energy source are widely distributed and have huge reserves with great potential for commercial development [1–4]. The superiority of the depressurization method has been confirmed by recent offshore NGH testing production activities [5–8]. However, the production capacity of offshore NGH testing conducted by China and Japan is still far below the commercial standard of 50×10^4 m^3/d [1]. Due to the significant gap between the daily production capacity of offshore NGH testing production and the industrialization threshold, achieving low-cost and efficient NGH development becomes a key challenge [9]. After conducting a systematic analysis of the entire NGH development process, Wu et al. believe that the most promising development direction to break through the bottleneck of NGH industrialization is the composite production mode of complex structured well (such as horizontal wells, radial lateral wells, and fishbone wells, etc.,) or group wells (well network) mainly consisting of multiple vertical/horizontal wells for depressurization mining combined with auxiliary heating (cable heating, microwave heating, and electromagnetic heating, etc.,) whilst simultaneously adopting stimulation techniques that are suitable for the target reservoir, such as CO_2 cap reconstruction, near wellbore reservoir hydraulic jet grouting, hydraulic fracturing, steam or brine injection [9]. Among them, the main approach for stimulation is to construct complex structured wells such as horizontal wells and radial lateral wells, etc., with the main mechanism for stimulation enlarging the drainage area [9]. Ye H et al. observed that a directional well and a multilateral well may significantly boost productivity, particularly in cluster wells, which can increase gas productivity by up to 2.2 times that of a single well [10]. Mao et al. investigated the impact of various helical multilateral well parameters on production capacity and concluded that it has the potential to achieve commercial exploitation of NGHs [11]. Xin X et al. discovered that the depth of laterals in a multilateral well is a critical factor determining production capacity [12]. Ye H et al. investigated the effect of various parameter settings of various well types, and the findings revealed that branch parameters had the greatest influence on the productivity [13]. Hao Y et al. discovered that fishbone wells are the best well types for long-term development of NGHs [14]. Jin G et al. discovered that interference at the multilateral well intersections can increase hydrate dissociation [15]. According to research by He J et al., the single horizontal well's production capacity was only around 59.3% lower than that of the six-branch fishbone well [16]. Cao X et al. discovered that well interference of a multibranch well is adverse to gas production [17]. Previous research has substantially prompted the application of complex structured wells in NGH development.

The coiled tubing drilling technique is widely used in conventional oil and gas extraction, due to its strong technical feasibility and low-cost advantages [18–22]. In recent years, the application of this technique in the mining of NGHs has received increasing attention. The primary research focus is applying the continuous tubing drilling technique to complete the drilling of two types of complex structured wells: horizontal serpentine wells and radial horizontal wells. For example, Wan et al. explored the technical feasibility of using the coiled tubing drilling technique for HSW drilling in NGH reservoirs. The research results verified the feasibility of the technology and the HSW can effectively improve the production capacity, reduce wellbore collapse problems, and has a relatively low cost [23]. Li et al. proposed a new method of using radial jet drilling to extract hydrates provided by the corresponding process flow and studied the extension limit and monitoring of the borehole trajectory [24]. Mahmood et al., using analytical models, investigated the gas production of RLW and HSW in extracting hydrates and found that the production capacity of RLW is positively correlated with the laterals's quantity, length, and radius, while the production capacity of HSW is positively correlated with the length and radius of the wellbore [25]. Zhang et al. found that radial wells can significantly increase production capacity in the early stage of hydrate depressurization mining, and the lateral length is the main controlling factor for overall production capacity [26,27]. Zhang et al.'s experiment simulated the extraction of hydrates in water-rich hydrate samples via vertical and radial wells, and found that the gas and water production of the radial well was approximately 120% and 139% of the vertical well, respectively [28]. Wan et al. conducted a numerical evaluation of the gas production capacity of different radial lateral wellbore deployment schemes in the Shenhu Sea area hydrate reservoir. The results indicate that radial lateral wellbore can effectively improve production efficiency [29]. According to the progress of continuous tubing drilling technology in hydrate development in the past decade (Table 1), it can be seen that there is currently limited research on the RLW and HSW, therefore this work was based on on-site data from China's first offshore NGH testing production and analyzed the gas and water production behavior of RLW and HSW with different completion layers and lateral lengths. The results provide a theoretical reference for the practical application of the above well types in the Shenhu Sea area.

Table 1. Progress of coiled tubing drilling technique in hydrate development.

Author	Year	Input	Work Summary	Output
Cinelli et al. [18]	2013	Technical review	Introduced the equipment and technical process of coiled tubing drilling, using a low permeability oilfield as an example to detail the completion process and production statistics for radial jet drilling.	Coiled tubing drilling technique is a low-cost and environmentally friendly method to improve productivity.
Kamel et al. [19,20]	2016, 2017	Technical review	Introduced the theoretical, and technological progress, procedures, applications, and challenges of coiled tubing drilling technique. Several global case studies were discussed.	Coiled tubing drilling technique is effective for increasing production, and is a feasible and cost-effective alternative to marginal oilfield fracturing.
Huang et al. [22]	2019	Technical review	Introduced the technical characteristics, advantages, and limitations of coiled tubing drilling technique. Discussed the drilling performance and trajectory measurement methods.	Coiled tubing drilling technique is a flexible new geo-energy development technology.
Wan et al. [23]	2019	Analytical model	Explored the feasibility of developing NGHs v HSW. Provided an analytical model to calculate the maximum achievable wellbore length (MAWL). Predict the production capacity based on the Furui equation.	The HSW mining technology is feasible for offshore NGHs, and demonstrates how to evaluate its productivity and economy.
Li et al. [24]	2020	Analytical model, experiment, and numerical simulation	Introduced the process flow of developing NGHs via radial lateral wells, the ability of jet rock breaking drilling, the feeding method and extension limit of jet drill bits, wellbore trajectory measurement, and control.	A new approach was proposed to develop marine NGHs using an integrated method of cavitation jet drilling radial horizontal wells and screen tube completion.
Mahmood et al. [25]	2021	Analytical model	Compared the production potential of HSW and RLW using the developed analytical model and on-site case data of NGH reservoirs in the South China Sea.	RLW produce slightly higher gas productivity than HSW.
Zhang et al. [26,27]	2020, 2021	Numerical simulation	Studied the performance of gas hydrate development by combining radial lateral wells with depressurization.	Radial lateral wells can improve gas recovery rates in the early stages of production and slow down secondary hydrates generation. The recovery rate of hydrates is linearly related to the lateral length.
Zhang et al. [28]	2022	Experimental	Experimental studies were conducted on the depressurization and extraction of hydrated sediments in both gas-rich and water-rich using vertical and radial lateral wells, respectively.	When extracting rich gas hydrate sediments, the gas production behavior of vertical and radial lateral wells is almost the same. When extracting rich water hydrate sediments, the cumulative gas production of radial lateral well increased by 20.16% compared to vertical well.
Wan et al. [29]	2024	Numerical simulation	A numerical evaluation was conducted on the gas production capacity of Class-1 type hydrate reservoirs using different radial lateral well deployment schemes.	Compared to a single vertical well, the cumulative gas production of a radial lateral well increased by approximately 208.53%.

2. Methodology

2.1. Method and Process

Taking China's first offshore NGH testing production as an example, the NGH development simulation software TOUGH + HYDRATE V1.0 was adopted to establish an ideal interlayer heterogeneity model based on SHSC4 well logging curve data. The gas production data of the site was fitted to verify the reliability of the numerical model. This work predicted and compared the gas and water production behavior of RLW and HSW with different completion lengths when they deployed at different layers, with a fixed production pressure difference of 6 MPa and continuous production for 360 days. The methodology flow chart is shown in Figure 1.

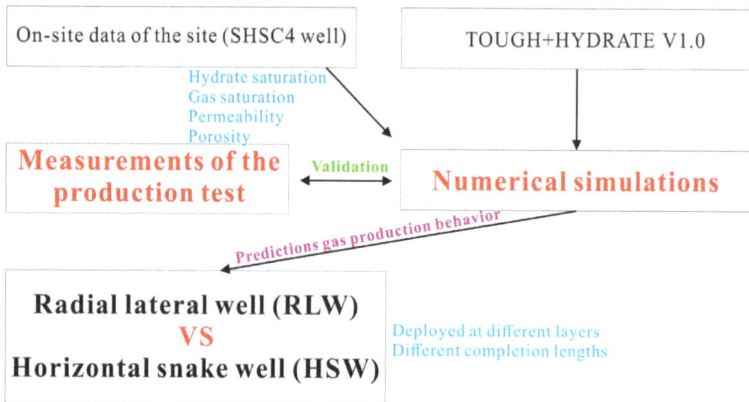

Figure 1. Methodology flow chart.

2.2. Geological Background

The SHSC4 well is located in the Baiyun sag (Figure 2). The water depths at this site are about 1266 m, and the seabed temperature is around 3.33–3.73 °C, with the geothermal gradient ranging from 45 to 67 °C/km [7,30]. The hydrate reservoir consists of three parts: the first layer is the natural gas hydrate layer rich in hydrates and water (GHBL, 201–236 mbsf); the second layer is the three-phase layer containing hydrates, high saturation free gas, and water (TPL, 236–251 mbsf); and the third layer is the free gas layer composed of water and low saturation free gas (FGL, 251–278 mbsf) [7].

2.3. Simulator Code

TOUGH + HYDRATE V1.0 is a well-known natural gas hydrate simulation code which considers the interactions between hydrate phases, multiphase flow, and heat transfer. It can accurately describe the dynamic changes in temperature, pressure, and saturation during the formation or dissociation process of hydrates [31]. The parallel version of this code was used for this work and adopted the equilibrium model for simulating hydrate extraction [32,33]. The main governing equation of this code is briefly introduced as follows [31]:

1. Mass conservation equation

The definition of the flow control equation for multicomponent fluid that follow mass conservation is as follows:

$$\frac{d}{dt}\int_{V_n} M^{\kappa} dV = \int_{\Gamma_n} F^{\kappa} \cdot \mathrm{n} d\Gamma + \int_{V_n} q^{\kappa} dV \tag{1}$$

In this equation, M^{κ} is the mass accumulation of components, F^{κ} is the flux, and q^{κ} is the source/sink ratio.

2. Energy conservation equation

The definition of the heat flow control equation follows energy conservation as follows:

$$\frac{d}{dt}\int V_n M^\theta dV = \int \Gamma_n F^\theta \cdot \mathrm{n} d\Gamma + \int V_n q^\theta dV \tag{2}$$

In this equation, θ is the heat component, M^θ is the heat accumulation, F^θ is the flux, and q^θ is the source/sink ratio.

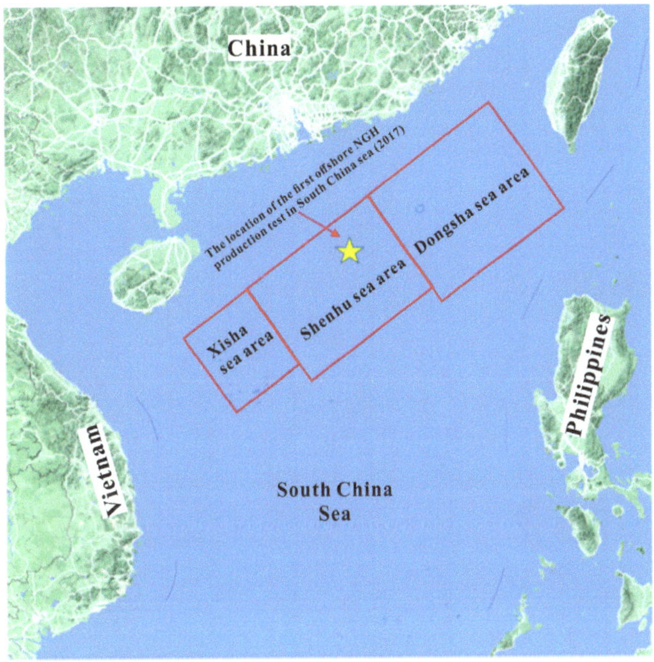

Figure 2. SHSC well site location [14]. (Adapted with permission from Hao, et al. Dynamic analysis of exploitation of different types of multilateral wells of a hydrate reservoir in the South China sea. *Energy & Fuels* 2022, *36*, 6083–6095., Copyright 2022 American Chemical Society.).

2.4. Model Discretization and Simulation Scenarios

A schematic diagram of the model is shown in Figure 3a. The x-y plane domain was discretized into 13,221 grids, and the model's z-axis was divided into 81 layers, with a total of 1,070,901 grids (Figure 3b). Hydrate dissociation is active near the wellbore and local refinement grids facilitate the capture of dynamic variations of temperature, pressure, and hydrate saturation. The minimum grid around the wellbore was set to x = 2.0 m, y = 2.0 m, and z = 1.0 m. This work established a total of nineteen simulation cases: (1) Single vertical well: the single vertical well with a length of 70 m was placed at the center of the model (Figure 3c). (2) Radial lateral well (RLW): Three simulation schemes: RLW-4 laterals, RLW-6 laterals, and RLW-8 laterals were established, each radial laterals well with a length of 357.05 m, 467.47 m, and 639.67 m respectively; RLW-4 laterals, RLW-6 laterals, and RLW-8 laterals were deployed at the middle of the three layers, respectively, (Figure 3d). (3) Horizontal snake well (HSW): Three simulation schemes: HSW-1 circle, HSW-1.5 circles, and HSW-2 circles were established, each horizontal snake well with a length of 357.05 m, 467.47 m, and 639.67 m respectively; HSW-1 circle, HSW-1.5 circles, HSW-2 circles were deployed at the middle of the three layers respectively (Figure 3e). Table 2 lists the detailed settings of the simulation scheme.

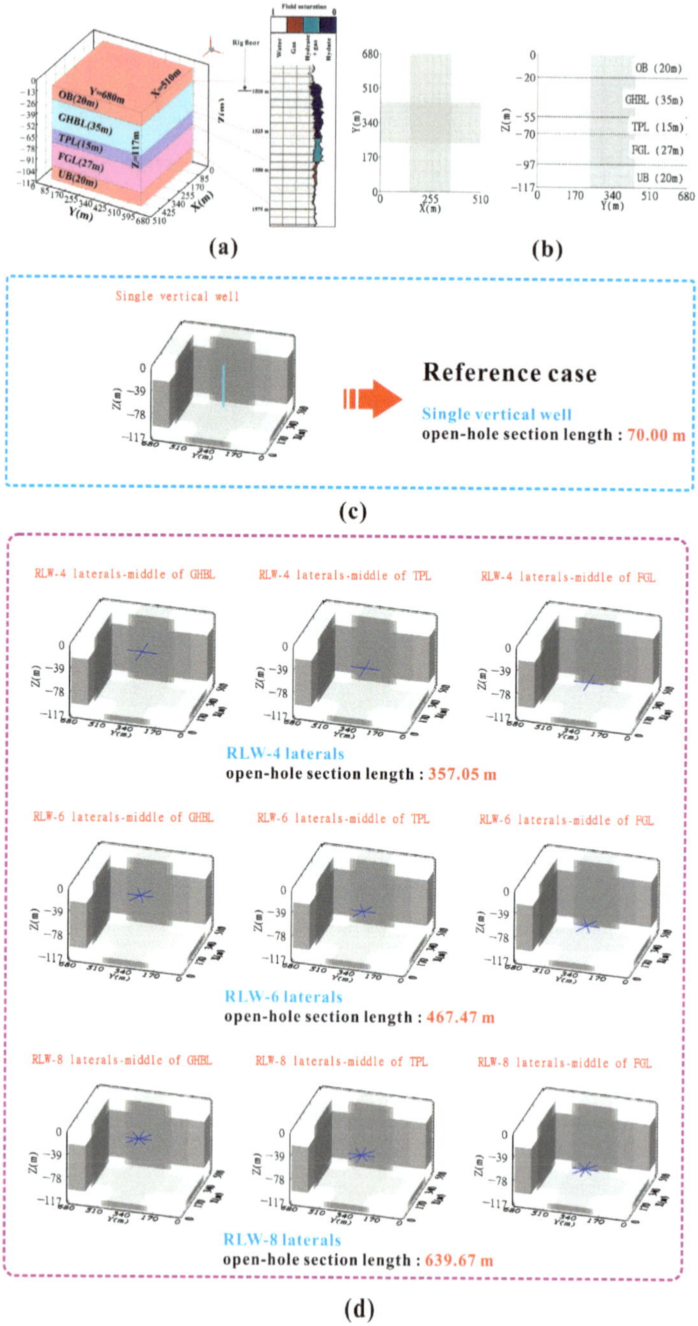

Figure 3. *Cont.*

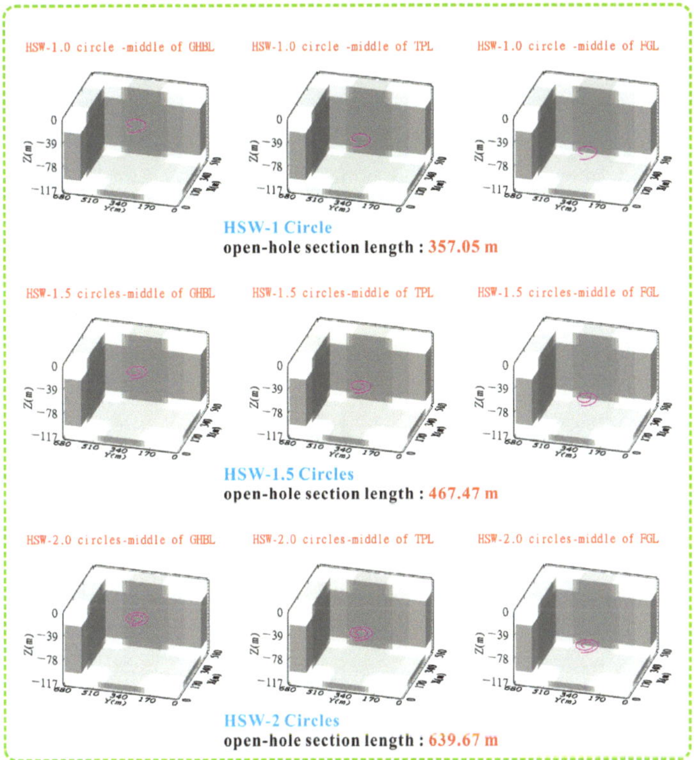

(e)

Figure 3. Schematic diagram of the model and well types: (**a**) geological model and Logging curve of SHSC-4 well. (**b**) Model mesh. (**c**) Schematic diagram of vertical well. (**d**) Schematic diagram of radial lateral well. (**e**) Schematic diagram of horizontal snake well.

Table 2. Detailed settings of the simulation scheme.

Case	Parameter Settings			
	L/(m)	*l*/(m)	*n*	Wellbore Location
Single vertical well	70	-	-	-
RLW-4 laterals	357.05	89.26	4	Middle of GHBL/TPL/FGL
RLW-6 laterals	467.47	77.91	6	
RLW-8 laterals	639.67	79.95	8	
HSW-1 circle	357.05	-	-	
HSW-1.5 circles	467.47	-	-	
HSW-2 circles	639.67	-	-	

Note: *L* is the open-hole completion length of wellbore; *l* is the length of each lateral wellbore; *n* is the quantity of lateral wellbore.

2.5. Model Initialization

GHBL, TPL, and FGL were initialized as individual subdomains, and the key was to maintain consistent heat flux between the contact surfaces of the subdomains. Finally, we combined the initialized subdomains as shown in Figure 4 [34–37] and set fixed temperatures and pressures at the top and bottom of the model to establish Dirichlet boundary conditions [38]. When the RLW and HSW were deployed at the middle of three layers, respectively, the production pressure difference between the wellbore grids and the reservoir was set to 6 MPa. In this work, the wellbore radius of the single vertical well was set to be 0.1 m, and the RLW and HSW were set to be 0.05 m [25].

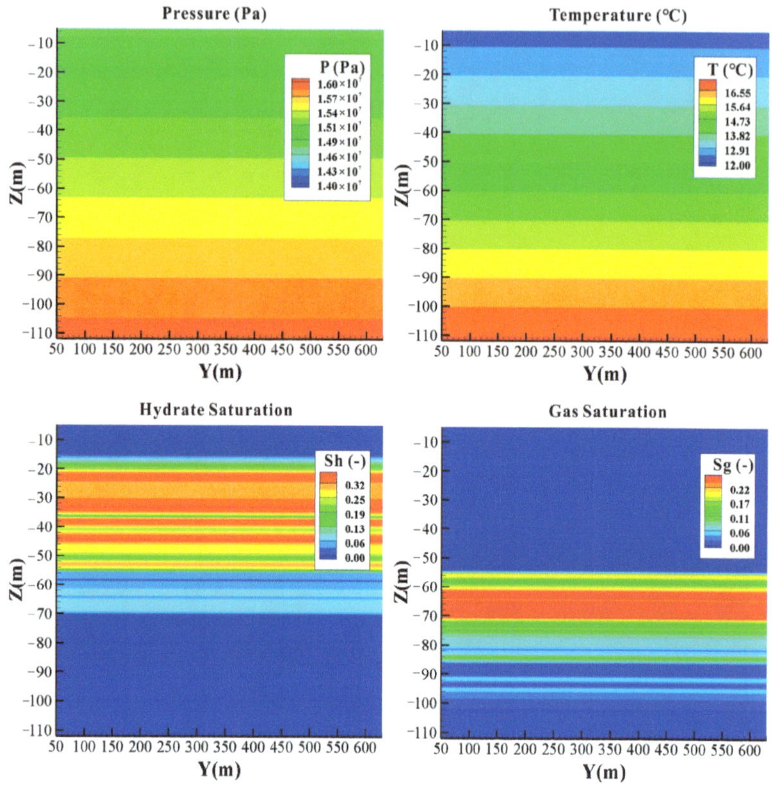

Figure 4. Model's initial conditions.

The physical properties of reservoirs, such as porosity, permeability, and saturation, were initialized based on the on-site data [7]. Since there was no information for the OB and UB, we assumed that their permeability was 2.0 mD and their porosity was 0.3. Table 3 provides the initial values of the main parameters.

Table 3. Initial values of the main parameters.

Parameter	Value and Unit	Data Sources
Overburden (OB) and Underburden (UB)'s thickness	20 m	[39]
GHBL's thickness	35 m	[40]
TPL's thickness	15 m	[40]
FGL's thickness	27 m	[40]
OB and UB's permeability	2.0 mD	-
GHBL's permeability	2.9 mD	[40]
TPL's permeability	1.5 mD	[40]
FGL's permeability	7.4 mD	[40]
OB and UB's porosity	0.30	[40]

Table 3. *Cont.*

Parameter	Value and Unit	Data Sources
GHBL's porosity	0.35	[40]
TPL's porosity	0.33	[40]
FGL's porosity	0.32	[40]
GHBL and TPL's hydrate saturation FGL's gas saturation	Extract from logging curve (Figure 3a)	[7]
Single vertical wellbore radius	0.1 m	[25]
Radial lateral wellbore radius	0.05 m	[25]
Horizontal snake wellbore radius	0.05 m	[25]
Production pressure difference	6.0 MPa	-
Salinity	3.0%	[40–42]
Grain density	2650 kg/m^3	[40–42]
Geothermal gradient	46 °C/km	[30]
Grain specific heat	1000 $J{\cdot}kg^{-1}{\cdot}K^{-1}$	[40–42]
Dry thermal conductivity	1.0 $W{\cdot}m^{-1}{\cdot}K^{-1}$	[40–42]
Wet thermal conductivity	3.1 $W{\cdot}m^{-1}{\cdot}K^{-1}$	[40–42]
Capillary pressure model	$P_{cap} = -P_0\left[(S^*)^{-1/\lambda} - 1\right]^{1-\lambda}$, $S^* = \frac{(S_A - S_{irA})}{S_{mxA} - S_{irA}}$	-
Maximum aqueous saturation S_{mxA}	1	[40–42]
Porosity distribution index λ	0.45	[40–42]
Initial capillary pressure P_0	10^4 Pa	[40–42]
Relative permeability model	$K_{rA} = [(S_A - S_{irA})/(1 - S_{irA})]^{nA}$, $K_{rG} = [(S_G - S_{irG})/(1 - S_{irA})]^{nG}$	-
Aqueous phase permeability reduction index n_A	3.5	[41]
Gas phase permeability reduction index n_G	2.5	[41]
Irreducible gas saturation S_{irG}	0.03	[41]
Irreducible aqueous saturation S_{irA}	0.30	[41]

2.6. Model Validation

Model validation is a crucial step in numerical simulation research. According to the data released by Li et al., the gas production of China's first offshore natural gas hydrate trial production is shown in Table 4 [43].

Table 4. Gas production of the first offshore NGH test production in China.

Duration/d	Cumulative Gas Volume/10^4 m^3	Gas Rate/10^3 $m^3{\cdot}d^{-1}$
0–8	12.80	16.00
9–16	3.20	4.00
17–22	2.37	3.95
23–31	2.71	2.98
32–42	2.42	2.20
43–60	7.40	4.11

The single vertical well was deployed at the center of the model with a length of 70 m, the completion interval was −201 to −271 mbsf (consistent with the model's −21 m to −91 m), and the wellbore grid had a production pressure difference of 3 MPa [44]. The position of the vertical well is shown in Figure 5.

Figure 6 shows the fitting results of gas production. It can be seen that the fitting results of simulated gas production and trial production data were within an acceptable range. According to the fitting results, this model can serve as the basic model for subsequent research.

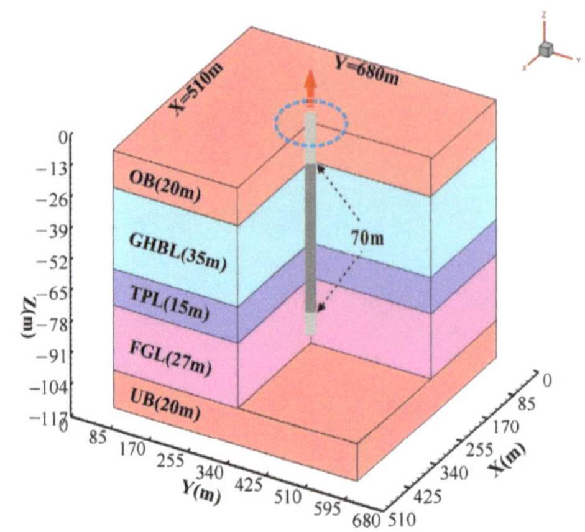

Figure 5. Vertical well position.

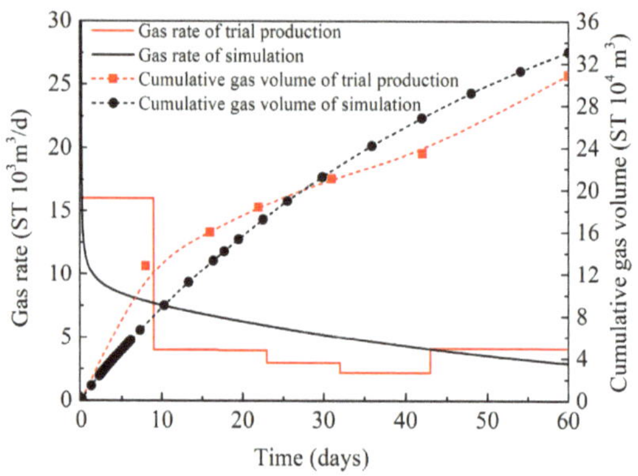

Figure 6. On-site gas production fitting.

3. Results and Discussion

3.1. RLW and HSW Deployed at GHBL

3.1.1. Evolution of Gas and Water Characteristics

Figure 7a,b shows the variation curves of gas production rate (Q_g) and cumulative gas production (V_g) with different RLW and HSW design deployment at the middle of GHBL. The gas production rate curves of these two well types can be divided into two stages. The existence of solid hydrates results in a lower effective permeability of the GHBL layer, therefore Q_g remains at a relatively low level in the early stages of production. After 90 days of depressurization, as the hydrates dissociation around the wellbore improves the seepage conditions, the free gas from TPL suddenly increases Q_g and V_g, leading to the second stage of production. Subsequently, they decrease as the driving force weakens. After 360 days of depressurization, the V_g of RLW-4 laterals, RLW-6 laterals, RLW-8 laterals, HSW-1 circle, HSW-1.5 circles, and HSW-2 circles were 453.83×10^4, 596.20×10^4, 731.84×10^4, 514.16×10^4, 644.57×10^4, and 849.53×10^4 ST m^3, compared to the single vertical well, and increased by 124.22%, 163.19%, 200.31%, 140.73%, 176.42%, and 232.53%, respectively. The results show that RLW and HSW can increase the drainage area and significantly improve production capacity. Figure 7c,d shows the variation curves of the water production rate (Q_w) and the gas-to-water ratio (R_{gw}). Compared with the single vertical well, the solid hydrates around the RLW and HSW's wellbore dissociation under the driving force, and the water produced via hydrates dissociation enters the wellbore, causing the Q_w to show a stable period before 90 days. With the free gas from TPL beginning to enter the wellbore, the Q_w suddenly decreases at 90 days. As a critical index for evaluating the efficiency of hydrate extraction, a higher R_{gw} (ST m^3 of CH_4/ST m^3 of water) implies better economically feasibility. When these two types of wells were deployed at GHBL, their R_{gw} was ultimately stable at around 100. Table 5 shows the gas production of these well types.

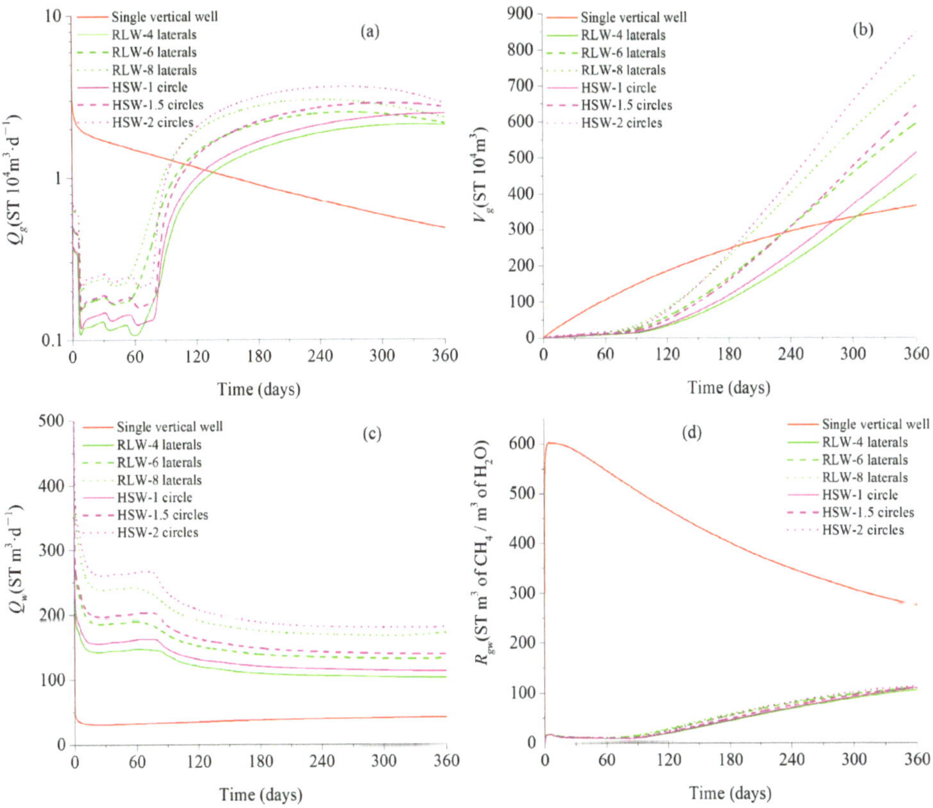

Figure 7. Gas and water production curves of RLW and HSW deployed at GHBL: (**a**) gas production rate Q_g. (**b**) Cumulative gas production V_g. (**c**) Water production rate Q_w. (**d**) Gas-to-water ratio R_{gw}.

Table 5. Gas production of RLW and HSW deployed at GHBL.

Case	Average Q_g (10^4 m^3/d)	V_g (10^4 m^3)	Compared to VW (Ref)
HSW-2 circles	2.36	849.53	232.53%
RLW-8 laterals	2.03	731.84	200.31%
HSW-1.5 circles	1.79	644.57	176.42%
RLW-6 laterals	1.66	596.20	163.19%
HSW-1 circle	1.43	514.16	140.73%
RLW-4 laterals	1.26	453.83	124.22%
Single vertical well	1.01	365.35	100.00%

3.1.2. Physical Characteristics of the Reservoir

The internal wellbore of HSW and the intersection of laterals in RLW had larger pressure drop areas (Figure 8a), which was due to the pressure superposition. This phenomenon was consistent with the findings of Jin et al. [15]. Compared with the well types deployed at TPL and FGL, the well types deployed at GHBL had a larger pressure gradient. This is because the presence of solid hydrates reduces the effective permeability of the reservoir, and allows for effective pressure propagation. The TPL and FGL contain free gas and the expansion effect of the gas limits the propagation of pressure, resulting in a smaller pressure gradient. Low-temperature areas were formed near the wellbore (Figure 8b) due to the heat absorption caused by the dissociation of hydrates (Figure 8c). Corresponding to the pressure field diagram, the internal wellbore of HSW and the intersection of laterals in RLW had a larger low-temperature area and hydrate dissociation range. A certain amount of gas was accumulated around the wellbore after 360 days of depressurization (Figure 8d).

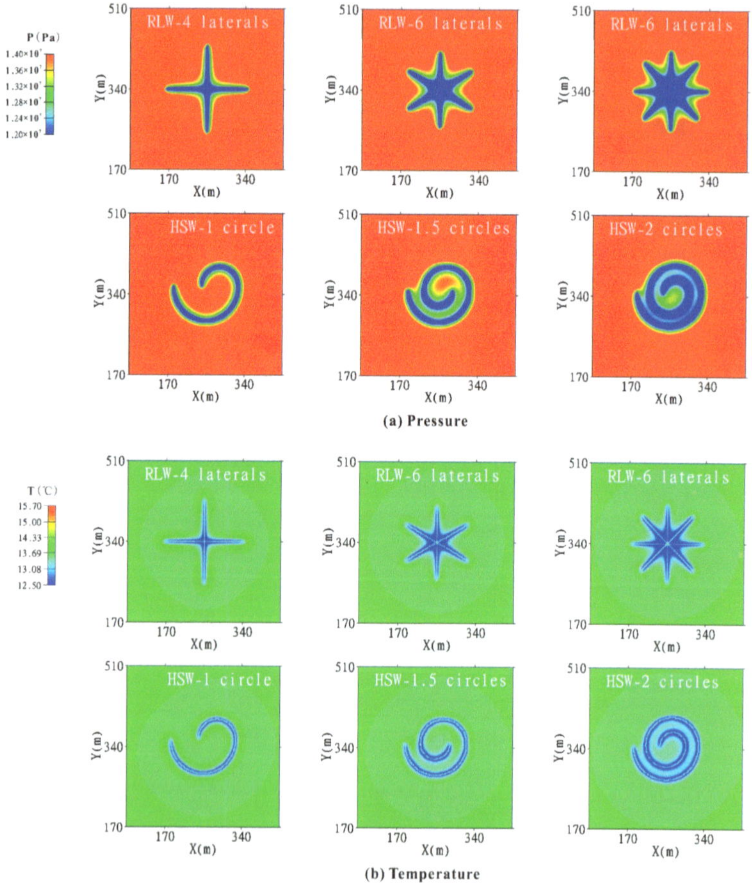

Figure 8. *Cont.*

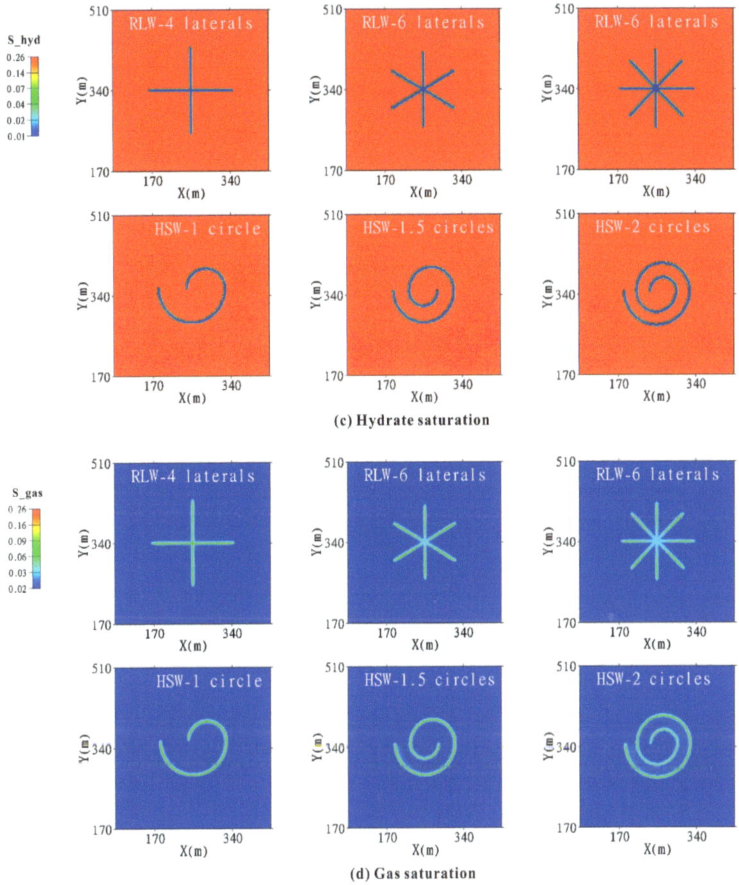

Figure 8. Physical characteristics distribution diagram of RLW and HSW deployed at GHBL.

3.2. RLW and HSW Deployed at TPL

3.2.1. Evolution of Gas and Water Characteristics

Figure 9a,b shows the variation curves of Q_g and V_g with different RLW and HSW design deployments at the middle of TPL. The Q_g of these two well types gradually decreased after reaching its peak value in the initial stage. Even so, its Q_g and V_g were the highest compared to the well types deployed at GHBL and FGL, which was because it can simultaneously recover hydrate dissociation gas from GHBL and free gas from TPL and FGL. Wan et al. also found the same results in previous studies [29]. After 360 days of depressurization, the V_g of RLW-4 laterals, RLW-6 laterals, RLW-8 laterals, HSW-1 circle, HSW-1.5 circles, and HSW-2 circles were 1215.12×10^4, 1294.38×10^4, 1356.88×10^4, 1305.72×10^4, 1463.54×10^4, and 1554.73×10^4 ST m^3, compared to the single vertical well, increased by 332.59%, 354.29%, 400.58%, 357.39%, 371.39%, and 425.54%, respectively. The results showed that the well types deployed at TPL had excellent production performance. It is worth noting that, similar to the wells deployed at GHBL, the overall production capacity of the HSW well was better than that of RLW, especially the production capacity of the HSW-1 circle was better than that of RLW-4 and RLW-6 laterals. This may be due to the smaller distance between the wellbore of HSW with spiral distribution, resulting in a larger range of pressure superposition and stronger synergistic production effects between wellbore. In this case, the reservoir at the root of the RLW laterals wellbore formed a certain amount of secondary hydrates, as shown in Figure 10c. Figure 9c,d shows the variation curves of Q_w and R_{gw}. When these well types were deployed at TPL, their Q_w was slightly lower overall compared to those deployed at GHBL and FGL. This was because a lot of free gas entered the wellbore, which affects water production; their R_{gw} was ultimately stable at around 200. Table 6 shows the gas production of these well types.

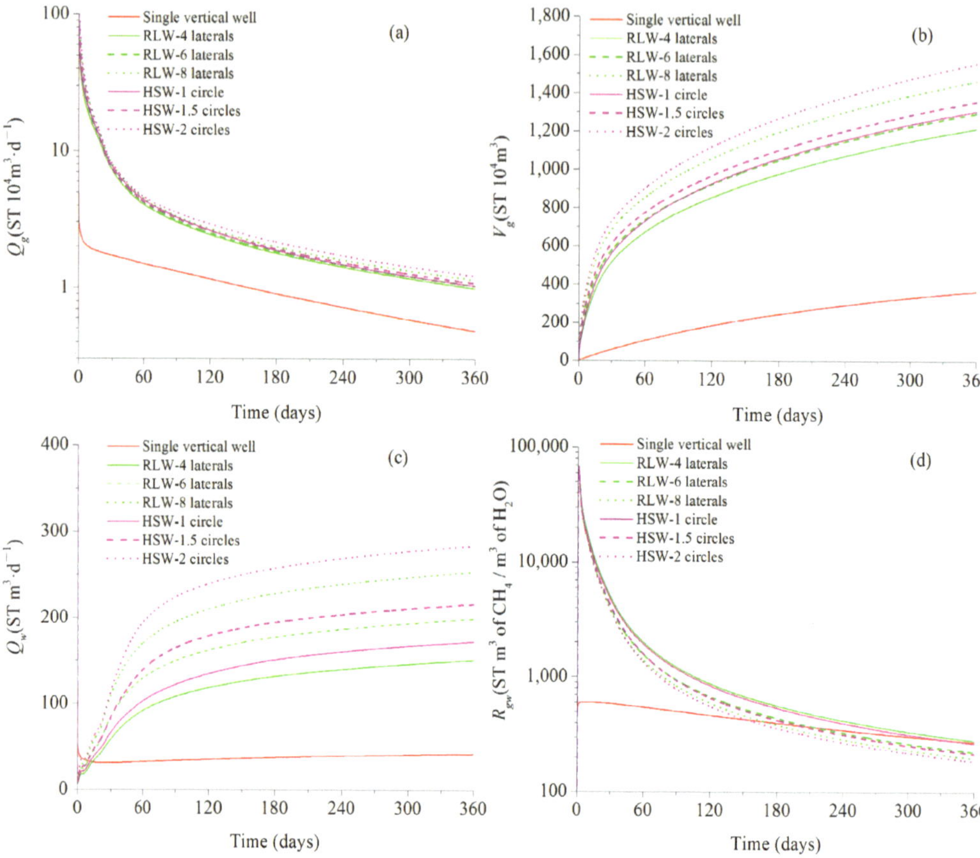

Figure 9. Gas and water production curves of RLW and HSW deployed at TPL: (**a**) gas production rate Q_g. (**b**) Cumulative gas production V_g. (**c**) Water production rate Q_w. (**d**) Gas-to-water ratio R_{gw}.

Table 6. Gas production of RLW and HSW deployed at TPL.

Case	Average Q_g (10^4 m^3/d)	V_g (10^4 m^3)	Compared to VW (Ref)
HSW-2 circles	4.32	1554.73	425.54%
RLW-8 laterals	4.07	1463.54	400.58%
HSW-1.5 circles	3.77	1356.88	371.39%
HSW-1 circle	3.63	1305.72	357.39%
RLW-6 laterals	3.60	1294.38	354.29%
RLW-4 laterals	3.38	1215.12	332.59%
Single vertical well	1.01	365.35	100.00%

3.2.2. Physical Characteristics of the Reservoir

The pressure superposition effect results in larger low-pressure area areas at the internal wellbore of HSW and the intersection of laterals in RLW (Figure 10a). The Joule–Thomson effect promotes the formation of low-temperature areas near wellbore reservoirs (Figure 10b). The reservoir at the root of the RLW laterals wellbore formed a certain amount of secondary hydrates after 360 days of depressurization (Figure 10c). Moreover, due to long-term mining, the surrounding areas of these well types formed corresponding low-saturation gas areas (Figure 10d).

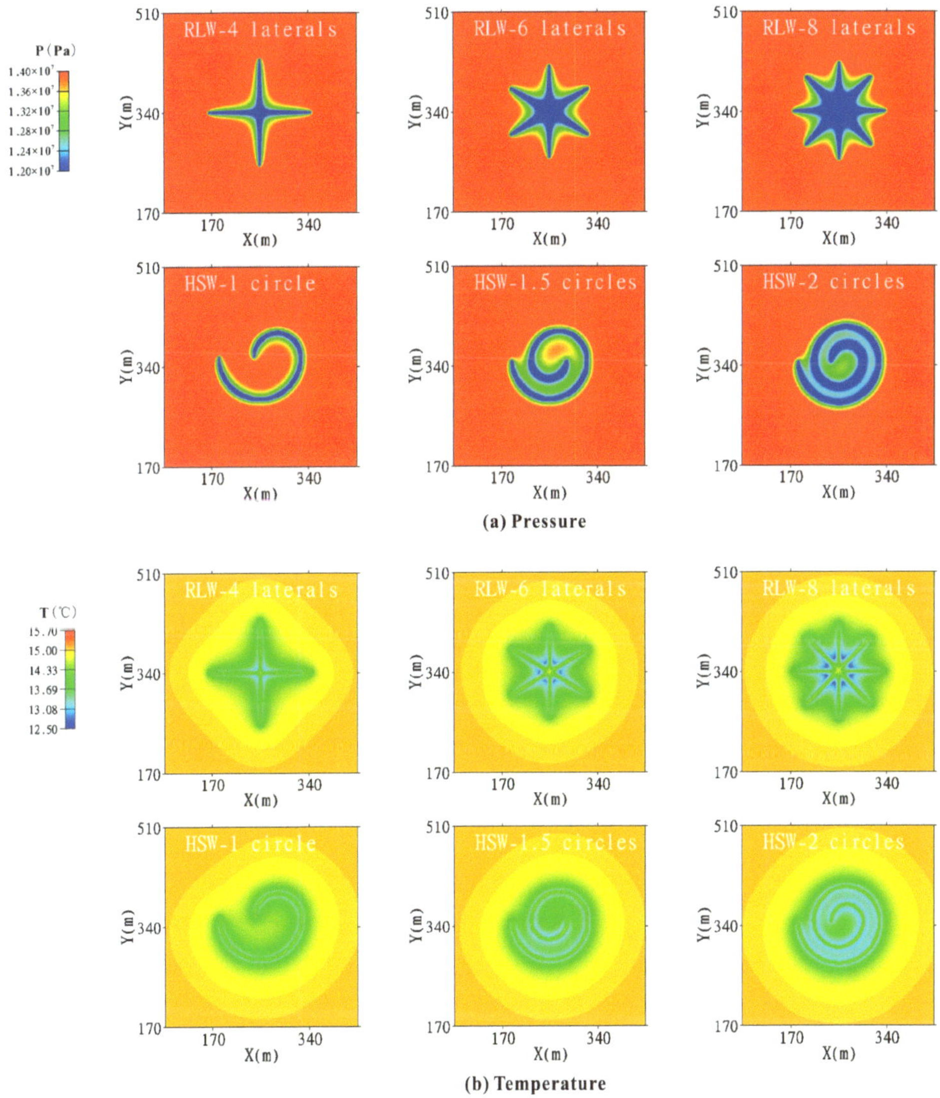

Figure 10. *Cont.*

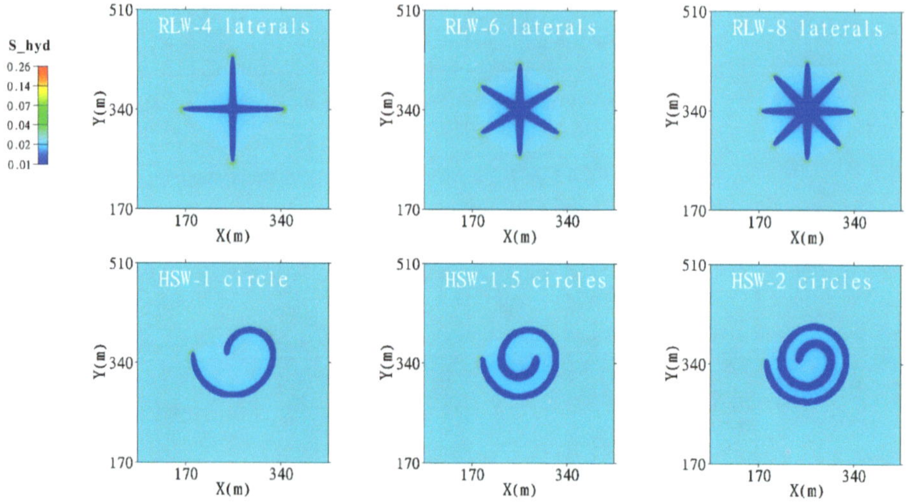

(c) Hydrate saturation

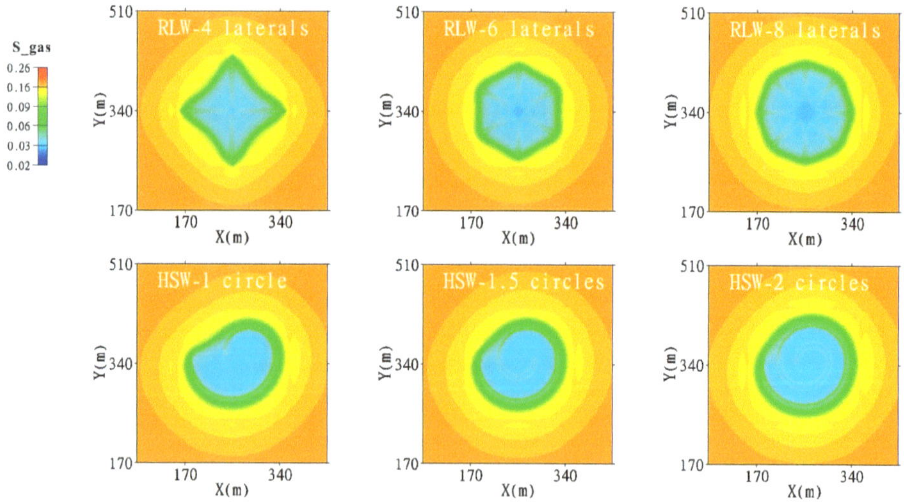

(d) Gas saturation

Figure 10. Physical characteristics distribution diagram of RLW and HSW deployed at TPL.

3.3. *RLW and HSW Deployed at FGL*

3.3.1. Evolution of Gas and Water Characteristics

Figure 11a,b shows the variation curves of Q_g and V_g with different RLW and HSW design deployments at the middle of FGL. After about eight days of depressurization, the Q_g of these two well types suddenly increased with the free gas from TPL entering the wellbore and gradually decreased with the weakening of the driving force. After 360 days of depressurization, the V_g of RLW-4 laterals, RLW-6 laterals, RLW-8 laterals, HSW-1 circle, HSW-1.5 circles, and HSW-2 circles were 1027.71×10^4, 1141.27×10^4, 1303.45×10^4, 1148.70×10^4, 1303.45×10^4, and 1396.74×10^4 ST m^3, compared to the single vertical well, increased by 281.29%, 312.38%, 356.77%, 314.41%, 330.46%, and 382.30%, respectively. Similar to the wells deployed at GHBL and TPL, the overall production capacity of HSW well was better than that of RLW and the production capacity of HSW-1 circle was better than that of RLW-4 and RLW-6 laterals again. Figure 11c,d shows the variation curves of Q_w and the R_{gw}. Compared with the well types deployed at GHBL and TPL, the well types deployed at FGL had a slightly higher water production rate because it had a higher water saturation of about 93%, and their R_{gw} was ultimately stable at around 100 to 200. Table 7 shows the gas production of these well types.

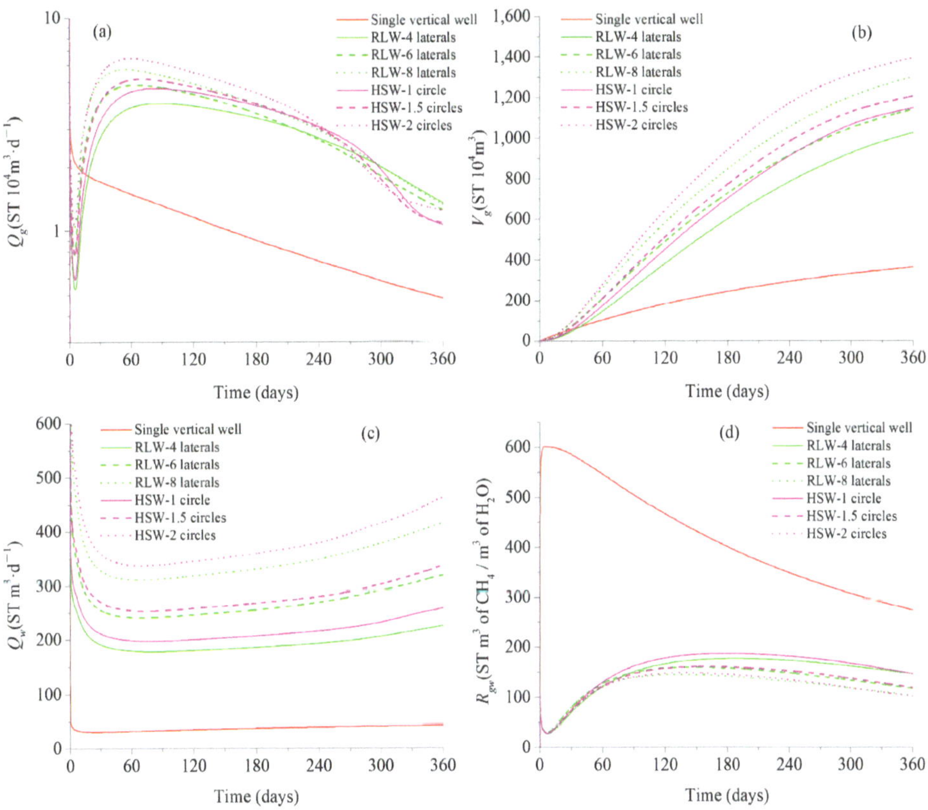

Figure 11. Gas and water production curves of RLW and HSW deployed at FGL: (**a**) gas production rate Q_g. (**b**) Cumulative gas production V_g. (**c**) Water production rate Q_w. (**d**) Gas-to-water ratio R_{gw}.

Table 7. Gas production of RLW and HSW deployed at FGL.

Case	Average Q_g (10^4 m^3/d)	V_g (10^4 m^3)	Compared to VW (Ref)
HSW-2 circles	3.88	1396.74	382.30%
RLW-8 laterals	3.62	1303.45	356.77%
HSW-1.5 circles	3.35	1207.33	330.46%
HSW-1 circle	3.19	1148.70	314.41%
RLW-6 laterals	3.17	1141.27	312.38%
RLW-4 laterals	2.85	1027.71	281.29%
Single vertical well	1.01	365.35	100.00%

3.3.2. Physical Characteristics of the Reservoir

Due to the superimposed pressure drop, the internal wellbore of HSW and the intersection of laterals in RLW had larger pressure drop areas (Figure 12a). Compared with the well types deployed at GHBL and TPL, the gas expansion effect weakened pressure propagation when the well types deployed at FGL. There were no low-temperature areas or secondary hydrates formed around the wellbore (Figure 12b,c), which was because FGL has a higher formation temperature. Additionally, a low saturation of free gas accumulated around these well type's wellbores (Figure 12d).

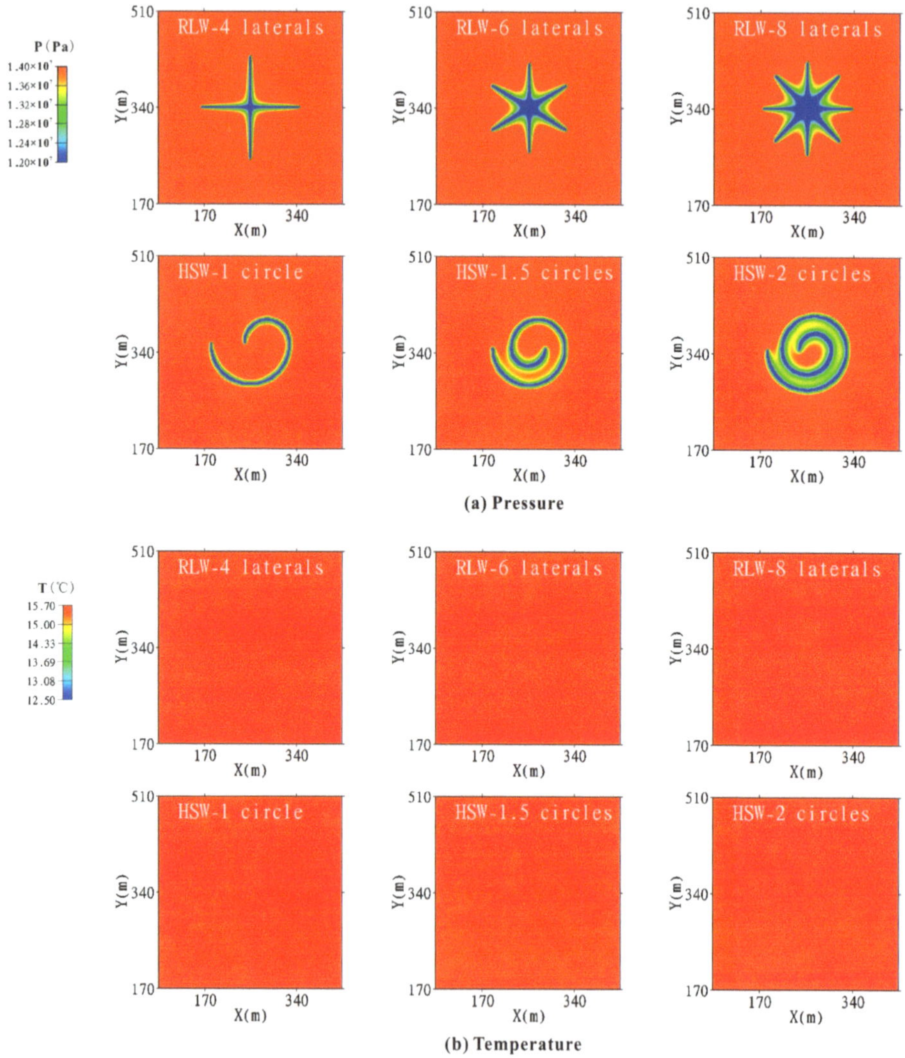

Figure 12. *Cont.*

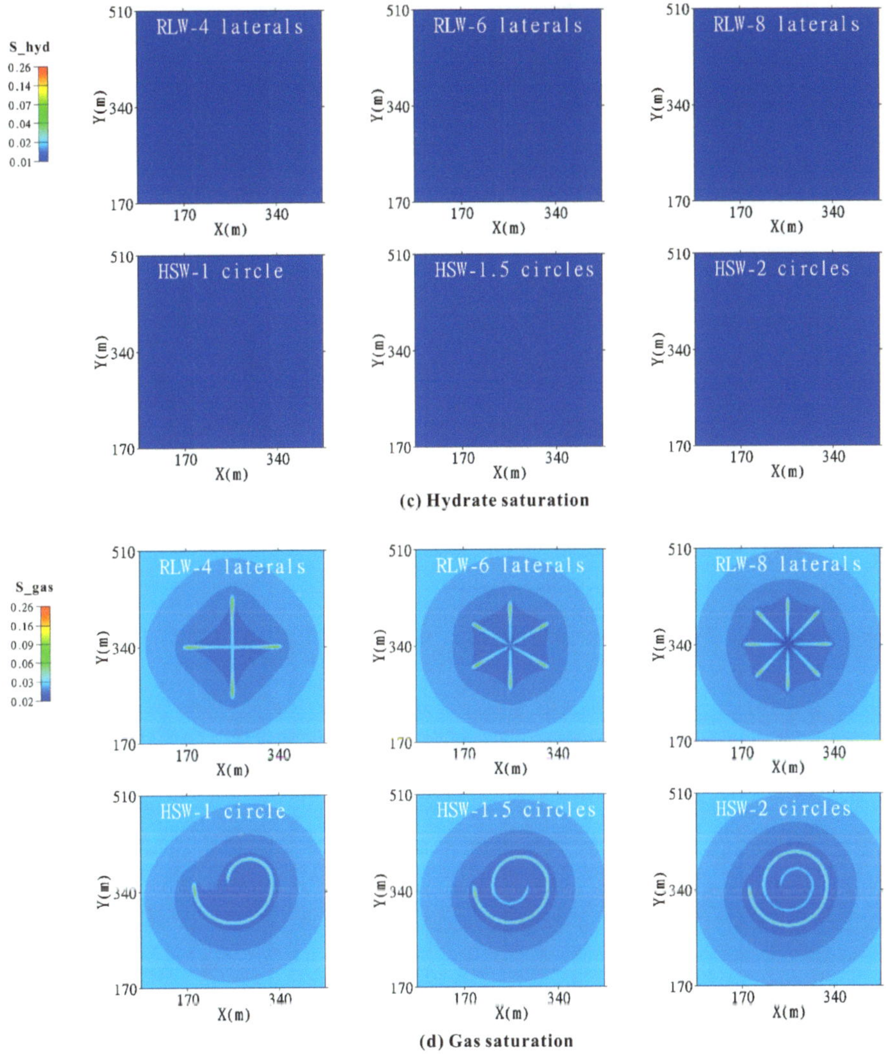

Figure 12. Physical characteristics distribution diagram of RLW and HSW deployed at FGL.

3.4. Discussion

3.4.1. Comparison of Production Capacity

The average Q_g and average R_{gw} are commonly used to evaluate production capacity. Figure 13 depicts the average Q_g and average R_{gw} of these well types during the 360-day production period. When these well types are deployed at GHBL, the average Q_g slowly increases with the dissociation of solid hydrates during production. Due to the synergistic pressure reduction effect between wellbores, the HSW well type performs better under the same completion length. When these well types are deployed at TPL or FGL, their average Q_g decreases with production as the driving force weakens. Similarly, due to the synergistic pressure reduction effect between wellbores, the HSW well type performs better. In addition, these well types have the best average R_{gw} performance when deployed at the TPL. Due to the production capacity, they may not be completely proportional to the well length. Therefore, the specific production index J is adopted as a supplementary indicator, which is mainly affected by the well types and the definition is as follows [9]:

$$J = Q_g / h\Delta P \tag{3}$$

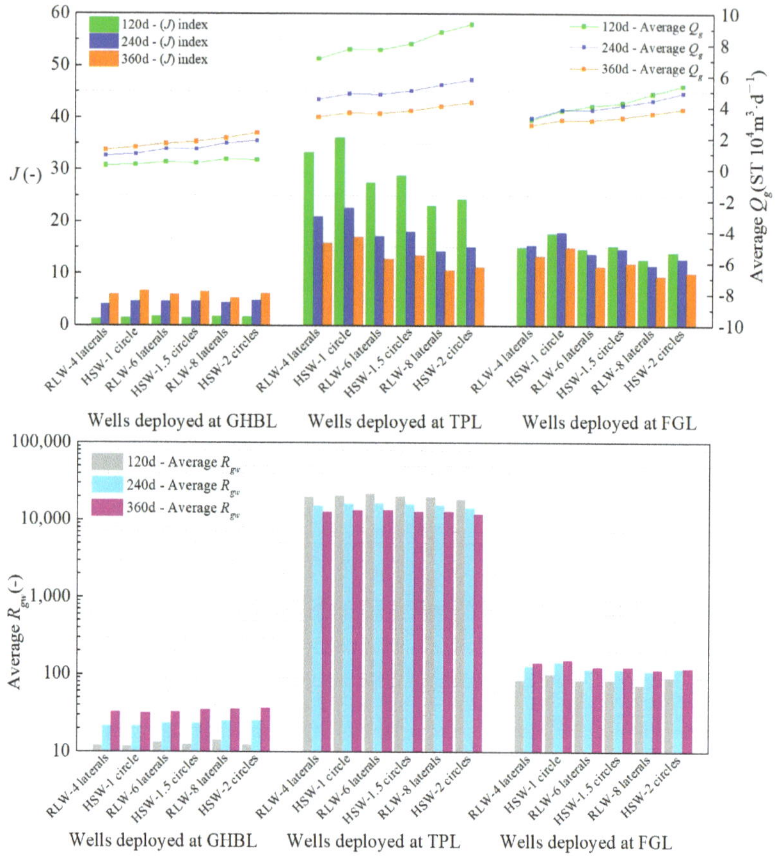

Figure 13. Histogram of average Q_g, average R_{gw} and J index, and t = 120, 240, 360 days.

Here, ΔP is the production pressure difference (MPa) and h is the well length (m). Figure 13 depicts J index of these well types during the 360-day production period. The productivity of these well types ranked as follows: TPL > FGL > GHBL. When these well types were deployed at the TPL, they had the best mining performance, and the HSW-1 circle well type stood out with an average Q_g of 3.63 × 10^4 ST m^3/d and a J-index of 16.93. Although the average Q_g of the HSW-1 circle well type was not the highest, its J-index was the highest among all well types, indicating that it had the best exploitation efficiency.

3.4.2. Summary and Recommendations

Unlike traditional drilling, coiled tubing drilling has a smaller wellbore size and turning radius, providing self-propulsion through hydraulic jetting. The axial and lateral forces generated on the wellhead during radial drilling are much lower, which can greatly improve the stability of the wellhead; this method has much lower drilling and production costs, and has great potential for application in future hydrate development, which is worth further study [24]. This work was based on on-site data from China's first offshore natural gas hydrate testing and production site and numerically analyzed the production behavior of RLW and HSW. Compared with the single vertical well production, RLW and HSW can effectively increase production capacity by enlarging the drainage area and the productivity is directly proportional to the total completion length, which is consistent with the results of many similar studies (e.g., Mahmood et al., 2021; Zhang, 2020, 2021) [25–27]. Different from the previous research results of Mahmood et al., during a short-term production period of 360 days, the overall production capacity of HSW was better than that of RLW, regardless of which layer they were deployed to [25]. This may be because previous research was based on analytical models, and factors such as the synergistic pressure reduction effect between wells could not be well considered. Meanwhile, RLW and HSW deployed at TPL had the highest production capacity during a 360-day production period. The total gas production of HSW-2 well type was about four times that of a single vertical well, reaching 1.554×10^7 ST m^3. It is worth noting that the HSW-1 circle well type stood out with an average Q_g of 3.63×10^4 ST m^3/d and a J-index of 16.93; it had the highest J-index among all well types, which means the best mining efficiency. It is recommended to choose the HSW-1 circle well type, if the coiled tubing drilling technique is used for on-site testing production of NGHs in the future. This work still has certain limitations. In the future, it is necessary to consider the real reservoir environment to establish a three-dimensional heterogeneous model, and further combine wellbore heating or reservoir reconstruction techniques to study the production behavior of RLW and HSW in-depth.

4. Conclusions

Based on the on-site data of China's first offshore NGH testing production site in the Shenhu Sea area, an ideal interlayer heterogeneity model of the SHSC4 well was established and the productivity of RLW and HSW were numerically evaluated with different completion layers and lateral lengths. The following results were obtained:

(1) RLW and HSW can effectively improve production capacity by expanding the drainage area, which is directly proportional to the number and length of laterals and the length of the horizontal wellbore. Different from previous research results, during a short-term production period of 360 days and due to the synergistic pressure reduction effect between wellbores, the overall production capacity of HSW was better than that of RLW, regardless of which layer they were deployed to.

(2) RLW and HSW deployed at the TPL had optimal mining performance within a 360-day production period due to their highest R_{gw} performance. The V_g of the HSW-2 circles well type was about four times that of a single vertical well, reaching 1.554×10^7 ST m^3. It is worth noting that the HSW-1 circle well type stood out with an average Q_g of 3.63×10^4 ST m^3/d and a J-index of 16.93 after 360-day production; it had the highest J-index among all well types, which meant the best mining efficiency. It is recommended to choose the HSW-1 circle well type, if the coiled tubing drilling technique is used for on-site testing production of NGHs in the future.

(3) Coiled tubing drilling has a smaller wellbore size and turning radius. With the advantages of strong technical feasibility and low-cost, it has great potential for application in hydrate development. In the future, it is necessary to consider the real reservoir environment, combined with stimulation methods such as wellbore-assisted heating and reservoir reconstruction, to further investigate the gas and water production behavior of RLW and HSW in different types of NGH reservoirs.

Author Contributions: T.W.: Conceptualization, Methodology, Software, Writing—Original Draft. M.W.: Writing—review and editing, Supervision. H.L.: Resources, Funding acquisition. Z.L.: Formal analysis, Investigation. Z.C.: Formal analysis, Investigation. L.T.: Resources. Q.L.: Data curation, Visualization. J.Q.: Data curation, Visualization. J.W.: Writing-review and editing, Supervision, Project administration. All authors have read and agreed to the published version of the manuscript.

Funding: National Key Research and Development Program of China (No. 2021YFB3401405 and No. SQ2023YFC2800361); Guangzhou Science and Technology Program (No. 202206050002); Guangdong Basic and Applied Basic Research Foundation (No. 2022A1515011902); and the Director General's Scientific Research Fund of Guangzhou Marine Geological Survey (No. 2023GMGSJZJJ00027).

Institutional Review Board Statement: Not applicable.

Informed Consent Statement: Not applicable.

Data Availability Statement: Data will be made available on request.

Conflicts of Interest: The authors declare that they do not have any commercial or associative interest that represent conflicts of interest in connection with the submitted work.

Nomenclature

L	Open hole completion length of wellbore (m)
l	Length of each lateral wellbore (m)
n	Quantity of lateral wellbore
M^{κ}	Mass accumulation of component κ, (kg/m^3)
F^{κ}	Mass flux of component κ, kg/(m^2·s)
q^{κ}	Sink/source of component κ, kg/(m^3·s)
M^{θ}	Energy accumulation (J/m^3)
F^{θ}	Energy flux, J/(m^2·s)
q^{θ}	Sink/source of heat, J/(m^3·s)
V	Volume (m^3)
Γ	Surface area (m^2)
t	Times (s)
φ	Porosity
Q_g	Gas production rates at well (m^3/d)
Q_w	Water production rates at well (m^3/d)
V_g	Cumulative gas production at well (m^3/d)
R_{gw}	Ratio of cumulative gas to cumulative gas (ST m^3 of CH_4/m^3 of H_2O)
J	Specific production index (-)
β	Phase, $\beta = A, G, H, I$ is aqueous, gas, hydrate and ice, respectively
κ	Component, $\kappa = w, m, i, h$ is water, methane, salt, and hydrate, respectively
S_{β}	Saturation of phase β
T	Temperature (°C)
P_{cap}cap	Capillary pressure (Pa)
P_0	Initial capillary pressure (Pa)
S^*	Saturation for capillary pressuremodel
S_{mxA}	Maximum aqueous saturation
S_{irA}	Irreducible saturation of aqueous phase
S_{irG}	Irreducible saturation of gas phase
n_A	Permeability reduction index for aqueous phase
n_G	Permeability reduction index for gas phase
λ	Porosity distribution index
k	Permeability (m^2)
g	Gravity acceleration (m/s^2)
$k_{r\beta}$	Relative permeability of phase β

Abbreviations

OB	Overburden layer
UB	Underburden layer
GHBL	Gas hydrate bearing layer
TPL	Three phase layer
FGL	Free gas layer
NGH	Natural gas hydrate
RLW	Radial lateral well
HSW	Horizontal snake well

References

1. Sloan, E. Fundamental principles and applications of natural gas hydrates. *Nature* **2003**, *426*, 353–359. [CrossRef] [PubMed]
2. Boswell, R. Is gas hydrate energy within reach? *Science* **2009**, *325*, 957–958. [CrossRef] [PubMed]
3. Chong, Z.; Yang, S.; Babu, P.; Linga, P.; Li, X. Review of natural gas hydrates as an energy resource: Prospects and challenges. *Appl. Energy* **2016**, *162*, 1633–1652. [CrossRef]
4. Boswell, R.; Collett, T. Current perspectives on gas hydrate resources. *Energy Environ. Sci.* **2011**, *4*, 1206–1215. [CrossRef]

5. Yamamoto, K.; Terao, Y.; Fujii, T.; Ikawa, T.; Seki, M.; Matsuzawa, M.; Kanno, T. Operational Overview of the First Offshore Production Test of Methane Hydrates in the Eastern Nankai Trough. In Proceedings of the Offshore Technology Conference, Houston, TX, USA, 5–8 May 2014; OnePetro: Richardson, TX, USA, 2014.
6. Yamamoto, K.; Wang, X.; Tamaki, M.; Suzuki, K. The second offshore production of methane hydrate in the Nankai Trough and gas production behavior from a heterogeneous methane hydrate reservoir. *RSC Adv.* **2019**, *9*, 25987–26013. [CrossRef] [PubMed]
7. Li, J.; Ye, J.; Qin, X.; Qiu, H.; Wu, N.; Lu, H.; Xie, W.; Lu, J.; Peng, F.; Xu, Z.; et al. The first offshore natural gas hydrate production test in South China Sea. *China Geol.* **2018**, *1*, 5–16. [CrossRef]
8. Ye, J.; Qin, X.; Xie, W.; Lu, H.; Ma, B.; Qiu, H.; Liang, J.; Lu, J.; Kuang, Z.; Lu, C.; et al. The second natural gas hydrate production test in the South China Sea. *China Geol.* **2020**, *3*, 197–209. [CrossRef]
9. Wu, N.; Li, Y.; Wan, Y.; Sun, J.; Huang, L.; Mao, P. Prospect of marine natural gas hydrate stimulation theory and technology system. *Nat. Gas Ind. B* **2021**, *40*, 173–187. [CrossRef]
10. Ye, H.; Wu, X.; Li, D. Numerical Simulation of Natural Gas Hydrate Exploitation in Complex Structure Wells: Productivity Improvement Analysis. *Mathematics* **2021**, *9*, 2184. [CrossRef]
11. Mao, P.; Wan, Y.; Sun, J.; Li, Y.; Hu, G.; Ning, F.; Wu, N. Numerical study of gas production from fine-grained hydrate reservoirs using a multilateral horizontal well system. *Appl. Energy* **2021**, *301*, 117450. [CrossRef]
12. Xin, X.; Li, S.; Xu, T.; Yuan, Y. Numerical investigation on gas production performance in methane hydrate of multilateral well under depressurization in Krishna-Godavari basin. *Geofluids* **2021**, *2021*, 9936872. [CrossRef]
13. Ye, H.; Wu, X.; Li, D.; Jiang, Y. Numerical simulation of productivity improvement of natural gas hydrate with various well types: Influence of branch parameters. *J. Nat. Gas Sci. Eng.* **2022**, *103*, 104630. [CrossRef]
14. Hao, Y.; Yang, F.; Wang, J.; Fan, M.; Li, S.; Yang, S.; Wang, C.; Xiao, X. Dynamic analysis of exploitation of different types of multilateral wells of a hydrate reservoir in the South China sea. *Energy Fuels* **2022**, *36*, 6083–6095. [CrossRef]
15. Jin, G.; Peng, Y.; Liu, L.; Su, Z.; Liu, J.; Li, T.; Wu, D. Enhancement of gas production from low-permeability hydrate by radially branched horizontal well: Shenhu Area, South China Sea. *Energy* **2022**, *253*, 124129. [CrossRef]
16. He, J. Numerical simulation of a Class I gas hydrate reservoir depressurized by a fishbone well. *Processes* **2023**, *11*, 771. [CrossRef]
17. Cao, X.; Sun, J.; Qin, F.; Ning, F.; Mao, P.; Gu, Y.; Li, Y.; Zhang, H.; Yu, Y.; Wu, N. Numerical analysis on gas production performance by using a multilateral well system at the first offshore hydrate production test site in the Shenhu area. *Energy* **2023**, *270*, 126690. [CrossRef]
18. Cinelli, S.; Kamel, A. Novel technique to drill horizontal laterals revitalizes aging field. In Proceedings of the SPE/IADC Drilling Conference and Exhibition, Amsterdam, The Netherlands, 5–7 March 2013.
19. Kamel, A. RJD: A cost effective frackless solution for production enhancement in marginal fields. In Proceedings of the SPE Eastern Regional Meeting, Canton, OH, USA, 13–15 September 2016.
20. Kamel, A. Radial Jet Drilling: A Technical Review. In Proceedings of the SPE Middle East Oil & Gas Show and Conference, Manama, UK, 8 March 2017.
21. Kamel, A. A technical review of radial jet drilling. *J. Pet. Gas Eng.* **2017**, *8*, 79–89.
22. Huang, Z.; Huang, Z. Review of Radial Jet Drilling and the key issues to be applied in new geo-energy exploitation. *Energy Procedia* **2019**, *158*, 5969–5974. [CrossRef]
23. Wan, L.; Shaibu, R.; Hou, X.; Guo, B. A feasibility study of producing natural gas from subsea hydrates with horizontal snake wells. In Proceedings of the Offshore Technology Conference Brasil. OTC, Rio de Janeiro, Brasil, 29–31 October 2019; D011S014R006.
24. Li, G.; Tian, S.; Zhang, Y. Research progress on key technologies of natural gas hydrate exploitation by cavitation jet drilling of radial wells. *Pet. Sci. Bull.* **2020**, *5*, 349–365.
25. Mahmood, M.; Guo, B. Productivity comparison of radial lateral wells and horizontal snake wells applied to marine gas hydrate reservoir development. *Petroleum* **2021**, *7*, 407–413. [CrossRef]
26. Zhang, P.; Tian, S.; Zhang, Y.; Li, G.; Zhang, W.; Khan, W.; Ma, L. Numerical simulation of gas recovery from natural gas hydrate using multi-branch wells: A three-dimensional model. *Energy* **2020**, *220*, 119549. [CrossRef]
27. Zhang, P.; Tian, S.; Zhang, Y.; Li, G.; Wu, X.; Wang, Y. Production simulation of natural gas hydrate using radial well depressurization. *Pet. Sci. Bull.* **2021**, *3*, 417–428.
28. Zhang, P.; Zhang, Y.; Wang, W.; Wang, T.; Tian, S. Experimental study on natural gas hydrate extraction with radial well depressurization. *Pet. Sci. Bull.* **2022**, *03*, 382–393.
29. Wan, T.; Yu, M.; Lu, H.; Chen, Z.; Li, Z.; Tian, L.; Li, K.; Huang, N.; Wang, J. Numerical Simulation of Vertical Well Depressurization with Different Deployments of Radial Laterals in Class 1-Type Hydrate Reservoir. *Energies* **2024**, *17*, 1139. [CrossRef]
30. Zhang, W.; Liang, J.; Lu, J.; Wei, J.; Su, P.; Fang, Y.; Guo, Y.; Yang, S.; Zang, G. Accumulation features and mechanisms of high saturation natural gas hydrate in shenhu area, northern south china sea. *Pet. Explor. Dev.* **2017**, *44*, 708–719. [CrossRef]
31. Moridis, G.; Kowalsky, M.; Pruess, K. *TOUGH+ Hydrate V1.0 User's Manual*; Report LBNL-0149E; Lawrence Berkeley National Laboratory: Berkeley, CA, USA, 2008.
32. Zhang, K.; Moridis, G.; Wu, Y.; Pruess, K. A domain decompositionapproach for large-scale simulations of flow processes in hydrate-bearing geologic media. In Proceedings of the 6th International Conference on Gas Hydrates, ICGH 2008, Vancouver, BC, Canada, 6–10 July 2008.
33. Kowalsky, M.; Moridis, G. Comparison of kinetic and equilibrium reaction models in simulating gas hydrate behavior in porous media. *Energy Convers. Manag.* **2007**, *48*, 1850–1863. [CrossRef]

34. Sun, J.; Zhang, L.; Ning, F.; Lei, H.; Liu, T.; Hu, G.; Lu, H.; Lu, J.; Liu, C.; Jiang, G.; et al. Production potential and stability of hydrate-bearing sediments at the site GMGS3-W19 in the South China Sea: A preliminary feasibility study. *Mar. Pet. Geol.* **2017**, *86*, 447–473. [CrossRef]
35. Yuan, Y.; Xu, T.; Xin, X.; Xia, Y. Multiphase Flow Behavior of Layered Methane Hydrate Reservoir Induced by Gas Production. *Geofluids* **2017**, *2017*, 7851031. [CrossRef]
36. Sun, J.; Ning, F.; Li, S.; Zhang, K.; Liu, T.; Zhang, L.; Jiang, G.; Wu, N. Numerical simulation of gas production from hydrate-bearing sediments in the Shenhu area by depressurising: The effect of burden permeability. *J. Unconv. Oil Gas Resour.* **2015**, *12*, 23–33. [CrossRef]
37. Moridis, G.; Kowalsky, M.; Pruess, K. Depressurization-induced gas production from class 1 hydrate deposits. *SPE Reserv. Eval. Eng.* **2007**, *10*, 458–481. [CrossRef]
38. Feng, Y.; Chen, L.; Suzuki, A.; Kogawa, T.; Okajima, J.; Komiya, A.; Maruyama, S. Enhancement of gas production from methane hydrate reservoirs by the combination of hydraulic fracturing and depressurization method. *Energy Convers. Manag.* **2019**, *184*, 194–204. [CrossRef]
39. Yuan, Y.; Xu, T.; Jin, C.; Zhu, H.; Gong, Y.; Wang, F. Multiphase flow and mechanical behaviors induced by gas production from clayey-silt hydrate reservoirs using horizontal well. *J. Clean. Prod.* **2021**, *328*, 129578. [CrossRef]
40. Sun, Y.; Ma, X.; Guo, W.; Jia, R.; Li, B. Numerical simulation of the short- and long-term production behavior of the first offshore gas hydrate production test in the South China Sea. *J. Pet. Sci. Eng.* **2019**, *181*, 106196. [CrossRef]
41. Ma, X.; Sun, Y.; Liu, B.; Guo, W.; Jia, R.; Li, B.; Li, S. Numerical study of depressurization and hot water injection for gas hydrate production in China's first offshore test site. *J. Nat. Gas Sci. Eng.* **2020**, *83*, 103530. [CrossRef]
42. Yu, T.; Guan, G.; Wang, D.; Song, Y.; Abudula, A. Numerical investigation on the long-term gas production behavior at the 2017 Shenhu methane hydrate production-site. *Appl. Energy* **2021**, *285*, 116466. [CrossRef]
43. Li, S.X.; Yu, X.; Li, S. Prediction of gas production of Shenhu hydrate reservoir by depressurization and its simulation treatment. *China Offshore Oil Gas* **2020**, *32*, 122–127.
44. Qin, X.; Liang, Q.; Ye, J.; Yang, L.; Qiu, H.; Xie, W.; Liang, J.; Lu, J.; Lu, C.; Lu, H.; et al. The response of temperature and pressure of hydrate reservoirs in the first gas hydrate production test in South China Sea. *Appl. Energy* **2020**, *278*, 115649. [CrossRef]

energies

MDPI

Article

Study on the Evolution Law of Temperature, Pressure, and Productivity near the Well for Gas Hydrate Exploitation by Depressurization

Rongrong Qi [1,2], Hongfeng Lu [1,2,*], Chenlu Xu [1,2], Lu Yu [1,2], Changwen Xiao [1,2], Jinwen Du [1,2] and Yan Li [1,2,*]

1 Guangzhou Marine Geological Survey, China Geological Survey, Guangzhou 511458, China; qrr_cup@163.com (R.Q.); gmgsxcl@126.com (C.X.); yulu@mail.cgs.gov.cn (L.Y.); xiaocw2024@126.com (C.X.); djw8067@163.com (J.D.)

2 National Engineering Research Center of Gas Hydrate Exploration and Development, Guangzhou 511458, China

* Correspondence: lhongfeng@mail.cgs.gov.cn (H.L.); liyangmgs@163.com (Y.L.)

Abstract: In this paper, a one-dimensional model of gas–water two-phase productivity for hydrate depressurization is established, which takes into account permeability variation and gas–water two-phase flow. By solving the coupled algebraic equations of dissociation front position, equilibrium temperature, and pressure in an iterative scheme, the movement law of the hydrate dissociation front and the evolution process of temperature and pressure near the well were obtained, and the effects of bottom hole pressure, reservoir temperature, and hydrate saturation on productivity were analyzed. The results show that the hydrate reservoir is divided into a decomposed zone and an undecomposed zone by the dissociation front, and the temperature and pressure gradients of the former are greater than those of the latter. Reducing bottom hole pressure, increasing reservoir temperature, and increasing hydrate saturation all lead to an increase in temperature and pressure gradient in the decomposed zone. Methane gas production is a sensitive function of bottom hole pressure, reservoir temperature, and hydrate saturation. The lower the bottom hole pressure, the higher the reservoir temperature, the lower the hydrate saturation (within a certain range), and the higher the gas production rate. The trend of the water production curve is the same as that of gas, but the value is 3–4 orders of magnitude smaller, which may be due to the large difference in the viscosity of gas and water, and the gas seepage speed is much larger than that of water.

Keywords: gas hydrate; depressurization; dissociation front; productivity

Citation: Qi, R.; Lu, H.; Xu, C.; Yu, L.; Xiao, C.; Du, J.; Li, Y. Study on the Evolution Law of Temperature, Pressure, and Productivity near the Well for Gas Hydrate Exploitation by Depressurization. *Energies* **2024**, *17*, 3728. https://doi.org/10.3390/en17153728

Academic Editors: Lunxiang Zhang and Hongsheng Dong

Received: 11 June 2024
Revised: 15 July 2024
Accepted: 26 July 2024
Published: 29 July 2024

1. Introduction

Natural gas hydrate (NGH) is a crystalline compound with a cage structure formed by hydrocarbon gas and water. NGH is a highly compressed natural gas resource, and 1 m^3 of NGH can be decomposed to obtain 160–180 m^3 (standard) of natural gas [1]. NGH exists widely in marine sediments and permafrost. The global methane carbon content in gas hydrates is estimated to be 10^{16} kg [2]. Therefore, NGH is recognized as a promising future energy source in the 21st century.

The principle of hydrate dissociation is based on changing the thermodynamic equilibrium of the three-phase system (water–hydrate–gas) [3,4]. In this study, we focus on the depressurization method for hydrate reservoirs. By reducing the bottom hole pressure, the equilibrium of the hydrate phase is broken, resulting in hydrate dissociation and the release of natural gas. Makogon (1997) used the classical Stefan problem to describe the process of hydrate dissociation and obtained a self-similar solution of the pressure distribution [5]. However, the model does not take into account the effect of water release. Based on the research of Makogon, Verigin et al. (1980) proposed a research model including the effect of water release, which considered the mass balance of gas and water at the dissociation front and assumed that the water produced by hydrate dissociation was stationary and did not affect the flow of gas [6]. Yousif and Sloan (1991) regarded the dissociation process in porous media as the Kim–Bishnoi dynamic isothermal process [7]. All the above models simplify the hydrate dissociation into an isothermal process, ignoring the thermal effect.

Holder and Angert (1982) evaluated the temperature distribution in the hydrate layer by using the conduction and heat transfer equation, considering the temperature change during the hydrate dissociation process [8]. Burshears et al. (1986) extended Holder and Angert's model, taking into account the influence of water transport in the formation [9]. In their model, however, only heat conduction is included in the energy equation. Selim and Sloan (1989) considered convective heat transfer in a one-dimensional model and obtained analytical expressions of reservoir temperature and pressure distribution under the assumption that reservoir water does not flow and the well temperature remains unchanged [10]. Makogon (1997) considered the influence of heat conduction, heat convection, and throttling processes in revealing the interface heat transfer process of hydrate dissociation, and finally obtained the analytical expression of temperature and pressure in a one-dimensional model [5]. Tsypkin (2000) described the movement of gas and water in hydrate reservoirs, respectively, and obtained self-similar solutions of temperature and pressure distributions by using the micro-perturbation method [11]. Zhao and Shang (2010) further considered the influence of hydrate dissociation heat and added the energy conservation equation on the dissociation front, proving that reducing bottom hole pressure can significantly increase gas production [12]. Hao et al. (2020) established a semi-analytical productivity model of NGH by depressurization and analyzed the factors affecting productivity [13]. Based on the classical analytical model of hydrate dissociation proposed by Selim and Sloan, Li et al. (2023) further considered geomechanics and established a new analytical model [14]. With the gradual deepening of research, scholars have gradually adopted numerical models to characterize hydrate dissociation and production. Yu et al. (2023) analyzed the production and dissociation characteristics of NGH by numerical method and considered the influence of reservoir stress in the model [15]. Liu et al. (2023) further used a multi-scale numerical model to simulate the dissociation process of methane hydrate in porous media, including two-phase flow, heat and mass transfer, dissociation kinetics, and hydrate structure evolution [16].

Both field data and experimental studies show that hydrate dissociation causes changes in reservoir permeability, and gas–water two-phase seepage after decomposition is an important factor affecting productivity [17,18]. Based on this, a one-dimensional gas–water, two-phase productivity model for hydrate depressurization is proposed in this paper. Considering hydrate saturation variation and water phase flow, a set of approximate self-similar formats is used to obtain the variational solutions of pressure, temperature, and productivity in the reservoir. This method is solved in an iterative format. The distribution of pressure and temperature near the well, the location of the front, and the evolution of gas/water production over time were obtained.

This paper mainly consists of three parts. Among them, Section 2 is the model building; Section 3 is the model solving; Section 4.1 is the dynamic analysis of hydrate dissociation; and Section 4.2 is the analysis of influencing factors in hydrate productivity.

2. Model Building

2.1. Physical Model and Basic Assumptions

Assume that there is a large pressurized methane hydrate reservoir underground, the reservoir pressure is *pe*, the reservoir temperature is T_e, and there are solid hydrates and natural gas in the reservoir pores. The hydrate is in a stable state at the initial temperature and pressure. When the bottom hole pressure p_G drops to a certain value, that is, $p_G < p_D < p_e$, where p_D is the dissociation pressure of hydrate at the dissociation temperature T_D. At this point, the balance of the hydrate phase near the well is disrupted, and it begins to decompose into natural gas and water. Over time, the process of hydrate dissociation spreads outward. It is assumed that hydrate dissociation in porous media occurs in a narrow region, which can be regarded as a surface, the so-called dissociation front. This moving front divides the volume of the reservoir into two distinct zones. The area near the well is the hydrate decomposed zone, where hydrate, water, and gas coexist in three phases, while the area far away from the well is the undecomposed zone, where only solid hydrate and natural gas exist. Because of the pressure gradient, the gas and water move inside the well, while the dissociation front moves in the opposite direction. The model adopts three important assumptions: (1) The pressure and temperature at any point of the dissociation front are the equilibrium pressure p_D and the equilibrium temperature T_D; (2) The hydrate reservoir is assumed to be porous and contain natural gas. When the dissociation front moves towards the hydrate zone, heat must be supplied to the front edge due to the endothermic property of the hydrate dissociation process. In this case, heat conduction is ignored compared with thermal convection. (3) The decomposed zone contains three phases of hydrate, methane gas and water, in which only two phases of gas and water flow.

2.2. Basic Mathematical Model

In this section, the basic mathematical model adopted by Makogon (1997) [5] and Ji et al. (2001) [19] is followed. The difference is that gas–water two-phase seepage is considered in the decomposed zone. Assuming a hydrate reservoir, the well location is shown in Figure 1. Using a one-dimensional model, for the decomposed zone, the gas–water two-phase pressure equation can be seen according to the mass conservation equation as follows:

$$\frac{2\phi_1 \mu_g}{K_1 k_{rg1}} \frac{\partial p_1}{\partial t} = \frac{\partial^2 p_1^2}{\partial x^2} \tag{1}$$

$$\frac{2\phi_1 \mu_w}{K_1 k_{rw1}} \frac{\partial p_1}{\partial t} = \frac{\partial^2 p_1^2}{\partial x^2} \tag{2}$$

where

$$\phi_1 = (1 - S_H - S_w)\phi \tag{3}$$

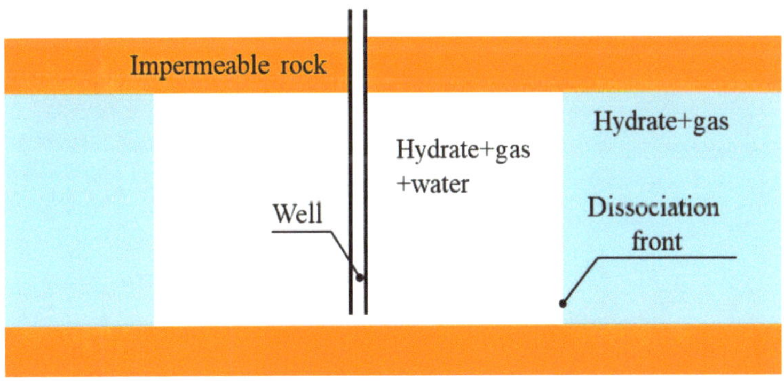

Figure 1. One-dimensional model of hydrate reservoir (modified from literature [19]).

Only free methane gas flows in the undecomposed zone. Similarly, the pressure expression in the undecomposed region can be expressed as:

$$\frac{2\phi_2 \mu_g}{K_2 k_{rg2}} \frac{\partial p_2}{\partial t} = \frac{\partial^2 p_2^2}{\partial x^2} \tag{4}$$

where

$$\phi_2 = (1 - S_{w0} - S_{H0})\phi \tag{5}$$

In the above equation, ϕ_1 and ϕ_2 are the porosities of the decomposed and undecomposed zones, respectively; μ_g and μ_w are the viscosities of gas and water, respectively; K_1 is the absolute permeability of the decomposed zone, and K_2 is the initial absolute permeability of the reservoir; k_{rg1} and k_{rg2} are gas relative permeability in the decomposed and undecomposed zones, respectively; k_{rw1} is the relative permeability of water in the decomposed zone; k_{rg0} is the initial gas relative permeability of the reservoir; p_1 and p_2 represent the pressure in the decomposed and undecomposed zones, respectively; S_H is the hydrate saturation in the decomposed zone; S_w is the water saturation of the decomposed zone; S_{w0} is the original water saturation of the reservoir; and S_{H0} is the initial hydrate saturation of the reservoir.

According to the hypothesis of the model, conduction heat transfer is much smaller than convective heat transfer in the case of natural gas in the hydrate reservoir. Therefore, the heat convection effect of fluid is considered in this paper, and the heat conduction effect is ignored. The temperature distribution equation of the hydrate reservoir is as follows:

$$\frac{\partial T_1}{\partial t} = \frac{c_v K_1}{c_1 \mu_1}\frac{\partial p_1}{\partial x}\left(\frac{\partial T_1}{\partial x} - \delta\frac{\partial p_1}{\partial x}\right) + \eta\frac{\phi_1 c_v}{c_1}\frac{\partial p_1}{\partial t} \tag{6}$$

$$\frac{\partial T_2}{\partial t} = \frac{c_v K_2}{c_2 \mu_2}\frac{\partial p_2}{\partial x}\left(\frac{\partial T_2}{\partial x} - \delta\frac{\partial p_2}{\partial x}\right) + \eta\frac{\phi_2 c_v}{c_2}\frac{\partial p_2}{\partial t} \tag{7}$$

where T_1 and T_2 are the temperatures of the decomposed and undecomposed zones, respectively; c_v is the volume heat capacity of gas; c_1 and c_2 are the comprehensive specific heat capacities of the decomposed and undecomposed zones, respectively; δ is the throttle coefficient of gas; η is the adiabatic coefficient of gas.

$$\begin{aligned} c_1 &= \phi(S_w c_w + S_g c_g + S_H c_H) + (1-\phi)c_r \\ c_2 &= \phi(S_{w0} c_w + S_{g0} c_g + S_{H0} c_H) + (1-\phi)c_r \\ \mu_1 &= S_w \mu_w + (1 - S_w)\mu_g \\ \mu_2 &= S_{w0}\mu_w + (1 - S_{w0})\mu_g \end{aligned} \tag{8}$$

where c_w, c_g, c_H, and c_r represent the specific heat capacities of water, gas, hydrate, and rocks, respectively; S_g is the gas saturation in the decomposed zone; S_{g0} is the original gas saturation of the reservoir.

The auxiliary equation is:

$$S_g + S_w + S_H = 1 \tag{9}$$

The initial and boundary conditions are:

$$\begin{aligned} p_1(0,t) &= p_G \\ p_2(x,0) &= p_2(\infty,t) = p_e \end{aligned} \tag{10}$$

$$p_1(l(t),t) = p_2(l(t),t) = p_D(T_D) \tag{11}$$

$$T_2(x,0) = T_2(\infty,t) = T_e \tag{12}$$

$$T_1(l(t),t) = T_2(l(t),t) = T_D \tag{13}$$

where p_G is the bottom hole pressure, p_e is the initial reservoir pressure, T_e is the initial reservoir temperature, p_D is the dissociation front pressure, and T_D is the dissociation front temperature. According to the hypothesis of the model, p_D and T_D are the equilibrium pressure and temperature. $l(t)$ is the distance between the dissociation front and the bottom hole.

Constant bottom hole pressure and constant reservoir pressure are, respectively, used to approximate the pressure treatment, as follows:

$$\begin{aligned} \frac{\partial p_1^2}{\partial t} &\approx 2p_G \frac{\partial p_1}{\partial t} \\ \frac{\partial p_2^2}{\partial t} &\approx 2p_e \frac{\partial p_2}{\partial t} \end{aligned} \tag{14}$$

Equations (1), (2), and (4) are linearized to:

$$\frac{\partial p_1^2}{\partial t} = \chi_{1i} \frac{\partial^2 p_1^2}{\partial x^2} \tag{15}$$

$$\frac{\partial p_2^2}{\partial t} = \chi_2 \frac{\partial^2 p_2^2}{\partial x^2} \tag{16}$$

where the subscripts $i = g, w$, respectively, represent the gas phase and the water phase.

$$\begin{aligned} \chi_{1i} &= \frac{K_1 k_{ri1} p_G}{\phi(1-S_w-S_H)\mu_i} \\ \chi_2 &= \frac{K_2 k_{rg2} p_e}{\phi(1-S_{w0}-S_{H0})\mu_g} \end{aligned} \tag{17}$$

From the boundary condition relation (10)–(13), self-similar solutions of Equations (15) and (16) can be obtained according to Makogon's (1997) method [5]:

$$p_1^2 = p_G^2 - (p_G^2 - p_D^2)\frac{erf\lambda_{1i}}{erf\alpha_{1i}} \tag{18}$$

$$p_2^2 = p_e^2 - (p_e^2 - p_D^2)\frac{erf\lambda_2}{erf\alpha_2} \tag{19}$$

where

$$\begin{aligned} \lambda_{1i} &= \frac{x}{2\sqrt{\chi_{1i}t}} \\ \lambda_2 &= \frac{x}{2\sqrt{\chi_2 t}} \end{aligned} \tag{20}$$

$$\begin{aligned} \alpha_{1i} &= \sqrt{\frac{\gamma}{4\chi_{1i}}} \\ \alpha_2 &= \sqrt{\frac{\gamma}{4\chi_2}} \end{aligned} \tag{21}$$

$$l(t) = \sqrt{\gamma t} \tag{22}$$

In the above equation, γ represents the constant of the dissociation front movement, which needs to be solved iteratively in this paper. The error function and complementary error function are defined as:

$$\begin{aligned} erf(\xi) &= \frac{2}{\sqrt{\pi}}\int_0^{\xi} e^{-\eta^2} d\eta \\ erfc(\xi) &= 1 - erf(\xi) \end{aligned} \tag{23}$$

Similarly, self-similar solutions to Equations (6) and (7) can be obtained:

$$T_1 = T_D + A_{1i}\delta\Big[erf\lambda_{1i} - erf\alpha_{1i} + (\frac{\eta}{\delta}B_1 - 1)(\Psi_1(\lambda_{1i}) - \Psi_1(\alpha_{1i}))\Big] \tag{24}$$

$$T_2 = T_e - A_2\delta\Big[erfc\lambda_2 + (\frac{\eta}{\delta}B_2 - 1)\Psi_2(\lambda_2)\Big] \tag{25}$$

where

$$\Psi_1(\xi_1) = \frac{2}{\sqrt{\pi}}\int_0^{\xi_1} \frac{\eta e^{-\eta^2}}{\eta + C_{1i}e^{-\eta^2}}d\eta$$
$$\Psi_2(\xi_2) = \frac{2}{\sqrt{\pi}}\int_{\xi_2}^{\infty} \frac{\eta e^{-\eta^2}}{\eta + C_2 e^{-\eta^2}}d\eta \tag{26}$$

$$A_{1i} = \frac{1}{2erf\alpha_{1i}}\frac{p_D^2 - p_G^2}{p_G}$$
$$A_2 = \frac{1}{2erfc\alpha_2}\frac{p_e^2 - p_D^2}{p_e} \tag{27}$$

$$B_1 = \frac{\phi_1 c_v}{c_1}$$
$$B_2 = \frac{\phi_2 c_v}{c_2} \tag{28}$$

$$C_{1i} = \frac{p_D^2 - p_G^2}{p_G}\frac{c_v}{c_1}\frac{1}{2\sqrt{\pi}erf\alpha_{1i}}\frac{K_1 k_{ri1}}{\mu_1 \chi_{1i}}$$
$$C_2 = \frac{p_e^2 - p_D^2}{p_e}\frac{c_v}{c_2}\frac{1}{2\sqrt{\pi}erfc\alpha_2}\frac{K_2 k_{rg2}}{\mu_2 \chi_2} \tag{29}$$

2.3. Dissociation Front Motion

At the dissociation front, according to the phase equilibrium of natural gas hydrates, the relationship between T_D and p_D can be expressed as:

$$\log_{10} p_D = a(T_D - T_0) + b(T_D - T_0)^2 + c \tag{30}$$

where T_0 is the reference temperature, which is 273.15 K in this paper; a, b, and c are empirical constants related to hydrate composition. Makogon (1997) [5] used the least squares method to obtain three values of 0.0342/K, 0.0005/K^2, and 6.4804 based on the equilibrium temperature and pressure data of methane hydrate.

The mass balance equation of gas in the dissociation front is obtained by Verigin et al. (1980) [6]:

$$\rho_1 v_1 - \rho_2 v_2 = -[S_H \varepsilon \rho_H - (1 - S_w)\rho_1 + (1 - S_H)\rho_2]\phi\frac{dl}{dt} \tag{31}$$

In the above equation, ρ_1 and ρ_2 are methane gas densities in the decomposed and undecomposed zones, respectively; ρ_H is the hydrate density; ε is the mass fraction of gas in methane hydrate; and v_1 and v_2 are methane gas velocities in the decomposed and undecomposed zones, respectively.

At the dissociation front,

$$\rho_1(l,t) = \rho_2(l,t) = \rho_0\frac{p_D T_0}{z p_0 T_D} \tag{32}$$

z is the compression factor of methane gas; ρ_0 is the density of methane gas in the standard state (pressure p_0, temperature T_0).

Substituting Equation (32) into Equation (31) yields:

$$v_1(l,t) - v_2(l,t) = -\left[\varepsilon S_H \frac{\rho_H}{\rho_0}\frac{p_0}{p_D}\frac{T_D}{T_0}z - (S_H - S_w)\right]\phi\frac{dl}{dt} \tag{33}$$

The mass balance equation of water is:

$$\rho_w \phi_1 S_w = (1 - \varepsilon)\rho_H \phi_2 S_H \tag{34}$$

where ρ_w is the density of water.

Substituting Equations (18) and (19) into Equation (33) yields an equation with the determinative constant γ:

$$K_1 k_{ri1} \frac{p_D^2 - p_G^2}{\sqrt{\pi \chi_{1i}}} \frac{e^{-\alpha_{1i}^2}}{erf\alpha_{1i}} - K_2 k_{rg2} \frac{p_e^2 - p_D^2}{\sqrt{\pi \chi_2}} \frac{e^{-\alpha_2^2}}{erf\alpha_2} = A\sqrt{\gamma} \tag{35}$$

$$A = \left[\varepsilon S_H \frac{\rho_H p_0 T_D}{\rho_0 T_0} z - (S_H - S_w) p_D\right] \phi \mu_1 \tag{36}$$

In addition, at the dissociation front, $\lambda_2 = \alpha_2$, Equation (25) becomes:

$$T_D = T_e - A_2 \delta \left[erfc\alpha_2 + (\frac{\eta}{\delta} B_2 - 1)\Psi_2(\alpha_2)\right] \tag{37}$$

2.4. Permeability Model of Decomposition Zone

The hydrate dissociation zone is a three-phase flow zone of gas–water–hydrate. With the dissociation of hydrate, the absolute permeability of the decomposed zone increases gradually, and the relative permeability of gas and water changes constantly. Masuda et al. (1999) [20] gave the empirical equation for the change in absolute permeability with hydrate saturation:

$$K = K_0 (1 - S_H)^N \tag{38}$$

In the above equation, K_0 is the absolute reservoir permeability when hydrate saturation equals 0; N is the permeability decline index. The Masuda model has been widely cited and improved by a large number of scholars, but different hydrate reservoirs have different N values, which need to be obtained by fitting the seepage experiment data. Minagawa et al. (2005) used an exponential relationship to fit the permeability data under different hydrate saturations and obtained a decline index N ranging from 2.5 to 9.8 [21]. Liang et al. (2011) believe that the value of N is between 2 and 15 [22]. In this paper, the N value is 5.

The gas–water relative permeability is based on the algorithm proposed by Hong and Pooladi-Darvish [23,24]:

$$k_{rg} = k_{rg0} \overline{S_g}^{1/2} (1 - \overline{S_{wH}}^{1/m})^{2m} \tag{39}$$

$$k_{rw} = k_{rw0} \overline{S_w}^{1/2} \left[1 - (1 - \overline{S_w}^{1/m})^m\right]^2 \tag{40}$$

where

$$\overline{S_g} = \frac{1 - S_w - S_H - S_{gr}}{1 - S_{wr} - S_{gr}}$$

$$\overline{S_{wH}} = \frac{S_w + S_H - S_{wr}}{1 - S_{wr} - S_{gr}}$$

$$\overline{S_w} = \frac{S_w - S_{wr}}{1 - S_{wr} - S_{gr}}$$

In the equation, m = 0.45, S_{wr} = 0.3, S_{gr} = 0.05, k_{rw0} = 0.5, and k_{rg0} = 1.0.

3. Model Solving

In the decomposed zone, given a hydrate saturation S_H, the corresponding gas saturation S_g and water saturation S_w can be obtained by combining Equations (9) and (34), and then the absolute permeability and relative permeability of gas and water under different hydrate saturations can be calculated. The nonlinear Equations (30), (35) and (37) are solved iteratively to obtain T_D, p_D, and γ values under different conditions. Methane gas production (per unit length) can be obtained by the following equation:

$$Q_g = \frac{K_1 k_{rg1}}{\mu_1} \frac{\partial p_1(0,t)}{\partial x} = \frac{K_1 k_{rg1}}{\mu_1} \frac{p_D^2 - p_G^2}{p_G} \frac{1}{erf\alpha_{1g}} \frac{1}{2\sqrt{\pi \chi_{1g} t}} \tag{41}$$

The water production per unit length is:

$$Q_w = \frac{K_1 k_{rw1}}{\mu_1}\frac{\partial p_1(0,t)}{\partial x} = \frac{K_1 k_{rw1}}{\mu_1}\frac{p_D^2 - p_G^2}{p_G}\frac{1}{erf\alpha_{1w}}\frac{1}{2\sqrt{\pi \chi_{1w} t}} \tag{42}$$

4. Result Analysis

A one-dimensional dissociation model is established, and the basic parameters of the reservoir are shown in Table 1. In this section, numerical results of the time evolution of hydrate reservoir pressure and temperature profiles under different conditions are given. In addition, changes in methane gas/water production over time and the location of the dissociation front were evaluated. The sensitivity of natural gas production to different reservoir parameters was discussed.

Table 1. Parameters of hydrate reservoir.

Reservoir Parameter	Value	Reservoir Parameter	Value
Initial reservoir pressure p_e (MPa)	15	Density of hydrate ρ_H (kg·m^{-3})	900
Porosity ϕ	0.4	Methane gas density under standard conditions ρ_0 (kg·m^{-3})	0.7
Gas compression factor z	0.8	Density of water ρ_w (kg·m^{-3})	1000
Volume heat capacity of gas c_v (J·kg^{-1}·K^{-1})	3000	Initial hydrate saturation S_{H0}	0.6
Specific heat capacity of water c_w (J·kg^{-1}·K^{-1})	4200	Initial water saturation S_{w0}	0.2
Specific heat capacity of gas c_g (J·kg^{-1}·K^{-1})	3000	Initial gas saturation S_{g0}	0.2
Specific heat capacity of hydrate c_H (J·kg^{-1}·K^{-1})	2200	Mass fraction of gas in hydrate ε	0.129
Specific heat capacity of rock c_r (J·kg^{-1}·K^{-1})	1800	Viscosity of water μ_w (mPa·s)	1.2
		Viscosity of gas μ_g (mPa·s)	0.02
		Throttling coefficient of gas δ (K·Pa^{-1})	8×10^{-7}
		Adiabatic coefficient of gas η (K·Pa^{-1})	3.2×10^{-7}

4.1. Dynamic Analysis of Hydrate Depressurization

Based on the given reservoir temperature T_e (15 °C, 12 °C, 10 °C), bottom hole pressure p_G (10 MPa, 8 MPa, 5 MPa), and hydrate saturation S_H (0.5, 0.3, 0.2), the Equations (30), (35) and (37) are solved iteratively. Hydrate dissociation temperature T_D, dissociation pressure p_D, and values representing the movement of the hydrate dissociation front were obtained (Table 2).

Table 2. Values of dissociation temperature, dissociation pressure, parameter γ, and the distance l from the dissociation front to the wellbore for a given reservoir and well condition.

T_e (°C)	p_G (MPa)	S_H (MPa)	T_D (°C)	p_D (MPa)	γ (m^2/s)	l (m)
15	10	0.5	14.92	12.62	1.26×10^{-4}	18.09
15	8	0.5	14.92	12.62	1.68×10^{-4}	20.88
15	5	0.5	14.92	12.62	2.14×10^{-4}	23.56
12	10	0.5	11.87	10.94	1.02×10^{-4}	16.25
10	10	0.5	9.85	10.44	7.69×10^{-5}	14.12
15	10	0.3	14.92	12.65	3.90×10^{-3}	100.09
15	10	0.2	14.92	12.65	1.64×10^{-2}	206.02

As can be seen from the table, when the bottom hole pressure gradually decreases from 10 MPa to 8 MPa and 5 MPa, the dissociation temperature and pressure remain unchanged, which are 14.92 °C and 12.62 MPa. Similarly, as the hydrate saturation decreases, the dissociation temperature and pressure remain almost constant. Reservoir pressure, however, has a great influence on the dissociation temperature and pressure. With the decrease in reservoir temperature, the dissociation temperature and pressure of hydrate decrease sharply. Temperature is particularly affected; for every 1 °C decrease in reservoir temperature, the dissociation temperature decreases by about 1 °C. The above conclusion, that hydrate dissociation temperature and pressure are sensitive functions of reservoir temperature, is consistent with the research of Ji et al. (2001) [19].

Table 2 also shows that under the premise of keeping other conditions unchanged, the lower the bottom hole pressure, the faster the hydrate dissociation front moves, and the farther the dissociation front spreads to the depth of the reservoir. Additionally, under the same conditions, the lower reservoir temperature provides less energy for hydrate dissociation, resulting in slower movement of the dissociation front. In addition, the decrease in hydrate saturation leads to a sharp increase in the movement velocity of the dissociation front. When the hydrate saturation gradually decreases from 0.5 to 0.3 and 0.2, the movement distance of the dissociation front rapidly increases from 14.12 m to 100.09 m and 206.02 m.

Figure 2 shows the variation curve of the dissociation front with time under different conditions. It can be seen that the front advances with time, and the advancing speed gradually slows down. It may be that in the early stage, hydrate dissociation occurs near the bottom of the well, the dissociation rate is fast, and the corresponding dissociation front moves faster. As the dissociation continues, the front moves deeper into the reservoir, the dissociation speed slows down, and the corresponding movement speed of the front decreases. The variation trend of the dissociation front with bottom hole pressure, reservoir temperature, and hydrate saturation is consistent with the conclusion in Table 2. With the decrease in bottom hole pressure and the increase in reservoir temperature, the motion speed of the dissociation front gradually increases (Figure 2a,b). The dissociation front, especially the sensitivity function of hydrate saturation, moves rapidly with the decrease in hydrate saturation (Figure 2c). It should be pointed out that when we discussed the relationship between the dissociation front and time, we assumed that the hydrate saturation remained unchanged, which is contrary to the fact, and this is also a limitation of the model in this paper. Consideration of dissociation dynamics will be added to the model in the subsequent research.

Figure 3 shows the pressure distribution curves near the well under different reservoir and well conditions. As shown in the figure, the hydrate reservoir is divided into two zones by the dissociation front, the decomposed zone and the undecomposed zone, and the pressure distribution in the two zones is quite different. In the decomposed zone, the pressure decreases rapidly from the dissociation pressure to the bottom hole pressure, and the pressure drop funnel is deep. In the undecomposed zone, the pressure slowly recovers to the original reservoir pressure, and the pressure drop funnel is relatively gentle. The pressure distribution curves of different reservoirs and well conditions have certain differences. The pressure drop funnel in the decomposed zone deepens as the bottom hole pressure decreases, while the pressure distribution curve in the undecomposed zone almost coincides (Figure 3a). This is due to the gradual increase in the production pressure differential as the bottom hole pressure decreases, which in turn increases the driving force of hydrate dissociation. As a result, the rate of hydrate dissociation is accelerated, more fluid is extracted from the reservoir, and the rate of pressure reduction in the decomposed zone is accelerated. Reservoir temperature and hydrate saturation have a significant influence on the pressure distribution of the hydrate reservoir. As the reservoir temperature increases, the pressure drop funnel in the decomposed zone deepens, while the pressure drop funnel in the undecomposed zone decreases (Figure 3b). This is because the reservoir temperature provides an energy supply. The higher the reservoir temperature, the faster the hydrate dissociation in the decomposed zone, and the deeper the pressure drop funnel formed. In addition, it can be seen from Table 2 that the dissociation pressure increases with an increase in reservoir temperature. Therefore, the higher the reservoir temperature, the closer the dissociation pressure is to the original reservoir pressure, and the shallower the pressure drop funnel in the undecomposed zone. On the contrary, as hydrate saturation decreases, less hydrate is available for dissociation, the pressure drop funnel in the decomposed zone decreases, and the corresponding pressure drop funnel in the undecomposed zone increases

(Figure 3c). This is because the lower the hydrate saturation in the decomposed zone, the farther the dissociation front moves (Table 2). At this time, under the same pressure difference (the difference between the dissociation pressure and the bottom hole pressure), a wider decomposed zone corresponds to a smaller pressure drop funnel. Correspondingly, when the hydrate saturation is lower, the disturbance in the undecomposed zone becomes larger, and the pressure drop funnel increases.

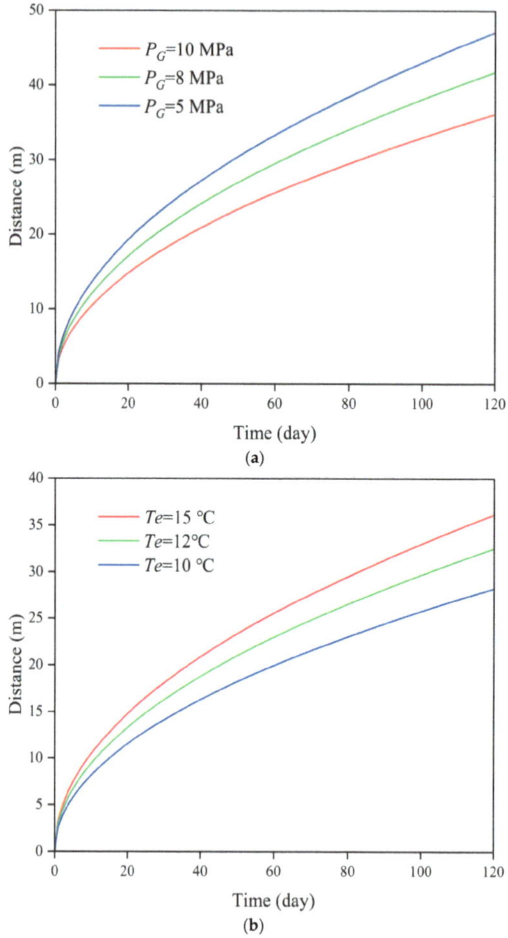

Figure 2. *Cont.*

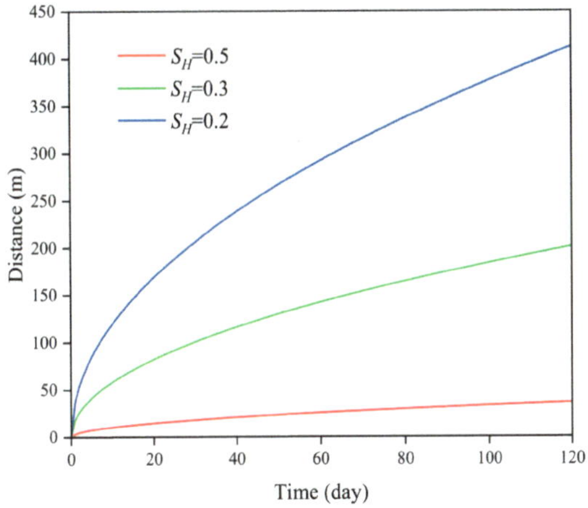

(c)

Figure 2. Variation in the dissociation front with time. (**a**) Different bottom hole pressures; (**b**) Different reservoir temperatures; (**c**) Different hydrate saturation.

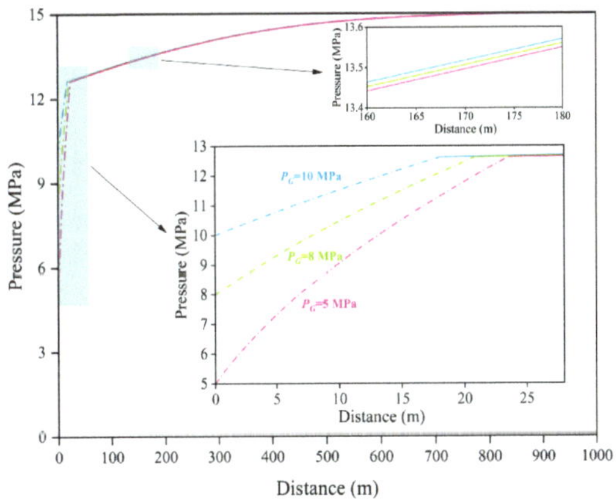

(a)

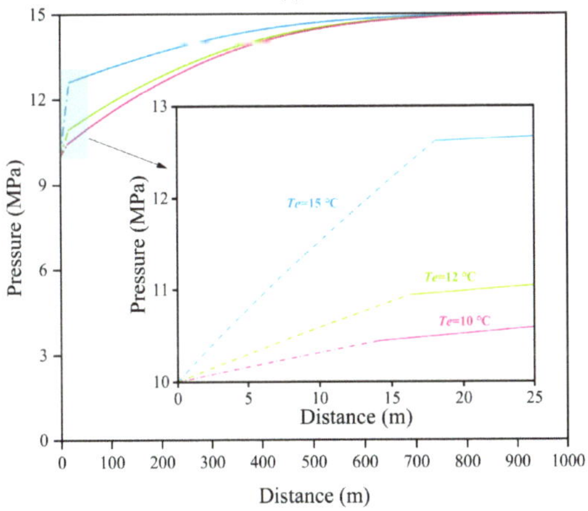

(b)

Figure 3. *Cont.*

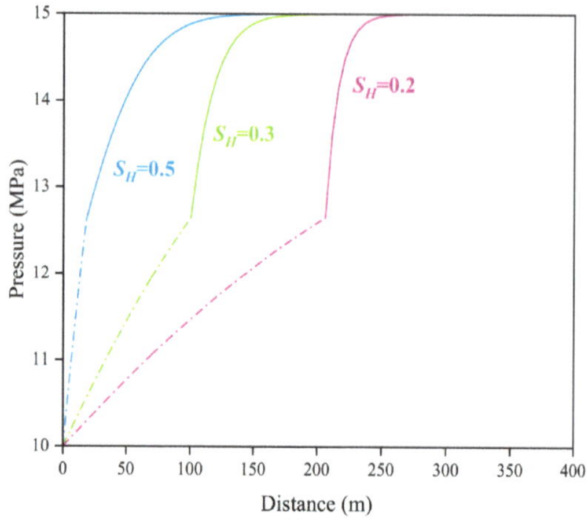

(c)

Figure 3. Pressure distribution near the well (dotted line represents decomposed zone, solid line represents undecomposed zone). (**a**) Different bottom hole pressures; (**b**) Different reservoir temperatures; (**c**) Different hydrate saturation.

Figure 4 shows the temperature distribution curve near the well under different conditions. As mentioned above, the hydrate reservoir is divided into two zones by the dissociation front, and the temperature changes in the two zones are quite different. In the decomposed zone, the lowest temperature appears at the wellbore, which may be due to the fact that only heat convection is considered in this paper and the heat from deep formation cannot be transferred to the well circumference in time. From the wellbore to the dissociation front, the temperature rises rapidly, and the temperature gradient is greatest at the dissociation front. This is because hydrate dissociation is an endothermic reaction, and the dissociation front is where the decomposition is most intense and absorbs a large amount of heat. In the undecomposed zone, the temperature slowly rises from the dissociation temperature to the original reservoir temperature.

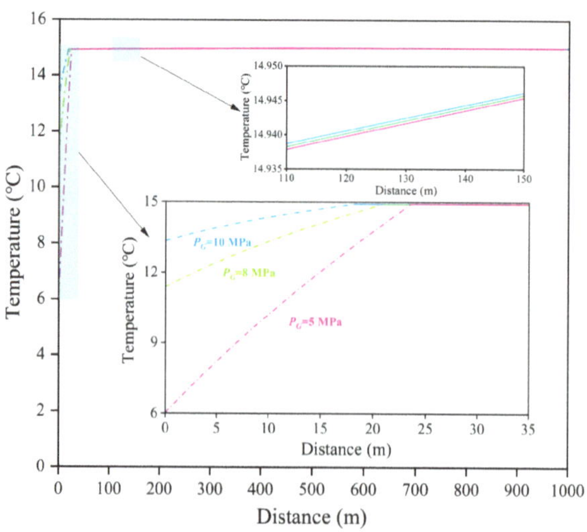

(a)

Figure 4. *Cont.*

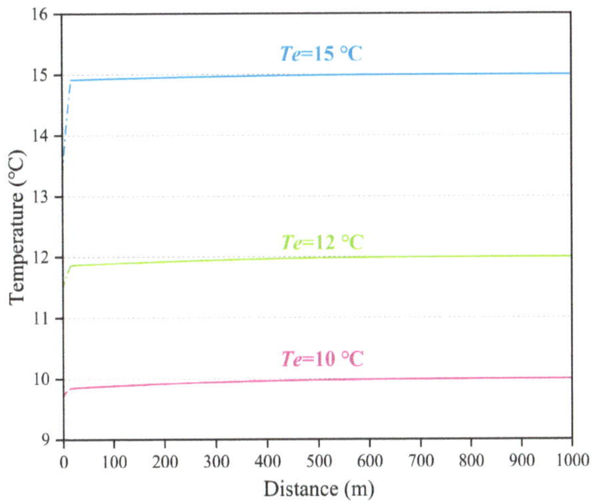

(**b**)

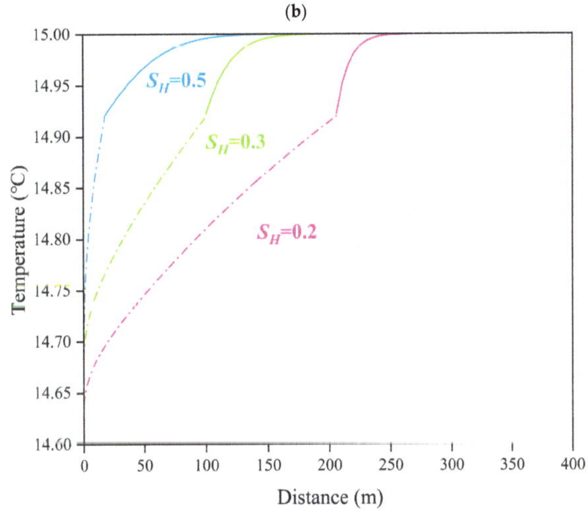

(**c**)

Figure 4. Temperature distribution curve near the well (dotted line represents decomposed zone, solid line represents undecomposed zone). (**a**) Different bottom hole pressures; (**b**) Different reservoir temperatures; (**c**) Different hydrate saturation.

As can be seen from Figure 4a, bottom hole pressure has a great influence on the temperature change in the decomposed zone. When the bottom hole pressure is 10 MPa, 8 MPa, and 5 MPa, the temperature gradient in the decomposition zone is about 0.087 °C/m, 0.169 °C/m, and 0.378 °C/m, respectively, showing a gradually increasing trend. As noted before, the lower bottom hole pressure causes more hydrates to reach equilibrium conditions and begin to decompose, thus absorbing more heat, resulting in a faster cooling of the decomposed zone. Similarly, the variation in reservoir temperature will also affect the temperature distribution in the decomposed zone. When the reservoir temperature is 15 °C, 12 °C, and 10 °C, respectively, the temperature gradient in the decomposition region shows a decreasing trend (Figure 4b). This may be due to the fact that when the initial reservoir temperature is low, hydrate dissociation absorbs a large amount of heat, resulting in a substantial decrease in local reservoir temperature, causing hydrate self-protection effects, secondary formation, and other problems, thus slowing down hydrate dissociation [13].

The change in hydrate saturation also affects the temperature distribution around the well. As hydrate saturation gradually decreases, less hydrate is available for dissociation, and the heat consumed gradually decreases, thus reducing the temperature gradient in the decomposed zone (Figure 4c).

For the undecomposed zone, the temperature distribution varies under different bottom hole pressures (Figure 4a). It can be seen from Table 2 that the change in bottom hole pressure has no effect on the dissociation temperature. Whereas, the lower the bottom hole pressure, the farther the dissociation front moves. This indicates that the lower the bottom hole pressure, the more dissociation heat is required from the reservoir and thus, the lower the temperature in the undecomposed zone. Under different reservoir temperatures, the temperature distribution curves in the undecomposed zone are almost parallel, with differences only in numerical values. All curves converge from dissociation temperature to reservoir temperature (Figure 4b). The effect of hydrate saturation on the temperature distribution in the undecomposed zone is similar to that of bottom hole pressure. The dissociation temperature does not change with hydrate saturation. The lower the hydrate saturation, the farther the dissociation front moves (Table 2). This indicates that when hydrate saturation is low, the undecomposed zone is disturbed more (more heat needs to be provided for hydrate dissociation), thus the temperature drop gradient is larger.

4.2. *Analysis of Factors Affecting Production Capacity of NGH Depressurization*

In this paper, the simulated natural gas production was verified by the data reported in the literature. Ye et al. reported that in the Shenhu area of the South China Sea, the natural gas production was about 2.87×10^4 m^3/day. The thickness of the gas-bearing zone in this area was about 89 m, and the length of the horizontal well was about 300 m [25]. As our research shows, the predicted gas production is of the order of 10^{-5} m^3/s per 1 m^2 of well. This is about 2.31×10^4 m^3/day, based on reservoir and well data in the Shenhu area. Considering the difference in reservoir permeability and mining conditions, the simulated gas production is within an acceptable range.

4.2.1. Influence of Bottom Hole Pressure on Productivity

The variation in methane gas production (unit length) with time under different bottom hole pressures is shown in Figure 5a. As expected, gas production decreases with the square root of time. Methane gas production is a sensitive function of bottom hole pressure. As bottom hole pressure decreases, the pressure gradient near the well increases (Figure 3a). Accordingly, the driving force of hydrate dissociation increases, and the movement speed of the hydrate dissociation front accelerates (Figure 2a). Therefore, the higher the peak gas production of production wells. Although lowering the bottom hole pressure can improve the gas production rate, too low a bottom hole pressure drop will lead to a large and rapid decomposition of hydrate, and the temperature near the wellbore will drop significantly (Figure 4a), resulting in problems such as the self-protection effect and secondary formation of hydrate, which hinder further decomposition of hydrate. Therefore, it is necessary to calculate and select the optimal bottom hole pressure in actual depression-mining so that gas production can meet industrial demand without causing the self-protection effect and secondary formation of hydrate due to excessive gas production [13]. Figure 5b shows the variation in the water production (unit length) with time under different bottom hole pressures. The trend of the water production curve is the same as that of gas, but it is 4 orders of magnitude less numerically. There may be two reasons for the low water yield. One is that there is a big difference in the viscosity between the two phases of gas and water, and the gas seepage is much higher than that of water. The other is that most of the hydrate decomposing water is trapped in the formation.

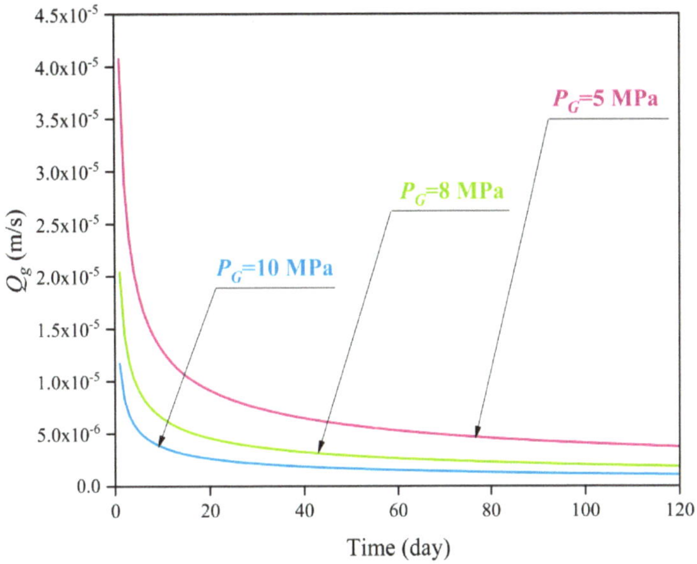

(**a**)

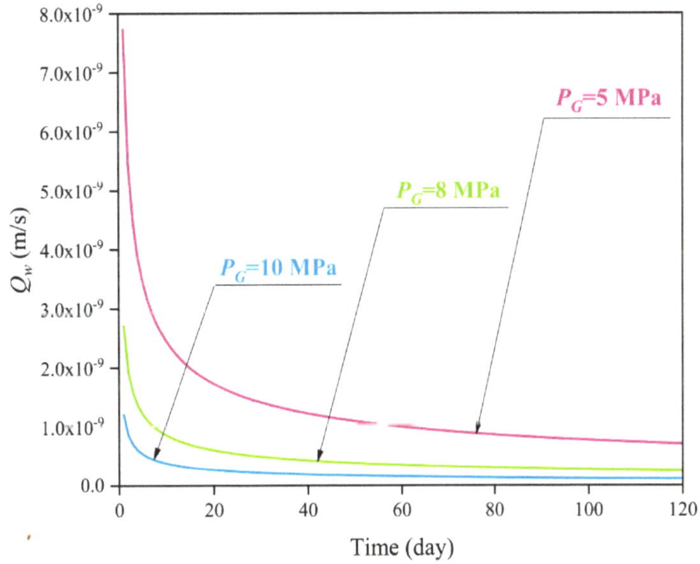

(**b**)

Figure 5. Variation curves of gas (**a**) and water (**b**) production with bottom hole pressure.

4.2.2. Influence of Reservoir Temperature on Productivity

Figure 6 shows the relationship between gas and water production (unit length) over time at different reservoir temperatures. As shown in the figure, both gas and water production decrease with the square root of time, and both are sensitive functions of reservoir temperature. The higher the reservoir temperature, the more gas and water production gradually increases. In the one-dimensional model presented here, the heat required for hydrate dissociation must be supplied by the hydrate reservoir. A higher initial temperature means that more heat is available for hydrate dissociation. Consequently, more hydrates begin to decompose, and thus more decomposition products are produced. Therefore, reservoir temperature becomes an important control parameter. It should be emphasized that for real gas hydrate reservoirs, heat can also be transferred from the side, which will significantly affect the production process of natural gas.

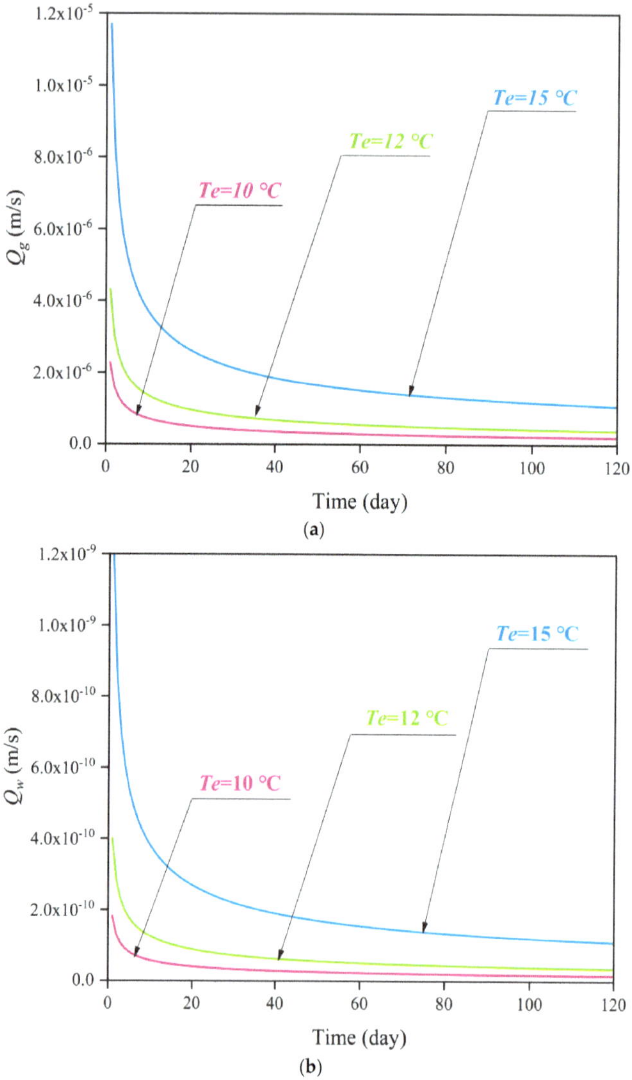

Figure 6. Variation curves of gas (**a**) and water (**b**) production with reservoir temperature.

4.2.3. Influence of Hydrate Saturation on Productivity

Figure 7a shows the variation curve of gas production (unit length) when hydrate saturation values are 0.5, 0.3, and 0.2. As can be seen from the figure, gas production increases with a decrease in hydrate saturation. This is because when the hydrate saturation is small, the absolute permeability of the reservoir is relatively large, and thus the corresponding gas production is large. Unlike gas production, water production does not always increase with a decrease in hydrate saturation. As can be seen from Figure 7b, as the hydrate saturation decreases, the water production increases first and then decreases. It should be pointed out that when the hydrate saturation is relatively high (S_H = 0.5), the water production is almost zero. As mentioned above, when the hydrate saturation reaches a certain value, such as 0.5, the absolute permeability of the reservoir is limited. Additionally, the seepage velocity of gas is much higher than that of water. Thus, in the limited permeability, gas preferentially produces, and most of the water remains in the reservoir, resulting in almost zero water production. With the decrease in hydrate saturation (S_H = 0.3), the absolute permeability of the reservoir increases, resulting in an increase in water production. When the hydrate saturation decreases further (S_H = 0.2), although the absolute permeability of the reservoir is relatively large, the water production, however, is reduced due to the relatively small amount of final decomposed water.

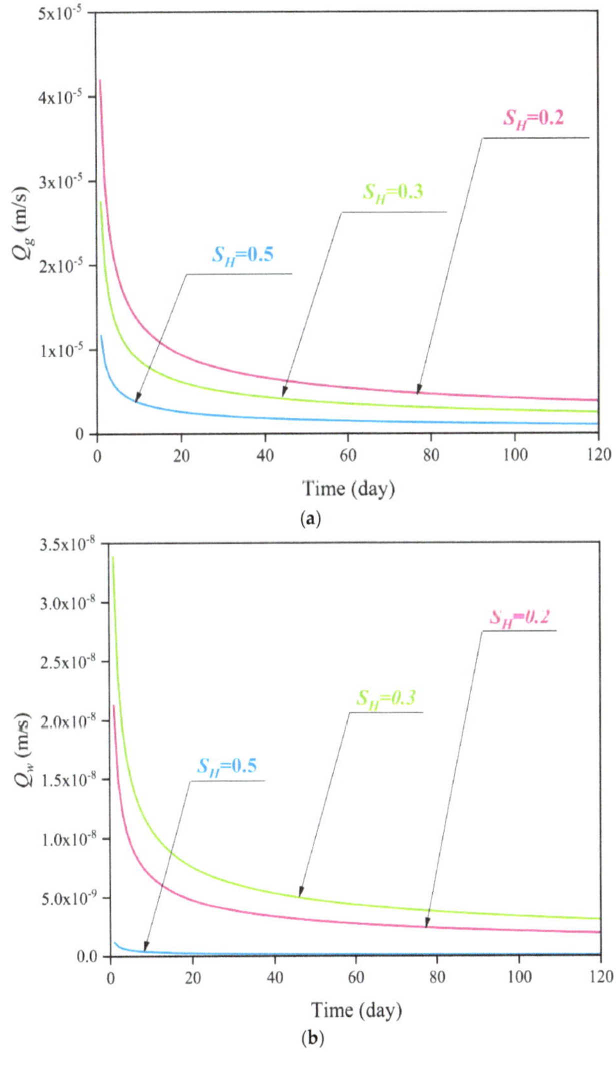

Figure 7. Variation curves of gas (**a**) and water (**b**) production with hydrate saturation.

In the exploitation of natural gas hydrates, methane gas production has always been the focus of attention. According to the above results, methane gas production is a sensitive function of bottom hole pressure, reservoir temperature, and hydrate saturation. By lowering the bottom hole pressure and increasing the reservoir temperature, the pressure and temperature gradient in the decomposed zone will increase. Both of these increases will improve the driving force of hydrate dissociation and push the dissociation front to move farther, thus promoting the development of methane gas. Nevertheless, with the increase in hydrate saturation in a certain range, the gas production of methane decreases, although the pressure and temperature gradient in the decomposed zone increase. This may be due to the fact that methane gas production is also affected by the absolute permeability of the reservoir. The increase in hydrate saturation will increase the driving force of hydrate dissociation, but it will also lead to a decrease in the absolute permeability of the reservoir, and the latter has a more obvious inhibitory effect on methane gas extraction. As a result, methane gas production is reduced.

5. Conclusions

In this paper, a one-dimensional numerical model of gas hydrate depressurization was established, in which convective heat transfer and gas–water two-phase seepage were taken into account. The movement law of the hydrate dissociation front and the evolution process of temperature and pressure near the well were analyzed, and the influence mechanisms of bottom hole pressure, reservoir temperature, and hydrate saturation on hydrate productivity were also discussed. According to the results obtained by the model, the following understandings were obtained:

(1) The dissociation front moves to the depth with time, and the moving speed slows down gradually. The dissociation front movement speed increases slowly with the decrease in bottom hole pressure and reservoir temperature, yet increases sharply with the decrease in hydrate saturation.

(2) The hydrate reservoir is divided into two zones by the dissociation front: the decomposed zone and the undecomposed zone. In the decomposed zone, the pressure decreases rapidly from the dissociation pressure to the bottom hole pressure, and the pressure drop funnel is deeper. In the undecomposed zone, the pressure slowly recovers from the dissociation pressure to the original reservoir pressure, and the pressure drop funnel is relatively gentle. Lower bottom hole pressure, higher reservoir temperature, and higher hydrate saturation all lead to an increased pressure gradient in the decomposed zone.

(3) Bottom hole pressure, reservoir temperature, and hydrate saturation all affect the temperature distribution in the decomposed zone. The temperature gradient in the decomposed zone is proportional to the reservoir temperature and hydrate saturation and inversely proportional to the bottom hole pressure.

(4) Gas and water production both increase with the decrease in bottom hole pressure, yet the latter is 4 orders of magnitude less than the former in numerical value, which may be due to the large difference in the viscosity of gas–water two phases, and the gas seepage speed is much larger than that of water.

(5) Both gas and water production are sensitive functions of reservoir temperature. A higher initial reservoir temperature can provide more heat for hydrate dissociation, and the corresponding decomposition products are higher.

(6) The gas production increases with the decrease in hydrate saturation, while the water production increases first and then decreases with the decrease in hydrate saturation. This indicates that when the hydrate saturation is low, although the absolute permeability of the reservoir is relatively large, the water production is still low even if the hydrate is completely decomposed due to the small hydrate content.

Based on the previous model, this paper has made some progress in considering the gas–water two-phase seepage in the decomposed zone, whereas the model itself still has some shortcomings. For instance, the model in this paper is a simple one-dimensional model, and only heat convection is considered in the thermal model. Additionally, the hydrate saturation in the decomposed zone is regarded as a fixed value when calculating the permeability, and its variation with time is not considered. These model defects need to be further improved in future studies.

Author Contributions: Writing—original, R.Q.; Funding acquisition, H.L.; Writing—review and editing, C.X. (Chenlu Xu); Project administration, L.Y.; Investigation, C.X. (Changwen Xiao); Software, J.D.; Supervision, Y.L. All authors have read and agreed to the published version of the manuscript.

Funding: This project was financially supported by the Guangzhou Science and Technology Program (No. 202206050001), Guangzhou Science and Technology Program (No. 2023A04J0305), Guangdong Basic and Applied Basic Research Foundation (No. 2022A1515011902), Guangzhou Science and Technology Program (No. 202206050002), the Youth Research Team Project of the National Engineering Research Center of Gas Hydrate Exploration and Development (No. NERC2024003).

Data Availability Statement: Data available in a publicly accessible repository.

Conflicts of Interest: The authors declare that they have no known competing financial interests or personal relationships that could have appeared to influence the work reported in this paper.

References

1. Collett, T.S. *Natural Gas Hydrate—Vast Resource, Uncertain Future*; USGS: Reston, VA, USA, 2001.
2. Sloan, E.D.; Koh, C.A. *Clathrate Hydrates of Natural Gases*, 3rd ed.; CRC Press: Boca Raton, FL, USA, 2007.
3. Li, Q.; Cheng, Y.; Li, Q.; Ansari, U.; Liu, Y.; Yan, C.; Lei, C. Development and verification of the comprehensive model for physical properties of hydrate sediment. *Arab. J. Geosci.* **2018**, *11*, 325. [CrossRef]
4. Li, Q.; Liu, J.; Wang, S.; Guo, Y.; Han, X.; Li, Q.; Cheng, Y.; Dong, Z.; Li, X.; Zhang, X. Numerical insights into factors affecting collapse behavior of horizontal wellbore in clayey silt hydrate-bearing sediments and the accompanying control strategy. *Ocean Eng.* **2024**, *297*, 117029. [CrossRef]
5. Makogon, Y.F. *Hydrates of Hydrocarbons*; Penn Well: Tulsa, OK, USA, 1997.
6. Verigin, N.N.; Khabibullin, I.L.; Khalikov, G.A. Linear problem of the dissociation of the hydrates of a gas in a porous medium. *Fluid Dyn.* **1980**, *15*, 144–147. [CrossRef]
7. Yousif, M.H.; Sloan, E.D. Experimental and theoretical investigation of methane-gas-hydrate dissociation in porous media. *SPE Reserv. Eval. Eng.* **1991**, *6*, 69–76.
8. Holder, G.D.; Angert, P.F. Simulation of gas production from a reservoir containing both gas hydrates and free natural gas. In Proceedings of the 57th Society of Petroleum Engineers Technology Conference, New Orleans, LA, USA, 26–29 September 1982.
9. Burshears, M.; Obrien, T.J.; Malone, R.D. A multi-phase, multi-dimensional, variable composition simulation of gas production from a conventional gas reservoir in contact with hydrates. In Proceedings of the SPE Unconventional Gas Technology Symposium, Louisville, KY, USA, 18–21 May 1986.
10. Selim, M.S.; Sloan, E.D. Heat and mass transfer during the dissociation of hydrates in porous media. *AIChE J.* **1989**, *35*, 1049–1052. [CrossRef]
11. Tsypkin, G.G. Mathematical Models of Gas Hydrates Dissociation in Porous Media. *Ann. N. Y. Acad. Sci.* **2000**, *912*, 428–436. [CrossRef]
12. Zhao, Z.; Shang, X. Analysis for temperature and pressure fields in process of hydrate dissociation by depressurization. *Int. J. Numer. Anal. Met.* **2010**, *34*, 1831–1845. [CrossRef]
13. Hao, Y.M.; Li, X.Z.; Li, S.X.; Zeng, X.Z.; Chen, Z.H. The semi-analytical two-phase productivity model of natural gas hydrate by depressurization. *Sci. Sin.-Phys. Mech. Astron.* **2020**, *50*, 064701. (In Chinese)
14. Li, X.; Lei, G.; Wan, Y.; Cheng, W.; Sun, J.X.; Ning, F.L. A new theoretical model incorporating geomechanical effects for hydrate dissociation in sediments under thermal stimulation. *Geoenergy Sci. Eng.* **2023**, *221*, 111220. [CrossRef]
15. Yu, Q.N.; Tang, H.M.; Li, C.L.; Chen, Z.J.; Zhang, K.; Yu, Y.; Liang, S. Numerical simulation of natural-gas-hydrate decomposition in process of heat-injection production. *Processes* **2023**, *11*, 2349. [CrossRef]
16. Liu, Z.; Xu, Q.; Yang, J.; Shi, L. Pore-Scale Modeling of Methane Hydrate Dissociation Using a Multiphase Micro-Continuum Framework. *Energies* **2023**, *16*, 7599. [CrossRef]
17. Kumar, A.; Maini, B.; Bishnoi, P.; Clarke, M.; Zatsepina, O.; Srinivasan, S. Experimental determination of permeability in the presence of hydrates and its effect on the dissociation characteristics of gas hydrates in porous media. *J. Pet. Sci. Eng.* **2010**, *70*, 114–122. [CrossRef]
18. Lee, J.; Park, S.; Sung, W. An experimental study on the productivity of dissociated gas from gas hydrate by depressurization scheme. *Energy Convers. Manag.* **2010**, *51*, 2510–2515. [CrossRef]
19. Ji, C.; Ahmadi, G.; Smith, D.H. Natural gas production from hydrate decomposition by depressurization. *Chem. Eng. Sci.* **2001**, *56*, 5801–5814. [CrossRef]
20. Masuda, Y.; Fujinaga, Y.; Naganawa, S.; Fujita, K.; Hayashi, Y. Modeling and experiment studies on dissociation of methane gas hydrates in Berea sandstone cores. In Proceedings of the Third International Gas Hydrate Conference, Salt Lake City, UT, USA, 18–22 July 1999.

21. Minagawa, H.; Ohmura, R.; Kamata, Y.; Ebinuma, T.; Narita, H.; Masuda, Y. Water permeability measurements of gas hydrate-bearing sediments. In Proceedings of the Fifth International Conference on Gas Hydrates, Trondheim, Norway, 13–16 June 2005.
22. Liang, H.; Song, Y.; Chen, Y.; Liu, Y. The Measurement of Permeability of Porous Media with Methane Hydrate. *Pet. Sci. Technol.* **2011**, *29*, 79–87. [CrossRef]
23. Hong, H.; Pooladi-Darvish, M. A numerical study on gas production from formations containing gas hydrates. In Proceedings of the Canadian International Petroleum Conference, Calgary, AB, Canada, 10–12 June 2003.
24. Hong, H.; Pooladi-Darvish, M. Simulation of Depressurization for Gas Production From Gas Hydrate Reservoirs. *J. Can. Pet. Technol.* **2005**, *44*, 39–46. [CrossRef]
25. Ye, J.L.; Qin, X.W.; Xie, W.W.; Lu, H.L.; Ma, B.J.; Qiu, H.J.; Liang, J.Q.; Lu, J.; Kuang, Z.G.; Lu, C.; et al. Main progress of the second gas hydrate trial production in the South China Sea. *Geol. China* **2020**, *47*, 557–568. (In Chinese with English Abstract)

energies

MDPI

Review

Gas Production and Storage Using Hydrates Through the Replacement of Multicomponent Gases: A Critical Review

Zhiyuan Zhu [1], Xiaoya Zhao [1], Sijia Wang [2], Lanlan Jiang [3], Hongsheng Dong [4,5,*] and Pengfei Lv [1,6,*]

1 School of Energy and Power Engineering, North University of China, Taiyuan 030051, China; sz202416028@st.nuc.edu.cn (Z.Z.); sz202416027@st.nuc.edu.cn (X.Z.)
2 Hebei Engineering Research Center of Advanced Energy Storage Technology and Equipment, School of Energy and Environmental Engineering, Hebei University of Technology, Tianjin 300401, China; 2023063@hebut.edu.cn
3 Key Laboratory of Ocean Energy Utilization and Energy Conservation of Ministry of Education, Dalian University of Technology, Dalian 116024, China; lanlan@dlut.edu.cn
4 School of Marine Science and Technology, Northwestern Polytechnical University, Xi'an 710072, China
5 National Innovation Platform (Center) for Industry-Education Integration of Energy Storage Technology, Xi'an Jiaotong University, Xi'an 710049, China
6 Key Laboratory of Shanxi Province for Solar Thermal Technology, Taiyuan 030051, China
* Correspondence: hsdong@nwpu.edu.cn (H.D.); pengfeilyu@nuc.edu.cn (P.L.)

Abstract: With the continuous growth of global energy demand and the gradual depletion of traditional fossil energy reserves, natural gas hydrates have attracted widespread attention as a potential clean energy source due to their vast reserves and wide distribution. Although various extraction methods, including depressurization, thermal stimulation, chemical inhibitors, and displacement methods, have been proposed, there are still challenges, such as low extraction efficiency, poor sustainability, and high costs, making it difficult to achieve large-scale engineering applications. Among these, the use of gases such as CO_2 for displacement extraction of natural gas hydrates can both develop hydrate resources and sequester CO_2, achieving a win–win situation for resource development and greenhouse gas reduction. This paper provides a detailed review of the multi-gas displacement extraction technology for natural gas hydrates, systematically summarizes the latest progress in thermodynamic and kinetic studies, analyzes the technical advantages and feasibility of combining displacement methods with traditional techniques, and explores the effects of multi-gas mixtures, such as N_2, CO_2, and H_2, and their ratios on hydrate extraction efficiency. Finally, this paper summarizes the technical challenges faced by displacement extraction methods for hydrates and offers future research directions to promote the development of multi-gas displacement technology for natural gas hydrates.

Keywords: natural gas hydrate; displacement method; thermodynamic properties; kinetic properties; multicomponent gas

Academic Editor: Reza Rezaee

Received: 24 January 2025
Revised: 12 February 2025
Accepted: 15 February 2025
Published: 18 February 2025

Citation: Zhu, Z.; Zhao, X.; Wang, S.; Jiang, L.; Dong, H.; Lv, P. Gas Production and Storage Using Hydrates Through the Replacement of Multicomponent Gases: A Critical Review. *Energies* **2025**, *18*, 975. https://doi.org/10.3390/en18040975

1. Introduction

1.1. Natural Gas Hydrate

Natural gas hydrate is a solid, ice-like substance formed by natural gas molecules and water molecules under low-temperature and high-pressure conditions. For every 1 cubic meter of solid hydrate that decomposes, approximately 163 cubic meters of natural gas are released. The total carbon stored in global hydrate resources is roughly twice the total carbon stored in all other conventional fossil fuels, including oil, coal, and conventional natural gas [1,2]. This compound is considered a clean energy source, as it produces lower carbon dioxide emissions during combustion compared to traditional fossil fuels, helping to reduce environmental pollution and mitigate global climate change issues [1]. Predictions suggest that there are vast natural gas hydrate resources on Earth, with reserves exceeding 15×10^{12} tons. If 17% to 20% of this resource could be exploited, it would be sufficient to provide ample energy supply for the next 200 years. Natural gas hydrate deposits are mainly distributed in two regions: one is the high-latitude land covered by permafrost, such as Siberia, Alaska, and the Canadian Archipelago; the other is the deep-sea sedimentary layers, particularly in the continental shelf and slope areas, especially in water depths ranging from 300 to 3000 m, such as the South China Sea depression in the western Pacific, the Blake Plateau in the western Atlantic, and the Arabian Sea Trench in the Indian Ocean. After nearly thirty years of global joint exploration, utilizing technologies such as seismic wave detection, seabed sampling, and deep-sea drilling, more than 230 natural gas hydrate deposits with industrial extraction potential have been identified worldwide. Figure 1 shows the geographical distribution, abundance levels, and the forms in which natural gas hydrates exist on land and in the ocean. In the figure, yellow circles (BSR) represent regions where natural gas hydrates have been discovered through the Bottom Simulating Reflector (BSR) method. BSR is an important seismic exploration marker for identifying natural gas hydrates. Red circles (by core) indicate areas where the presence of natural gas hydrates has been confirmed through core sampling. Core sampling is a direct method of obtaining underground material samples, which can definitively prove the existence of natural gas hydrates. Red squares (production) represent areas where natural gas hydrate extraction has already taken place. The figure marks several typical locations, such as Mallik in Canada, Messoyakha in Russia, and Nankai in Japan. From the map, it is evident that natural gas hydrate deposits are widely distributed across the globe. In the ocean, they are mainly concentrated along the continental margins, while on land, they are primarily found in high-latitude polar regions, such as those near the Arctic, demonstrating the immense market potential and sustainable development advantages of natural gas hydrates [3]. Natural gas hydrate can be used in various fields, such as civilian and industrial fuels, chemicals, and power generation, offering broad application prospects. It forms a complete industrial chain, from upstream exploration and development to midstream transportation and storage and downstream comprehensive utilization. Research into its extraction technology can not only accelerate energy technological innovation and industrialization but also provide significant support for global energy structure optimization, environmental protection, and climate change mitigation. With ongoing technological advancements, natural gas hydrate is expected to become an important component of the global energy system, playing a key role in the future energy transition [4].

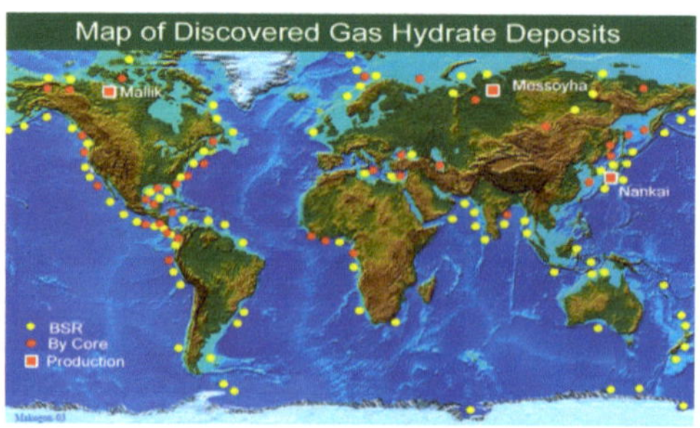

Figure 1. Distribution of discovered gas hydrate deposits. BSR stands for the deposit located by seismic refraction. By core refers to the areas where the presence of natural gas hydrates has been confirmed through core sampling. Production indicates the areas where natural gas hydrate extraction has already taken place [5].

1.2. Natural Gas Hydrate Extraction

1.2.1. Traditional Natural Gas Hydrate Extraction Methods

The extraction methods for natural gas hydrates are still in the research and testing stages. Despite facing numerous challenges, several technological solutions have been proposed. Most existing extraction methods are based on adjusting the pressure and temperature of the hydrates, shifting them from the stable hydrate zone to the dissociation zone, thereby enabling dissociation [6]. There are three main traditional methods for extracting natural gas hydrates: depressurization, thermal stimulation, and chemical inhibitor injection. The advantages and disadvantages of these three methods are shown in Table 1 [7–9].

Table 1. Advantages and disadvantages of different extraction methods.

Method	Advantages	Disadvantages
Depressurization Method	Technologically mature; simple to operate; no chemical additives; no negative impact on the environment	Requires prolonged depressurization, leading to increased extraction costs; relatively low extraction rate; rapid pressure changes may cause well leakage or collapse
Thermal Stimulation Method	Increase the decomposition rate of hydrates; applicable to low-temperature areas	High energy consumption; damages wellbore and reservoir; using thermal fluids may require treatment of the injected water; otherwise, it could introduce environmental contaminants
Chemical Inhibitor Injection Method	Low energy demand; adaptable to various reservoir conditions	High chemical costs; requires precise control of injection volume

The depressurization method used in the extraction of natural gas hydrates is a widely used technique. Its basic principle is to lower the pressure in the hydrate reservoir to disrupt the phase equilibrium of the hydrates, causing the dissociation of the hydrates into gas (such as CH_4) and water, thereby enabling the extraction of natural gas [10,11]. Currently, two main approaches are used to achieve depressurization: (1) the use of low-density drilling mud technology, which aims to achieve a depressurization effect; (2) the use of pump pressure technology to extract free gas and other fluids from beneath the natural gas hydrate reservoir to achieve the depressurization goal [12].

The thermal stimulation method is a technique that promotes natural gas release by heating the hydrate reservoir [13]. Its basic principle is to provide heat to the natural gas hydrate reservoir, increasing the temperature, thereby causing the dissociation of the hydrates into gas and water and releasing natural gas [14]. The implementation methods include thermal fluid injection, electric heating, and geothermal heating, among others [15]. Table 2 lists the advantages and disadvantages of different thermal extraction methods.

Table 2. Merit and demerit of different thermal recovery methods.

Method	Advantages	Disadvantages
Thermal fluid injection method	Increase recovery rate; widely applicable; high heat transfer efficiency	High energy consumption; significant heat loss; complex equipment maintenance
Electric heating method	High energy efficiency; precise control; minimal equipment requirements	High equipment cost; high power consumption; poor adaptability
Geothermal heating method	Low energy consumption; minimal heat loss; wide applicability	Slow heating effect; limited improvement in yield; less effective than high-temperature heating

The chemical inhibitor injection method prevents the formation of hydrates or promotes their dissociation by injecting chemical inhibitors, thereby improving natural gas recovery efficiency [16]. This method is particularly suitable for deep-sea or polar natural gas hydrate reservoirs, as hydrates in these regions are typically stable due to temperature and pressure conditions and are not easily dissociated naturally. Chemical inhibitors can effectively control the formation and dissociation of hydrates. Common chemical inhibitors, as shown in Table 3, include thermodynamic hydrate inhibitors, kinetic hydrate inhibitors, anti-agglomerants, and dual-function hydrate inhibitors [17].

Table 3. Advantages and disadvantages of different chemical inhibitors.

Inhibitor	Type	Advantages	Disadvantages
Thermodynamic Hydrate Inhibitors	Alcohols and electrolytes	Effectively reduce the formation temperature of hydrates, thereby preventing their formation	It usually requires a high dosage, leading to increased costs and environmental impact
Kinetic Hydrate Inhibitors	Polymeric compound	Effectively inhibit hydrate formation at a lower dosage, suitable for long-term flow assurance	In some cases, it may not be stable enough, and its effectiveness varies with different gas compositions
Anti-agglomerants	Compounds with various chemical structures	Prevent the agglomeration of hydrate particles, thereby reducing the risk of blockage	It needs to be used in combination with other inhibitors to enhance effectiveness
Dual-function Hydrate Inhibitors	Amino acids, ionic liquids, and nanoparticles	Combining the advantages of thermodynamic and kinetic inhibition provides a more comprehensive inhibitory effect	Relatively novel, and it requires further research to determine its long-term effects and cost-effectiveness

In the extraction process of natural gas hydrate, a combination of various traditional methods is often employed to enhance extraction efficiency. For example, the combination of depressurization and thermal stimulation. By reducing pressure, the stability of the hydrate is lowered, and the heating process accelerates the decomposition of the hydrate, thereby promoting the release of natural gas and shortening the extraction cycle. The advantage of this method lies in its ability to accelerate gas release and increase extraction rates. However, the thermal stimulation process requires a large amount of energy and may cause temperature fluctuations, which can affect the stability and continuity of extraction. Additionally, prolonged depressurization may lead to instability in the hydrate reservoir, affecting the gas recovery rate [12].

Furthermore, the combination of depressurization and chemical inhibitors can also significantly improve extraction efficiency. Chemical inhibitors can suppress the re-crystallization of the hydrate during depressurization, reducing instability and accelerating hydrate decomposition at lower temperatures, thereby maintaining high gas output [10,11]. However, chemical inhibitors are expensive, and their potential environmental impact needs to be evaluated and monitored, as long-term use may pose environmental risks.

Moreover, combining thermal stimulation with chemical inhibitors can prevent the re-crystallization of the hydrate at lower temperatures during the heating process, thus maintaining extraction efficiency. Heating helps accelerate the decomposition of the hydrate, while inhibitors further stabilize the decomposition process [16]. However, this method requires a large amount of energy input, and the use of chemical inhibitors is costly. Additionally, the ratio of heat energy to inhibitor use must be carefully controlled, as excessive use may lead to environmental issues. Therefore, while coupling methods can improve extraction efficiency, their implementation must carefully consider energy consumption, costs, and environmental impact.

1.2.2. CO_2 and CH_4 Displacement Method

Traditional methods promote the natural decomposition of hydrate into natural gas by breaking the phase equilibrium of hydrates in their initial state. However, when exploring the feasibility and economic viability of various extraction technologies, the stabilizing support role of natural gas hydrates in the surrounding formation cannot be overlooked. While traditional extraction methods promote hydrate decomposition, they may also weaken the formation's stability, potentially triggering geological disasters such as earthquakes and underwater landslides, which highlights the limitations of these extraction methods [18]. On the other hand, with the rapid development of human society and the surge in greenhouse gas emissions, this has become a major environmental issue that urgently needs to be addressed. In this context, CO_2 sequestration technology, as an effective means to promote CH_4 extraction, reduce CO_2 emissions, and alleviate the greenhouse effect, has become increasingly important. Therefore, in the exploration of natural gas hydrate extraction methods, it is also essential to actively consider how to integrate CO_2 sequestration technology to achieve a win–win situation for both energy development and environmental protection [19]. Figure 2 shows the development history of the displacement method research.

The principle of the displacement extraction method is to inject CO_2 into the submarine hydrate-bearing zone, replacing CH_4 in the hydrate with CO_2, thereby enabling the safe and efficient extraction of CH_4 hydrate while also achieving the long-term stable geological sequestration of CO_2 [20]. In addition, during the hydrate displacement extraction process, the generated CO_2 hydrate helps maintain the geological stability of the hydrate reservoir, effectively preventing formation collapse and instability caused by the phase transition and decomposition of CH_4 hydrate. In this process, CO_2 and CH_4 in the hydrate undergo a substitution reaction through physical and chemical processes, with CO_2 being fixed in the form of a hydrate, while the original CH_4 is released as gas [21]. The displacement method has strong technical adaptability and is capable of operating effectively under different temperature and pressure conditions, especially in low-temperature, high-pressure reservoirs. It offers good technical and economic feasibility and broad applicability, contributing to global emission reduction goals and the sustainable utilization of energy.

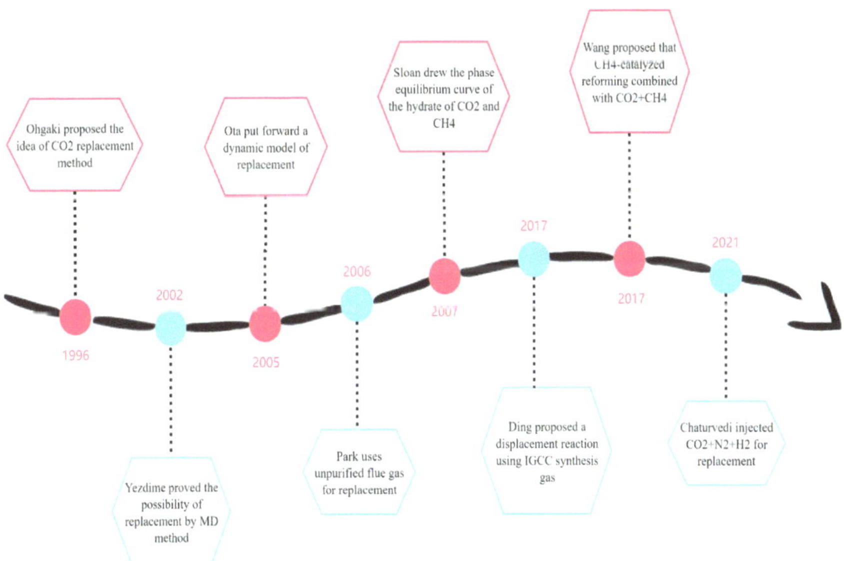

Figure 2. The development history of the research on replacement method [22–29].

2. Structure and Properties of Hydrates

2.1. Crystal Structure of Hydrates

Hydrates have a crystal structure similar to that of molecular sieves, where water molecules are connected by hydrogen bonds to form a regular lattice structure. These lattice units create a cage-like structure that can encapsulate and store gas molecules. Depending on the size and quantity of the gas molecules and their interaction with water molecules, the structure of hydrates is generally classified into three types: Type I, Type II, and Type H. Due to its simplicity and stability, the most commonly found hydrate in nature is Type I, composed of smaller gas molecules such as CH_4 (which makes up more than 90%). These are commonly referred to as methane hydrates [30,31].

The cage structure of hydrates is not always completely filled, and there are certain vacancies within the cages. These vacancies allow gas molecules to enter the cages and be "encapsulated." The stability of the hydrate and its gas storage capacity typically depend on the size, shape, and degree of matching between the vacancies in the cages and the gas molecules. During the formation of hydrates, gas molecules enter these vacancies and are surrounded by water molecules, forming a stable solid structure under specific temperature and pressure conditions [21,32]. For example, in Type I hydrates, the smaller 5^6-cage mainly accommodates one gas molecule, while the larger 5^{12}-cage can hold multiple small gas molecules. The presence of vacancies and the entry of gas molecules allow hydrates to store a large amount of gas, which is a key feature of hydrates as a medium for gas storage.

During the formation of hydrates, gas molecules are surrounded by water molecules and embedded into vacancies within the cages. Once the gas molecules enter the cages, they interact with the water molecules primarily through van der Waals forces and hydrogen bonding. Due to the small size and chemical inertness of the gas molecules, they can be stably surrounded by water molecules, forming a stable "cage-like" structure [32]. The size, type, temperature, and pressure of the gas molecules collectively determine the formation, stability, and storage capacity of hydrates. Suitable gas size and shape, low temperature, and high-pressure conditions, as well as the appropriate type and concentration of gas, facilitate the stable formation and efficient storage of hydrates. With changes in temperature and pressure, hydrates can undergo thawing or gas release, leading to the decomposition of the hydrate and the release of gas [33].

In the process of replacing methane with carbon dioxide for the extraction of natural gas hydrates, the maximum theoretical displacement efficiency is 75%. This phenomenon is primarily related to the crystalline structure of the hydrate and the interactions between gas molecules. Natural gas hydrates typically have a cage-like crystalline structure, which is mainly classified into two types: Structure I (S-I) and Structure II (S-II). In these hydrate structures, carbon dioxide molecules usually replace methane molecules. However, since carbon dioxide molecules are larger than methane molecules, carbon dioxide cannot completely replace all the methane molecules, resulting in a maximum displacement efficiency of about 75%.

In Structure I hydrates, methane molecules occupy the small cages, while carbon dioxide molecules preferentially fill the larger large cages. In Structure II hydrates, carbon dioxide molecules also preferentially occupy the larger cages, similar to the large cages in Structure I. These structural differences and the size disparity between carbon dioxide and methane molecules limit the complete replacement of methane by carbon dioxide [30,31,33].

2.2. Thermodynamic Study

Thermodynamic studies mainly focus on the formation phases of hydrates, stability regions, and the changes in free energy during the replacement process. Through thermodynamic analysis, the stability of CO_2 and CH_4 in hydrates and their replacement behavior under different temperature and pressure conditions can be predicted [34]. Additionally, research on phase equilibrium helps to determine stable operating conditions, control the thermodynamic driving forces of replacement reactions, study the dynamic processes of gas diffusion and exchange, ensure thermodynamic and kinetic stability during extraction, and provide a scientific basis for the long-term sequestration of CO_2 and environmental safety [26].

2.2.1. Thermodynamic Characteristics of Replacement Reactions

According to the basic principles of chemical thermodynamics, spontaneous reactions always proceed in the direction of decreasing Gibbs free energy. Yezdimer et al. [23] demonstrated through molecular dynamics simulations that the Gibbs free energy of CO_2 hydrate is lower than that of CH_4 hydrate. Additionally, the Gibbs free energy of the replacement reaction is negative, indicating that the reaction can proceed spontaneously [24]. From the perspective of phase change heat, the heat released during the decomposition of CO_2 hydrate (57.98 kJ/mol) is greater than the heat absorbed during the decomposition of CH_4 hydrate (54.94 kJ/mol). Therefore, the heat required for the decomposition of CH_4 hydrate can be provided by the heat released during the formation of CO_2 hydrate, and the replacement reaction does not require additional heat input from external sources [35,36].

Akihiro [37], in his study of the phase equilibrium temperature and pressure of CH_4 and CO_2 hydrates below the freezing point, found that they follow the same pattern. During the replacement process, a mixed gas of CO_2 and CH_4 is generated. Goel [38], through experimental research on the formation process of CH_4 and CO_2 hydrates, plotted phase equilibrium curves for mixtures of CO_2 and CH_4 at different ratios. He found that as the CO_2 content increased, the phase equilibrium pressure of the mixed gas decreased while the phase equilibrium temperature increased. CO_2 hydrate remains stable within the temperature range of 260–270 K and the pressure range of 0.6–1.7 MPa, while CH_4 hydrate is not easily formed. Circone et al. [39] further demonstrated through experiments that, within a specific temperature and pressure range (0.1 MPa, 168–218 K), CH_4 hydrate can decompose while CO_2 hydrate remains stable. Xie et al. [40], by fitting experimental data using 10 thermodynamic models, showed that from the perspective of adsorption thermodynamics, the adsorption amount of CO_2 is larger than that of CH_4 under the same conditions, with an adsorption strength range of 16.71–35.65 kJ/mol, which is higher than CH_4's range of 12.48–28.90 kJ/mol. This indicates that shale adsorbs CO_2 more strongly. Gas adsorption is a spontaneous and enthalpy-driven process, with a negative ΔG indicating spontaneous adsorption. Furthermore, the absolute value of ΔG for CO_2 is smaller than that for CH_4, meaning the CO_2 adsorption process is more likely to occur.

2.2.2. Stability of Hydrates and Gas Distribution Behavior

The stability and formation process of hydrates is closely related to the occupancy state of gas molecules within the hydrate crystal structure. Methane hydrate is typically stable at lower temperatures and higher pressures, while carbon dioxide hydrate has a relatively wider range of stable temperatures and pressures, and its formation conditions are more easily achieved compared to methane hydrate [26]. Additionally, the decomposition process of hydrates plays a significant role in gas replacement behavior and energy release. Therefore, a thorough study of the phase diagram and stability of hydrates is crucial for the development of these applications [41].

Ohgaki et al. [22] proposed a method for extracting natural gas hydrates using CO_2 substitution. This method involves introducing CO_2 gas into the gas–liquid hydrate three-phase system of CH_4, forming a gas–liquid hydrate equilibrium system containing both CO_2 and CH_4. Figure 3 shows the thermodynamic equilibrium curves for pure CH_4 and pure CO_2 hydrates, as well as the gas–liquid phase equilibrium curve for CO_2. The thermodynamic trends for the two natural gas hydrates can be summarized as follows: under the presence of gaseous CO_2, the formation conditions for pure CO_2 hydrate are milder than those for pure CH_4 hydrate. When gaseous CO_2 is injected into the methane hydrate layer, the unstable methane hydrate dissociates, releasing methane gas and liquid water. Subsequently, the melted water reacts with the injected CO_2 and spontaneously forms a more stable CO_2 hydrate as the product of the entire gas exchange reaction [42]. In the figure, areas A and B are located above the H_2O-hydrate-CO_2 equilibrium curve and below the H_2O-hydrate-CH_4 equilibrium curve, respectively. Thermodynamic analysis indicates that these regions suggest that, under specific temperature and pressure conditions, gaseous methane (CH_4) and CO_2 hydrate can coexist. Therefore, when CO_2 is injected into the CH_4 hydrate layer, CO_2 can be stored as a hydrate. This phenomenon reveals that CO_2 hydrate ($CO_2{\cdot}nH_2O$) is more stable than methane hydrate ($CH_4{\cdot}nH_2O$) [43]. Under the same environmental conditions, CO_2 can form and stabilize hydrates at lower temperatures and higher pressures, while CH_4 hydrate requires higher temperatures or lower pressures to remain stable.

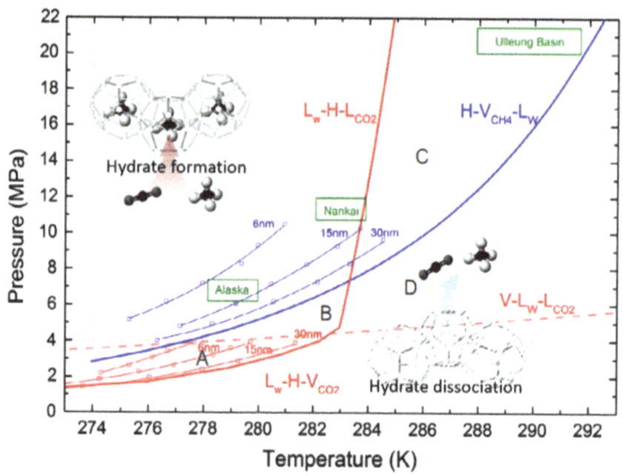

Figure 3. Phase diagram of CO_2 replacement hydrate. A–D represent the different regions enclosed by the thermodynamic equilibrium curves of pure CH_4 and pure CO_2 hydrates, as well as the gas-liquid phase equilibrium curve of CO_2 [42].

The process of CO_2 replacing methane occurs above the methane hydrate phase equilibrium line to ensure that CO_2 can effectively replace methane. The reaction must take place under conditions where methane hydrate remains stable. Only above the methane hydrate phase equilibrium line does methane hydrate maintain a stable solid structure, providing a basis for CO_2 molecules to substitute into the lattice. Below the phase equilibrium line, methane hydrate will spontaneously decompose, and there will be no stable methane hydrate structure available for CO_2 to carry out the replacement reaction. At the pressure–temperature conditions above the phase equilibrium line, the process of CO_2 and water molecules forming a hydrate is thermodynamically more favorable. Higher pressure and appropriate temperature conditions help CO_2 molecules overcome energy barriers to enter the hydrate lattice and replace methane molecules. This thermodynamic advantage drives the replacement reaction towards the formation of CO_2 hydrate and the release of methane [42]. In the process of CO_2 replacing methane, if the pressure–temperature conditions fluctuate near the methane hydrate phase equilibrium line, methane hydrate may re-form, encapsulating unreacted CO_2 or blocking pore channels, thus reducing the replacement efficiency. To minimize this risk, stable conditions above the phase equilibrium line should be maintained to ensure the continuous progress of the replacement reaction [41]. Additionally, if the CO_2 concentration is too high or poorly controlled, excessive formation of CO_2 hydrate at the sediment surface may occur, creating a dense shell that hinders CO_2 diffusion and methane release, thereby reducing replacement efficiency. By properly controlling the CO_2 injection rate and reaction conditions, this issue can be avoided, maintaining a high replacement efficiency. Therefore, controlling the replacement process above the methane hydrate phase equilibrium line ensures the stable existence of methane hydrate while reducing the risks of reformation of methane hydrate and excessive formation of CO_2 hydrate.

This conclusion indicates that CO_2 hydrate has a clear thermodynamic advantage, as it not only forms more easily under natural sediment conditions but also has greater potential to replace methane hydrate [44]. The figure also presents the equilibrium data between the liquid and gas phases of CO_2, which is crucial for understanding the behavior of CO_2 in hydrates. These data can help researchers better predict the conversion efficiency of CO_2 during the replacement process of natural gas hydrates, thereby optimizing the overall process of CO_2 sequestration and natural gas extraction.

At the same time, during the exchange process, because of their smaller size and stronger polarity, CO_2 molecules can more easily be embedded into the hydrate cages, especially in Structure I hydrates, where CO_2 can effectively replace CH_4 molecules. However, the competitive entry of CO_2 and CH_4 for entry may result in changes to the stability of the hydrates. This leads to the hindrance of the displacement process, with the main limiting factors including thermodynamics, competitive adsorption, kinetics, and hydrate stability. CO_2 replacement of methane requires specific temperature and pressure conditions, and unfavorable temperature and pressure may result in low replacement efficiency. At the same time, the competitive adsorption of CO_2 and methane in the hydrate structure may affect the replacement effect, especially when methane has already been adsorbed. Kinetically, the diffusion rate of CO_2 is relatively slow, and the size and structure of the hydrate pores also influence the rate of the displacement reaction. Furthermore, excessive CO_2 displacement may alter the stability of the hydrate, causing hydrate decomposition or instability. Finally, the entire displacement process often requires a long time, especially under low temperatures and high pressure, which limits large-scale applications. Therefore, the distribution and phase behavior of gas molecules in hydrates are not only central to the formation, stability, and dissociation processes of hydrates but are also key factors in the study of hydrate utilization and its environmental impacts [45].

2.3. Kinetic Studies

Kinetic studies primarily focus on several key factors in the gas displacement process, including reaction rates, diffusion mechanisms, and dynamic changes at the interface. Although thermodynamically favorable conditions for gas displacement may exist, the actual rate and efficiency of the displacement process are still constrained by various kinetic factors. Therefore, even when thermodynamic conditions are met, the actual occurrence of the gas displacement process is still limited by these kinetic factors, requiring further experimental studies and modeling for optimization [46].

2.3.1. Gas Diffusion and Exchange Rate

The diffusion process of gas within the hydrate is crucial to the gas displacement rate. In natural gas hydrates, gas molecules diffuse into the hydrate's cage structure, displacing the original gas molecules inside the hydrate [44]. This process involves several key steps, including the entry of gas molecules from the external environment into the hydrate interface, penetration through the solid structure of the hydrate, and exchange with the existing gas molecules.

The diffusion rate of CO_2 is typically higher than that of CH_4, which may be due to the larger size of CO_2 molecules, resulting in a more complex spatial occupation within the hydrate's cage structure. The structure of the hydrate is typically formed by water molecules arranged in a cage-like pattern, creating pore structures to trap gas molecules [18]. Due to the larger molecular volume of CO_2, CO_2 molecules may be more likely to aggregate and interact with each other within the hydrate's cage compared to methane molecules, thereby affecting its diffusion behavior [23].

In the gas displacement process, gas exchange primarily occurs at the gas–hydrate interface. This process refers to the interaction between gas molecules and water molecules within the hydrate, causing the gas molecules to enter the hydrate structure and potentially displace the original gas molecules [47]. During this process, a larger gas–hydrate interface area helps accelerate gas exchange. If the hydrate particles are smaller in size, the surface area per unit volume increases, leading to more contact between gas molecules and the hydrate, which in turn enhances the exchange rate. Additionally, if the hydrate structure is more loosely packed or has larger pore sizes, the diffusion rate of gas molecules within the hydrate is faster, improving the efficiency of gas exchange. Conversely, if the hydrate structure is more compact, the diffusion rate of gas molecules is restricted, and the displacement process may slow down [44].

2.3.2. Reaction Mechanisms in the Displacement Process

The process of CO_2 displacing methane hydrate is a complex and dynamic multi-step reaction that involves competition, diffusion, and the reorganization of the hydrate structure between gas molecules. In this process, CO_2 molecules compete with methane molecules for occupancy of the hydrate's empty cages. This process is not merely a simple gas exchange but also involves a series of changes related to the hydrate structure. A schematic diagram of the two CH_4 hydrate replacement mechanisms is shown in Figure 4 [46].

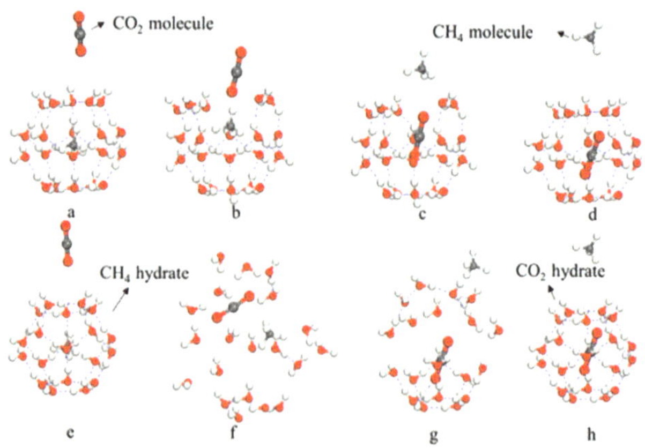

Figure 4. Schematic diagram of two CH_4 hydrate replacement mechanisms: (**a**,**e**) contact between CO_2 molecules and the CH_4 hydrate; (**b**) partial hydrogen bond fracture of the CH_4 hydrate cage (not necessary); (**c**) CH_4 molecules leave the hydrate cage, and CO_2 molecules enter hydrate cage; (**d**,**h**) CO_2 hydrate formation; the replacement process is complete; (**f**) complete dissociation of the CH_4 hydrate cage; (**g**) the two objects are interchangeable [46].

The CO_2-CH_4 hydrate replacement process can be divided into three steps: First, CO_2 contacts the surface layer of the CH_4 hydrate. As CO_2 enters the hydrate surface, the cage-like structure of the methane hydrate begins to break down, causing a portion of the methane gas to rapidly decompose and be released. At this point, methane gas escapes from the hydrate, and the structure of the surface layer of the hydrate is disrupted. As the displacement process progresses, CO_2 and CH_4 quickly combine with surrounding free water molecules, forming a mixed hydrate layer. Finally, CO_2 gas slowly diffuses into the interior of the hydrate and continues to displace the methane molecules within the hydrate structure. As CO_2 penetrates deeper into the hydrate, methane molecules originally occupying the internal cages are gradually replaced, and methane gas escapes outward, completing the displacement process. This stage of diffusion is relatively slow and depends on factors such as temperature, pressure, and the pore structure of the hydrate [20].

2.3.3. Kinetic Characteristics of the Displacement Reaction

Studies have shown that the CO_2-CH_4 displacement process is a complex dual mechanism significantly influenced by the occupancy of the hydrate cages. Bai et al. [48] conducted an in-depth study of this process using molecular dynamics (MD) simulations, revealing that the chemical potential changes of CO_2 molecules lead to the decomposition of methane hydrate and, by disturbing the structure of the hydrate, cause CH_4 molecules to be released from the cage-like structure. Nakano et al. [49] found experimentally that CO_2 diffuses faster than methane (CH_4) in the hydrate, proving that the CO_2-CH_4 hydrate displacement method is kinetically feasible. Tung et al. [50] pointed out that liquid CO_2 can directly replace methane (CH_4) in the hydrate without the need for the decomposition process. Other studies [51] explored the microscopic behavior of CO_2 and methane hydrates during the displacement reaction by using high-pressure reactors under different temperature and pressure conditions combined with molecular dynamics simulations, as shown in Figure 5. Figure 5A,B show the process of CO_2 replacing methane hydrate with and without free water. Figure 5C demonstrates that during the free water replacement process, the number of hydrogen bonds gradually decreases, and the decomposition rate of methane hydrate and the formation rate of CO_2 hydrate show different slopes at each stage. The results indicate that within a certain temperature and pressure range, as temperature and pressure increase, the displacement efficiency gradually improves.

Researchers have also conducted in-depth studies on the formation and decomposition mechanisms of hydrates using visualization techniques such as neutron diffraction, Raman spectroscopy, and nuclear magnetic resonance (NMR). For example, Uchida et al. [52] used Raman spectroscopy to confirm that the CO_2-CH_4 hydrate displacement reaction primarily occurs at the surface layer of the CH_4 hydrate. The results showed that during the displacement process, the concentration of methane (CH_4) gas in the gas phase gradually increased over time, but the rate of increase slowed down as the reaction progressed. This indicates that as the displacement reaction advances, CH_4 molecules in the hydrate are gradually replaced by CO_2, leading to a stabilization in the concentration of CH_4 gas. Kuang et al. [53] studied the microstructural evolution of CO_2 hydrates in porous media using NMR technology, revealing the structural changes of CO_2 hydrates in the pore space and their interactions with the surrounding environment under different conditions. Additionally, by analyzing the variation in pore water volume distribution with depth in sand layers under different storage conditions (as shown in Figure 6), they found differences in the formation locations of hydrates in porous media, confirming that CO_2 hydrates nucleate randomly in porous media and preferentially form in pore spaces. Their research provides an important theoretical basis for understanding the behavior of CO_2 hydrates in underground storage and the development of natural gas hydrates. Farahani et al. [54] compared the formation and decomposition of CH_4 hydrates in synthetic and natural sediment samples. Through experimental analysis, they found that CH_4 hydrates in natural sediments exhibited different kinetic characteristics during formation and decomposition compared to synthetic samples. This discovery provides new insights into the stability of natural gas hydrates, storage conditions, and gas displacement phenomena, presenting both challenges and opportunities for the exploitation and utilization of hydrates.

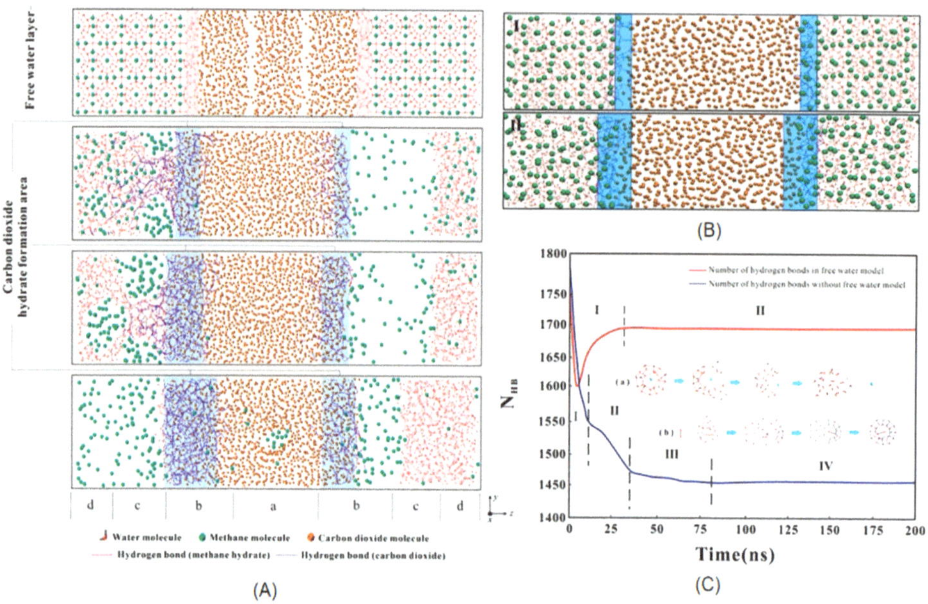

Figure 5. Carbon dioxide replacement of methane hydrate process. (**A**) Snapshot of carbon dioxide replacement of methane hydrate in the model with free water. (**B**) Snapshot of carbon dioxide replacement of methane hydrate without the free water model. (**C**) NHB variation with time during the replacement process of the free water model: (I) the destruction of the methane hydrate cage structure by the edge of carbon dioxide and (II) the generation of the carbon dioxide hydrate cage structure along the source direction [51].

In the practical process of CO_2 displacement for natural gas hydrate exploitation, thermodynamics and kinetics work together in a complementary manner. Thermodynamic studies provide the phase equilibrium conditions and driving forces for the displacement process, while kinetic studies reveal the reaction rates, diffusion mechanisms, and influencing factors of the process. By combining thermodynamic and kinetic models, researchers can optimize the various operating conditions involved in the extraction process to achieve more efficient gas displacement. In terms of temperature and pressure control, thermodynamic models can determine the optimal temperature and pressure ranges to promote the CO_2 displacement of methane. Kinetic simulations, on the other hand, can help researchers predict the reaction rates and efficiency under different temperature and pressure conditions, thereby identifying the most suitable operating conditions.

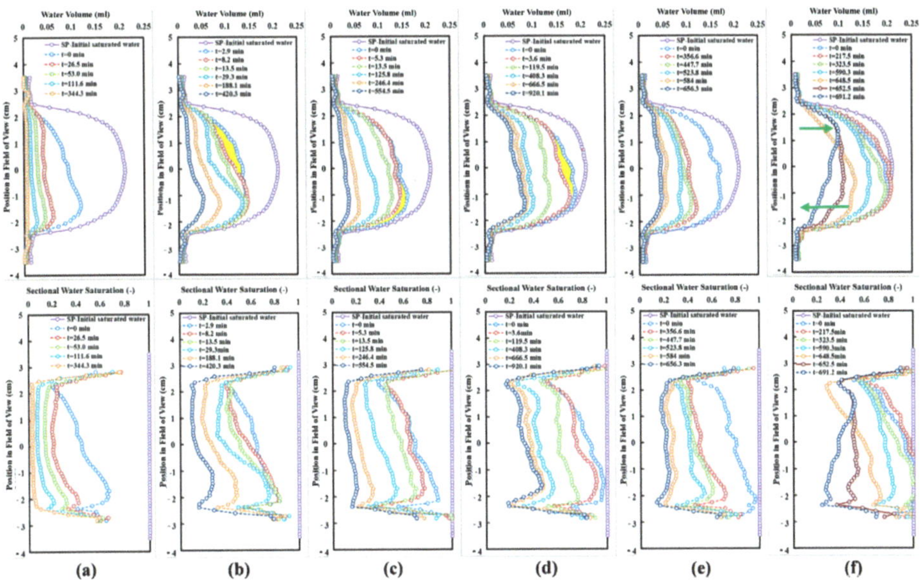

Figure 6. Variations in water volume (top) and sectional water saturation (bottom) distributions at vertical locations in porous media during the hydrate formation process [53]. Figure (**a**–**f**) show the variation in the distribution of pore water volume at different vertical positions in the sand layer under varying pressure, temperature, initial water saturation, and initial gas saturation conditions.

3. Combined Enhancement of Displacement Method and Traditional Methods

The displacement method, as a novel natural gas extraction technology, holds significant potential in improving extraction efficiency, reducing greenhouse gas emissions, and promoting sustainable extraction. However, its high cost, technical complexity, and environmental risks are key challenges that need to be addressed for its commercialization [46]. Therefore, researchers worldwide have conducted extensive studies on enhancing the CO_2-CH_4 displacement process, exploring various aspects from the macro-level displacement process to the micro-level displacement mechanisms. The aim is to find an efficient enhancement method that comprehensively considers factors such as economic cost, environmental protection, and process complexity. Table 4 presents a comparison of the impacts of the displacement method combined with various traditional methods of natural gas extraction.

3.1. Combined Extraction of Displacement Method and Depressurization Method

The traditional depressurization method has been applied in natural gas hydrate extraction to some extent. However, during the depressurization process, the dissociation of hydrates is an endothermic reaction, leading to a gradual decrease in methane production and a continuous drop in reservoir temperature. At the same time, a large amount of water is produced during depressurization, which not only further reduces gas production efficiency but may also cause a series of safety issues. This presents numerous challenges for the practical application of this method [21]. In contrast, the displacement method promotes hydrate dissociation by injecting external gases (such as CO_2) to replace methane molecules in the hydrate. Combining depressurization with the displacement method can fully leverage the synergistic effects of both, significantly improving methane release efficiency [48,55].

Table 4. A comparison of the impacts of different methods on the combined natural gas extraction.

Extraction Method Combination	Advantages	Disadvantages	Suitable Geological Conditions	Key Factors Affecting Extraction
Displacement Method Combined with Pressure Reduction Method	Increase extraction rate; sustainable gas release; reduce the risk of hydrate re-crystallization	High cost of gas injection; extraction efficiency limited by reservoir characteristics	Higher bottom pressure, better permeability, and porosity	Reservoir temperature and pressure; gas injection rate and gas selection
Displacement Method Combined with Thermal Stimulation Method	Thermal energy promotes hydrate dissociation, enhancing displacement effectiveness	Thermal stimulation method may cause potential damage to the reservoir; high energy consumption and relatively high cost	The hydrate layer at a lower temperature (0 °C to 10 °C) is relatively thick and evenly distributed	Reservoir temperature and thermal response characteristics; heat injection methods and temperature control; thermal stability and structure of the reservoir
Displacement Method Combined with Chemical Inhibitor Method	Improve the long-term stability of natural gas production; prevent hydrate recrystallization	Inhibitors may increase environmental risks; chemical inhibitors are expensive and could negatively impact extraction costs	Mid- to high-saturation hydrate reservoirs under low-temperature and high-pressure conditions	Selection and injection concentration of chemical inhibitors; cost and environmental friendliness of inhibitors; synergistic effect of inhibitors and displacement gases

The advantage of this combined method is that the depressurization process creates the necessary low-pressure environment for the injection of displacement gases, ensuring that the displacement gas can effectively penetrate the hydrate layer. Meanwhile, the displacement method accelerates the dissociation of the hydrate through gas exchange, greatly enhancing the release rate of CH_4 [47]. Specifically, when CO_2 is used as the displacement gas, it can undergo a substitution reaction with methane in the hydrate, which not only helps improve methane recovery but also enables the sequestration of greenhouse gases, thus offering significant environmental benefits.

Zhao et al. [56] proposed an alternative method for depressurization extraction. Experiments showed that after 2 h of depressurization dissociation, CO_2 was used to displace the hydrates. Compared to non-depressurization extraction, the total displacement rate after combined depressurization increased to over 30%. Lee et al. [57], using a triaxial compression testing apparatus and applying the method of depressurization followed by displacement, found that the methane recovery rate could be increased from 35% with pure CO_2 replacement to around 60%. They also studied the impact of the depressurization process and CO_2 injection on the stability of natural gas hydrate reservoir structure. The results indicated that, during the dissociation of hydrates, as the dissociation ratio of CH_4 hydrates gradually increased, the structural strength of the hydrates significantly decreased. This phenomenon suggests that the stability of the hydrates weakens with methane release, causing the reservoir structure to gradually become looser, which increases the risk of potential reservoir collapse or gas leakage during extraction. Further analysis revealed that when the dissociation ratio of CH_4 hydrates reaches 20%, subsequent CO_2 displacement can achieve the optimal displacement efficiency.

Chen et al. [58] employed a "cross-flow method" for depressurization combined with displacement using two wells. The experimental results indicated that the inlet pressure, outlet pressure, and confining pressure all had significant effects on production efficiency. Higher inlet pressure facilitated the effective injection of CO_2 into the porous medium (as shown in Figure 7a), promoting the formation of CO_2 hydrates and enhancing displacement extraction. Additionally, when the outlet pressure is higher than the equilibrium pressure of CO_2, depressurization can disrupt the structure of methane hydrates, thus increasing the hydrate displacement rate. The experiments also showed that the impact of outlet pressure was greater than that of confining pressure (Figure 7b). In this study, the utilization efficiency ranged from 27.2% to 46.6% (Figure 7c), reflecting that improving CO_2 utilization efficiency through depressurization is feasible, with outlet pressure being one of the most important parameters determining utilization efficiency.

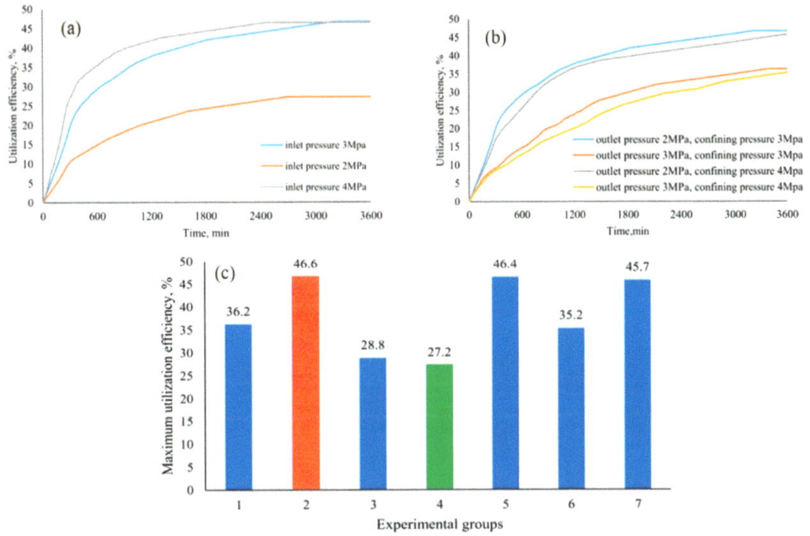

Figure 7. (**a**) Utilization efficiency vs. time curves for different inlet pressures. (**b**) Utilization efficiency vs. time curves for different outlet and confining pressures. (**c**) Maximum utilization efficiency in different experimental groups [58].

In contrast to the extraction sequence mentioned above, Yang et al. [59] adopted a method where CO_2 displacement is carried out for a period of time before depressurization. In the initial stage, CO_2 is injected to displace the hydrates, which increases the concentration of CO_2 in the hydrates. Then, by reducing the pressure, the driving force for hydrate dissociation is enhanced, further promoting the CH_4/CO_2 displacement reaction, ultimately achieving a total displacement rate of 80 mol%. The advantage of this extraction sequence is that it can utilize CO_2 more efficiently, not only increasing gas production but also effectively reducing the energy required for methane hydrate dissociation, making the overall extraction process more energy efficient and environmentally friendly.

Moreover, the combined method of depressurization and displacement can address the issues of slow methane release and uneven gas diffusion that occur in single depressurization methods. By properly adjusting the depressurization rate and the amount of displacement gas injected, the methane release process can be optimized, avoiding the reformation of hydrates or localized freezing caused by excessive depressurization or improper gas injection, thereby improving the stability and safety of the extraction process. This makes it a safer method for natural gas hydrate extraction [60].

3.2. Combined Extraction of Displacement and Thermal Stimulation Methods

The thermal stimulation method is simple to operate, technically mature, and widely applicable. However, when used alone for natural gas extraction, the thermal stimulation method can be costly and may lead to reservoir instability, potentially triggering geological hazards. Therefore, in practical applications, thermal stimulation is often combined with the displacement method. This combination can improve methane recovery rates while reducing the energy consumption and reservoir instability risks associated with thermal stimulation [61,62].

Zhang et al. [62] studied the CH_4 substitution rate from the perspective of combined thermal stimulation and displacement methods for natural gas extraction. They found that the methane recovery efficiency (CRE) during the hydrate substitution process significantly increased from the original 10–50% to 21–63%. Tupsakhare et al. [63] conducted heating stimulation experiments using a gas mixture of 85% CO_2 and 15% N_2 with heating powers of 100 W, 50 W, and 20 W. Under a 100 W heating power, about 60% of the methane gas could be extracted from the hydrate. In another study, Tupsakhare et al. combined CO_2/N_2 displacement with thermal stimulation to enhance hydrate recovery. The results showed that, compared to the recovery rate of 26.5% from thermal stimulation alone, the recovery efficiency of natural gas using the combined thermal stimulation and displacement method (68.8%) was significantly higher.

The efficiency of heat transfer varies significantly depending on the choice of heat source. When using an external heat source for hydrate extraction, its efficiency is generally lower than that of internal heat sources. The main reason for this is that heat tends to dissipate in the non-hydrate regions outside the hydrate layer [21]. Fan et al. [64] conducted experiments, as shown in Figure 8a, to explore three different heat injection modes: Incremental Heating Injection for Enhanced Replacement Recovery (IHIR), Decremental Heating Injection for Enhanced Replacement Recovery (DHIR), and Constant Heating Injection for Enhanced Replacement Recovery (CHIR). The changes in CH_4 recovery rates under these modes are presented in Figure 8b. The results show that incremental heat injection can lead to more CH_4 dissociation and replacement, achieving the highest CH_4 recovery rate (17.0%) and the highest gas mole fraction (13.5%) while also reducing the energy consumption for gas separation and purification.

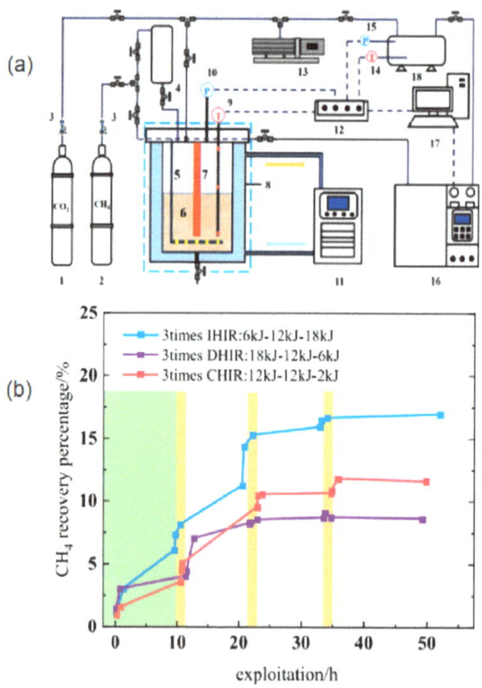

Figure 8. (**a**) Schematic diagram of hydrate exploitation simulation experiment. (**b**) The variation of CH_4 recovery percentage during the replacement recovery processes of three enhanced modes [64].

The combined use of thermal stimulation and displacement methods can play a crucial role in enhancing CH_4 recovery and CO_2 sequestration. However, ensuring the stability and safety of the operation through effective monitoring and adjustments remains a challenge for technical implementation. Additionally, in practical applications, optimizing the parameters of thermal stimulation and displacement to balance methane recovery and CO_2 sequestration is still an important area of research.

3.3. Combined Extraction of Displacement and Chemical Inhibitor Methods

The chemical inhibitor method has the advantages of accelerating the decomposition of natural gas hydrates and improving methane recovery efficiency. However, its drawbacks cannot be overlooked. The main issues include poor selectivity of the inhibitors, environmental pollution risks, high costs, and potential adverse effects on the formation and stability of hydrates, which is why it is rarely used alone in natural gas hydrate extraction. In recent years, some researchers have combined chemical additives with the CO_2-CH_4 hydrate displacement method to explore joint extraction technologies. Tetra-n-butylammonium bromide (TBAB), a commonly used hydrate promoter, has been applied to enhance CO_2-CH_4 displacement research. It effectively lowers the phase equilibrium conditions of CO_2 hydrate, thereby improving the displacement capacity of CO_2 for CH_4 hydrate and significantly enhancing the displacement efficiency [65]. Some studies also indicate that [66] chemical inhibitors can effectively improve CO_2's displacement efficiency for CH_4 hydrate, optimizing the extraction process. However, simultaneous decomposition and displacement significantly reduce displacement efficiency. Therefore, the optimal approach is to first add chemical inhibitors to decompose CH_4 hydrate, releasing most of the CH_4 gas, and then inject CO_2 for displacement, which can significantly improve displacement efficiency. In addition to thermodynamic inhibitors, researchers have also explored hydrate promoters and anti-agglomerants. Heydari and Peyvandi's study [67], which used biological surfactants to replace natural gas hydrates, showed that biological surfactants not only effectively promoted the formation of methane hydrates but also significantly improved CO_2 displacement efficiency. Specifically, the displacement ratio increased by 72.6%, and displacement kinetics improved by 39%. Although the biological surfactants were added during the hydrate formation phase, the study also indicated that their addition altered the hydrate morphology and had a positive impact on continuous displacement reactions.

Currently, this method is still in the early stages of research, with many scholars, both domestically and internationally, conducting related studies. The displacement potential and economic benefits remain unclear. In addition, issues such as the selection and optimization of inhibitors, environmental impacts, and their long-term effectiveness and stability continue to be major constraints for large-scale applications. Therefore, how to better simulate and understand these coupled processes, optimize process parameters, and achieve the best extraction results remains a significant challenge in technological progress.

Overall, the combined application of the displacement method and traditional extraction methods provides an innovative solution for natural gas extraction. This combination not only significantly optimizes extraction efficiency but also provides strong support for the sustainable development of energy. As shown in Table 5, the combination of the displacement method with three traditional methods demonstrates complementary advantages in several aspects, further enhancing the overall extraction efficiency.

Table 5. Combination of the displacement method with three traditional methods.

Method	Experimental Conditions	Advantages	CH4 Recovery Rate	Literature Source
Combined Pressure Reduction and Substitution Method	Experiments conducted using a customized high-pressure flow-through apparatus at different methane hydrate dissociation levels (0%, 20%, 40%, 60%, 80%, 100%)	The mechanical properties of methane hydrate-bearing sediments were considered to provide a basis for economically safe extraction; experimental studies were conducted to investigate the effects of various factors on mechanical properties and methane recovery rate	35.4–63.3%	Lee et al. [57]
Pressure Reduction-Assisted CO_2 Substitution Method	Design of a one-dimensional experimental setup to simulate the interface between horizontal wells, investigating the impact of different pressures (inlet pressure, outlet pressure) on CO_2 substitution behavior. The experimental temperature is 275 K, and the methane hydrate saturation is 32%	By combining the advantages of CO_2 substitution and pressure reduction, production efficiency is improved, and risks are reduced; the impact of pressure parameters on natural gas extraction was studied, providing theoretical support for further research and application	27.2–46.6%	Chen et al. [58]
Pressure Reduction-Assisted CO_2 Substitution Method	Study of the depressurization-assisted CO_2 substitution process by varying initial hydrate dissociation ratio (0%, 50%, 100%), substitution period (1, 4, 7 days), and CO_2 injection flow rate	The issues of weakened geomechanical strength of methane hydrate-bearing sediments caused by pressure reduction alone and the slow production rate during substitution were addressed; methane production and CO_2 sequestration efficiency were improved	Through depressurization-assisted substitution, the amount of CO_2 stored in the sediment can be greater than the amount of CH_4 produced, with approximately 92% of the initial methane being replaced by CO_2	Choi et al. [68]
Combined CH_4/CO_2 Substitution and Thermal Stimulation Method	Experiments conducted under different methane hydrate saturations, substitution zones, and freezing point conditions	The diffusion rate of CO_2 was increased through thermal stimulation, overcoming the diffusion limitation in the CO_2 substitution process alone; the methane substitution percentage, CO_2 storage efficiency, and energy efficiency under different conditions were analyzed and discussed	64.63%	Zhang et al. [62]
Combined CH_4/CO_2 Substitution and Thermal Stimulation Method	Experiments conducted in a large-scale hydrate vessel (LSHV) with heating rates of 20, 50, and 100 W	The effect of temperature on N_2 capture was studied, and it was found that N_2 is selectively captured in hydrate cages at temperatures below 12 °C	At a heating rate of 100 W, the mole number of methane during thermal stimulation is 8.5; during thermal stimulation with CO_2 substitution, it is 16; and during thermal stimulation with CO_2 + N_2 substitution, it is 20	Tupsakhare et al. [63]
Combined Thermal Stimulation and CH_4/CO_2 Substitution Method with Nanoparticle Addition	Experiments conducted in a high-pressure stainless steel reactor under different pressures (40 bar, 45 bar) and temperatures (5.5 °C, 8 °C, 10 °C)	Without the need for vacuum extraction, this method can effectively increase methane recovery and CO_2 storage efficiency; the optimal experimental conditions (45 bar and 8 °C) were determined	The recovery rate of CH_4 increased from 19.8% to 51.9%	Adibi et al. [69]
Inhibitor-Assisted Substitution Method (Using Methanol Solution)	Using an automated core flooding system to simulate and monitor fluid flow and studying the effects of different inhibitors on CH_4 hydrate dissociation and CO_2 substitution by varying the injected fluid	The CO_2 substitution method and thermodynamic hydrate inhibitor technology were combined to replace the simple CH_4 hydrate substitution process	Under the experimental conditions, the methane recovery rate exceeds 92%	Khlebnikov et al. [66]
Bio-Surfactant-Assisted Method (Using Rhamnolipid)	Studying the effect of different concentrations of rhamnolipids on the kinetics of methane hydrate formation and comparing it with the chemical surfactant SDS	Compared to the chemical surfactant SDS, it significantly reduces the induction time and total time; improves gas consumption and increases the kinetic growth rate of the hydrate; it can enhance the substitution rate and CO_2 storage capacity	Injecting rhamnolipid increased the substitution percentage by approximately 72.6%	Heydari et al. [67]

The combination of CO_2 injection, depressurization, thermal stimulation, and chemical inhibitors has shown significant potential in natural gas hydrate extraction, improving recovery rates (such as sustained methane production in the South China Sea pilot project) and carbon sequestration efficiency, contributing to the low-carbon transition. However, its technical depth and scaling face multiple challenges: the CO_2 injection and depressurization combination is limited by reservoir permeability differences, with low-permeability gas fields prone to insufficient fluidity and potential underground water contamination risks; combined with thermal stimulation, although it enhances gas flow (as seen in the Russia Mezoyakha gas field), the high-temperature environment accelerates equipment aging, increases maintenance costs, and irreversible changes in reservoir porosity threaten long-term stability [1]; when used with chemical inhibitors (such as ethylene glycol), although it can control hydrate dissociation risks (as seen in the South China Sea basin project in Japan), long-term use poses underground water contamination risks and relies on costly monitoring systems. On the economic side, high initial investments (equipment and reservoir modification) and specialized operation and maintenance costs far exceed the financial capacity of small and medium-sized enterprises, while the imperfect carbon trading market and fragmented policies further weaken revenue stability. Despite significant technical advantages, the large-scale application still needs to overcome core challenges such as geological adaptability, equipment durability, and environmental risks. In the future, AI should be used to optimize technical parameters, low-cost materials should be developed, and a global carbon market should be promoted, along with tax incentives and international standards coordination at the policy level, to balance energy efficiency and climate goals [4].

4. Multicomponent Gas Displacement Process

The CO_2-CH_4 hydrate replacement has become a widely studied field in academia. Previous studies have shown that up to 68% of the methane (CH_4) captured in hydrates can be replaced by carbon dioxide (CO_2) for recovery [20,70]. The main reasons for the low efficiency of CO_2 replacement in natural gas hydrate extraction are as follows: (1) The methane (CH_4) molecules in natural gas hydrates are relatively small and can be accommodated in the "large cages" and "small cages" of the hydrate structure. In contrast, CO_2 molecules, which are slightly larger and have a different molecular structure compared to CH_4, can only replace methane molecules in the large cages and are unable to effectively replace methane in the small cages. (2) There are strong van der Waals forces between methane molecules and hydrate cages, which makes the methane molecules tightly bound within the hydrate. (3) During the CO_2 replacement process, when CO_2 molecules enter the hydrate structure, especially at the hydrate surface, CO_2 forms a mixed hydrate layer with some methane molecules. This leads to difficulty in CO_2 entering the internal structure of the hydrate, limiting the exchange between CO_2 and methane and significantly reducing the replacement efficiency. Therefore, finding small molecular gases that can replace CH_4 in the small cages, increasing the hydrate phase equilibrium pressure, and enhancing gas transfer are key factors in improving the efficiency of CO_2 replacement for natural gas hydrate extraction.

4.1. CO_2-H_2 Mixture Replacement to Improve CH_4 Recovery Rate

The use of a CO_2 and H_2 mixture can improve the replacement efficiency [28]. The addition of H_2 does not lead to additional occupation of the hydrate cages. The reason hydrogen can facilitate the replacement process is primarily because H_2 molecules, being smaller, diffuse easily within the hydrate system. The introduction of H_2 reduces the partial pressure of methane in the gas phase, which helps to destabilize the methane hydrate, promoting its dissociation rather than directly replacing methane through gas exchange. Additionally, H_2 can enhance the mass transfer process, aiding in the diffusion of methane gas within the medium [71]. The CO_2-H_2 methane extraction technology uses mixed gas injection to enhance recovery and carbon sequestration. CO_2 displaces methane, while H_2 boosts reservoir permeability [28]. Benefits include hydrogen's high diffusivity for improved gas flow and catalytic carbon utilization (e.g., methane dry reforming). Key challenges include the following: hydrogen flammability risks, complex gas separation (membrane/cryogenic methods), and hydrogen embrittlement under high pressure. No large-scale applications exist yet, with insights drawn from hydrogen storage and CO_2-EOR. Heterogeneous reservoirs risk uneven gas distribution and efficiency drops. High costs dominate, alongside safety expenses and uncertain carbon revenue. Future success hinges on affordable green hydrogen, hydrogen-resistant materials, and integrated "production-extraction-sequestration" systems to balance feasibility and economics in low-carbon transitions.

This finding aligns with the research by Ding et al. [27], who used a simulated CO_2/H_2 mixed gas integrated gasification combined cycle (IGCC) to replace CH_4 in methane hydrates and monitored the changes in CH_4, CO_2, and H_2 gas concentrations using a gas chromatograph (GC). Three key points (hydrate phase "A" point, gas–hydrate interface "B" point, and gas phase "C" point) were selected for Raman detection (as shown in Figure 9). As shown in Figure 9a, the experiment formed sI CH_4 hydrate. During the continuous injection of the CO_2/H_2 gas mixture into the reactor, the Raman spectra changes at the hydrate phase point (Figure 9c,d) indicated the formation of CO_2 and CH_4 hydrates, suggesting that the replacement occurs only between CO_2 and CH_4 molecules. No H_2 signal was detected in the figures, as shown in Figure 9b, indicating that H_2 was not encapsulated in the hydrate cages. The experiment demonstrated that the replacement process occurs in two steps: first, the dissociation of CH_4 hydrate, followed by the formation of CO_2 hydrate. Furthermore, the CH_4 recovery rate through CH_4-CO_2/H_2 replacement was greater than 71%, significantly higher than the CH_4-CO_2 replacement (50%). Notably, H_2 does not compete with CH_4 for occupying hydrate cages but rather plays a facilitative role in the CO_2-CH_4 replacement process.

Figure 10a shows the phase equilibrium conditions of hydrates formed by mixtures of $H_2 + CH_4 + CO_2$ at different ratios. The injection ratio of H_2 has a significant impact on the phase equilibrium of the $H_2 + CH_4 + CO_2$ ternary hydrate. As the H_2 ratio increases, the dissociation kinetics of the hydrate become more favorable, but the separation barrier for CO_2 also increases. In contrast, a lower H_2 ratio favors the formation of the hydrate but may lead to a lower CH_4 yield [72].

The study by Sun et al. [73] shows that when the mole fraction of CO_2 in the gas is low, the impact of the mixed gas on the hydrate structure is more significant. Additionally, as the hydrogen content increases, the dissociation rate of CH_4 hydrates also shows an increasing trend. Under appropriate temperature and pressure conditions, the methane replacement rate can reach over 90% [74]. Wang et al. [28] conducted experiments with different gas compositions (mole ratios of CO_2 and H_2), during which they monitored parameters such as the mole ratio of CO_2 and H_2 during the gas exchange process, methane replacement rate, and CO_2 sequestration ratio. The study indicates that there is a balance point between the mole fractions of CO_2 and H_2: too high a mole fraction of CO_2 promotes CO_2 sequestration but sacrifices gas yield, while a higher mole fraction of H_2 favors gas yield but is not conducive to CO_2 sequestration. A balance in CO_2 substitution can be achieved within a CO_2 mole fraction range of 55% to 72%, as shown in Figure 10b. Xu et al. [75] conducted replacement experiments under specific temperature and pressure conditions and analyzed the composition of released gases using gas chromatography

(GC), measuring the CH_4 replacement efficiency during the process. The results show that there are significant differences in the final composition and replacement efficiency of CH_4 in experiments with different gas mixtures. The order of replacement efficiency is as follows: $CO_2/H_2 > CO_2/N_2 >$ pure CO_2. Xie et al. [76] clarified the impact of gas partial pressure on hydrate replacement in CH_4-rich systems. The experimental results, for the first time, show that in a $CO_2/CH_4/H_2$ ternary gas system, even if the partial pressure of CH_4 exceeds the phase equilibrium pressure of CH_4 hydrates, H_2 can promote the premature dissociation of CH_4 hydrates. The effects of different H_2 and CO_2 gas mixing ratios on the extraction efficiency of pure hydrate phases and hydrate-sediment mixed phases are detailed in Table 6.

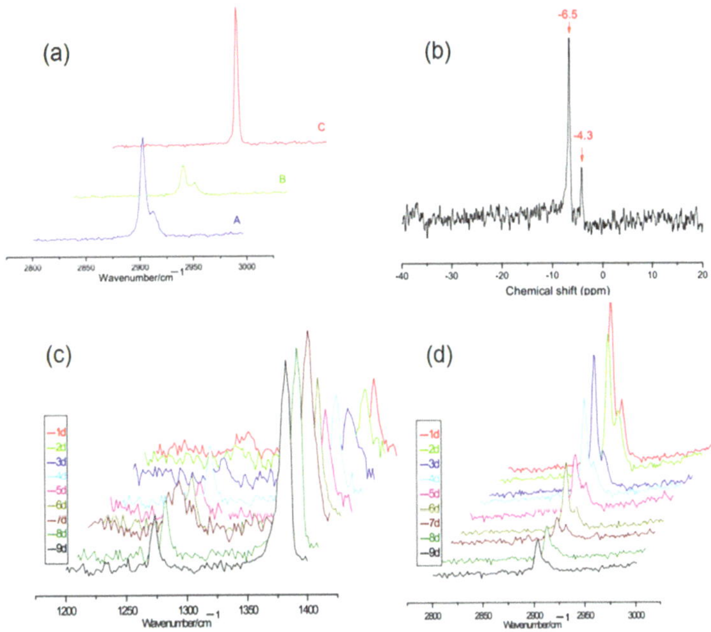

Figure 9. (**a**) Raman spectroscopy of the points A (hydrate phase point), B (gas hydrate interface point), and C (gas phase point) after CH_4 hydrate formation. (**b**) ^1H NMR spectrum of the hydrate phase after replacement. (**c**) The changes of Raman spectroscopy for CO_2 hydrate (**d**) The changes of Raman spectroscopy for CH_4 hydrate [27].

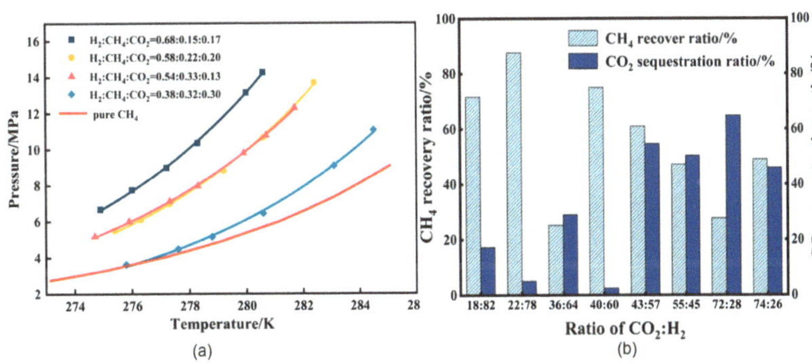

Figure 10. (**a**) Hydrate phase equilibrium conditions for $H_2 + CH_4 + CO_2$ mixtures. (**b**) Comparison of the effect of CO_2 and H_2 at different ratios on CH_4 recovery and CO_2 sequestration rates [72].

Table 6. The effect of the $CO_2 + H_2$ mixed gas replacement method on the enhancement of the natural gas hydrate recovery rate.

Mixed Gas Ratio (CO_2/H_2)	Hydrate Medium	Temperature	Pressure	CH_4 Recovery Rate	Literature Source
0.72/0.28	Sandstone + brine	275.6 K	5.0 MPa	28.0%	Wang et al. [28]
0.55/0.45				47.0%	
0.36/0.64				25.0%	
0.18/0.82				70.0%	Xu et al. [75]
0.4/0.6	Pure water	274.0 K	4.5 MPa	78%	
0.601/0.399		274.2 K	6.0 MPa	32%	Sun et al. [73]
0.74/0.26	Quartzite + brine	276.0 K	3.6 MPa	41.4~52.4%	
0.74/0.26	Quartz sand + brine	276.0 K	3.7 MPa	30.0~50.0%	
0.4/0.6	Quartz sand + brine	276.0 K	3.7 MPa	40.0~75.0%	Sun et al. [73]
0.22/0.78		276.0 K	3.7 MPa	12.0~88.0%	

4.2. CO_2 and N_2 Mixture Displacement to Improve CH_4 Recovery Rate

The application of CO_2 and N_2 mixed gas displacement in methane (CH_4) extraction is considered an effective method to significantly improve the methane recovery rate from natural gas hydrates. The CO_2 and nitrogen gas mixed methane extraction technology involves the simultaneous injection of CO_2 and N_2, which work synergistically to displace adsorbed methane and reduce pressure, achieving both enhanced recovery and carbon sequestration. Its advantage lies in the ability to adjust the gas mixture ratio to suit geological conditions (such as permeability and temperature) while reducing the risk of secondary hydrate formation. However, there are significant technical challenges: the gas mixture ratio needs precise control to avoid pore blockage or decreased efficiency, adsorption separation technologies (such as pressure swing adsorption) have limited adaptability to complex formations, and high-pressure corrosive environments exacerbate equipment wear and increase maintenance costs. In practical applications, while experiences from CO_2-enhanced oil recovery (such as Shengli Oilfield) and nitrogen-assisted extraction can be referenced, large-scale cases remain scarce, and heterogeneous formations are prone to gas retention or micro-seismic risks. Economically, the high initial investment and operational energy consumption depend on carbon trading revenues. However, small and medium-sized enterprises face challenges due to financial pressure and fragmented policies, especially in regions with underdeveloped carbon markets. Future progress will require technological optimization (such as AI control and low-cost catalysts), policy coordination (global carbon pricing), and industry chain integration (gas fields—chemicals—sequestration) to help balance resource development and emission reduction goals in the energy transition. Research by Park et al. [25] shows that when flue gas is used to replace methane (CH_4) in Type I hydrates, the displacement efficiency can reach 85%, while the displacement efficiency of pure CO_2 gas is only 64%. Raman spectroscopy results indicate that approximately 23% of CH_4 in the hydrate cages is replaced by N_2, while 62% is replaced by CO_2. This suggests that CH_4 can be displaced by N_2, and the occupation of N_2 effectively improves the replacement efficiency. Koh et al. [42] summarized that N_2, as a smaller guest molecule, can form type II hydrates on its own. Different concentrations of mixed gases, pure CO_2, and N_2 lead to the formation of different Type I and Type II hydrate structures. Since CO_2 molecules are larger, the addition of N_2 gas can significantly improve the displacement efficiency (as shown in Figure 11). When pure CO_2 gas is used to replace methane in hydrates, CO_2 molecules are thermodynamically unstable in the small Type I cages (5^{12}).

Therefore, displacement mainly occurs in the larger Type I cages ($5^{12}6^2$), and methane molecules in the small cages are not replaced.

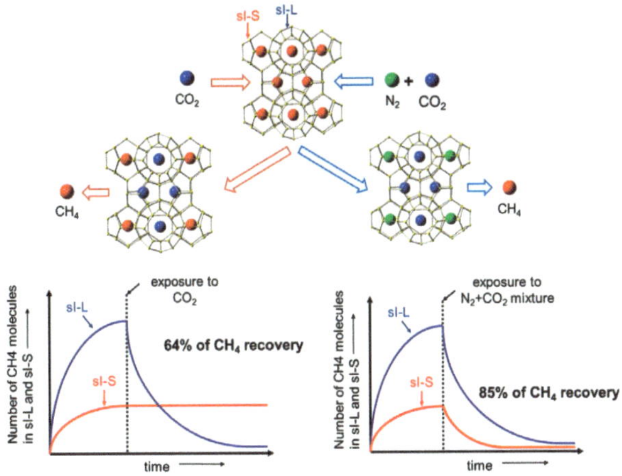

Figure 11. Comparison of sI CH_4 hydrate recovery yield by using pure CO_2 (left) and CO_2 + N_2 mixture (right). sI-L represents $5^{12}6^2$ cages, and sI-S represents 5^{12} cages [30].

After the introduction of N_2, N_2, as a smaller molecule, tends to occupy the small cages, which allows N_2 and CO_2 to work together to achieve the displacement process in both large and small cages. This interaction not only enhances methane recovery but also effectively improves production efficiency [77]. Research by Pandey et al. [78] also indicates that the use of a CO_2 and N_2 mixed gas can further increase methane recovery. Furthermore, as the concentration of CO_2 in the CO_2 + N_2 mixture increases, the methane recovery rate also rises, while the stability of the mixed CH_4-CO_2 hydrate is enhanced.

Specifically, the essential mechanism of this process lies in the competition between N_2 and CO_2 gases for entry into the molecular cage structures of the hydrate. Due to the different molecular behaviors of N_2 and CO_2 during hydrate formation, N_2 gas can influence the dissolution of CO_2 and the conditions for hydrate formation. When N_2 competes with CO_2 for entry into the hydrate cages, it alters the thermodynamic stability of the hydrate, making the temperature and pressure conditions for hydrate formation more stringent than those for pure CO_2 hydrates.

Various researchers have conducted studies on the impact of different CO_2 and N_2 mixing ratios on methane (CH_4) hydrate recovery rates, and the results are shown in Table 7. Lee et al. [79] primarily used Differential Scanning Calorimetry (DSC) and Pressure–Volume–Temperature (PVT) experimental methods to investigate the replacement process of methane hydrate ($CH_4 \cdot nH_2O$) under different gas mixing systems, particularly the displacement process with CO_2 and N_2 mixed gases. The study found that the injection of mixed gases effectively promotes the dissociation of methane hydrates, enabling methane recovery while also trapping CO_2 as CO_2 hydrates or mixed hydrates. Additionally, the gas mixture ratio significantly influences the displacement efficiency, with different gas combinations exhibiting varying replacement effects. During the displacement process, the original pure methane hydrate gradually transforms into a mixed hydrate composed of CH_4, CO_2, and N_2. This change leads to increased fluctuations in heat flow, thereby affecting the thermodynamic and kinetic properties of the hydrate.

In a CO_2 and N_2 mixed gas environment, when the nitrogen (N_2) proportion is relatively low, the methane (CH_4) displacement recovery rate is very low. However, when the N_2 proportion exceeds 50%, the methane displacement recovery rate increases to 12.2%, while the CO_2 hydrate sequestration rate reaches 42.8%. In contrast, when pure liquid CO_2 is used for natural gas hydrate extraction, the CO_2 sequestration rate is only 22.1%, significantly lower than the effectiveness of the CO_2 and N_2 mixed gas displacement method [59,75,79–81].

Table 7. The effect of the CO_2 + N_2 mixed gas displacement method on the improvement of the natural gas hydrate recovery rate.

Gas Mixture Ratio (CO_2/N_2)	Hydrate Medium	Temperature	Pressure	CH_4 Recovery Rate	Literature Source
0.1/0.9	Porous silica + water	274.0 K	11.5/14.6/18.6 MPa	77%/80%/79%	Lee et al. [79]
0.2/0.8			13.7 MPa	80%	
0.6/0.4	Pure water		4.5 MPa	73.4%	Xu et al. [75]
0.146/0.854	Silica sand + water	273.3 K	4.2 MPa	53.3%	Yang et al. [59]
0.28/0.72	Pure water + SDS solution	284.3K	9.0 MPa	13.2%	Niu et al. [80]
0.5/0.5	Pure water	273.9 K	5.0/6.7 MPa	8.3%/17.7%	Zhou et al. [81]
0.75/0.25		274.0 K	2.6/3.2/3.5 MPa	9.5%/12.6%/17.9%	

Experimental results and the equilibrium pressure data of the hydrate at different N_2 ratios (as shown in Figure 12) indicate that when CO_2 and N_2 mixed gases are used for natural gas hydrate extraction, the equilibrium pressure of the hydrate increases with the increase in the N_2 proportion. At higher N_2 concentrations, the hydrate requires higher pressure to maintain its stability. The stability of CH_4 hydrate decreases, making it more prone to replacement reactions, and the recovery rate significantly improves [72].

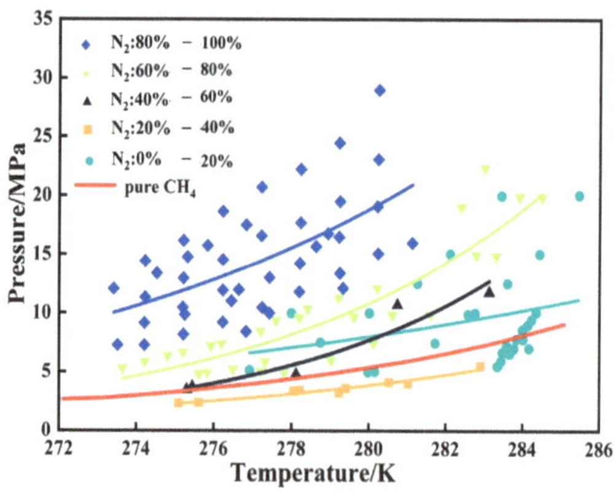

Figure 12. Images of hydrate phase equilibrium at different N_2 ratios [72].

Since sI hydrates are widely present in natural gas hydrates, most studies focus on the replacement reactions in sI hydrates. However, sII hydrates have also been found in nature in certain regions, and the replacement mechanism of sII hydrates is significantly different from that of sI hydrates. Seo et al. promoted the replacement reaction of sII (C_3H_8 + CH_4) hydrate by externally injecting a CO_2/N_2 (50:50) mixed gas. The study showed that the replacement degree of C_3H_8 + CH_4 hydrate was about 54%. This method not only achieved a higher gas recovery rate than the pure CO_2 replacement method but also maintained the crystal structure of the hydrate after the replacement reaction (as shown in Figure 13b), indicating that the replacement process did not cause any structural changes in the sII hydrate. At the same time, according to the Raman spectrum of the CO_2 and N_2 molecular vibration modes in the C_3H_8 + CH_4 hydrate (Figure 13a), it was observed that both CO_2 and N_2 molecules occupied the cages of the sII hydrate. In Figure 13c, the reduction in C_3H_8 and CH_4 amounts in the initial C3H8 + CH_4 hydrate and the increase in CO_2 and N_2 amounts in the replaced C_3H_8 + CH_4 hydrate are similar, further confirming that CH_4 in the small cages is replaced by N_2, while C_3H_8 in the large cages is mainly replaced by CO_2 [82].

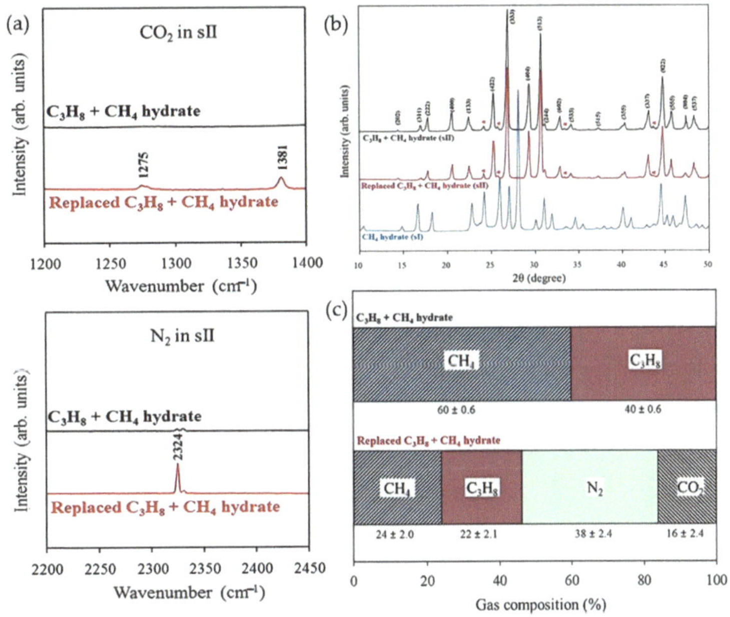

Figure 13. (**a**) Raman spectra of the initial C_3H_8 + CH_4 hydrate and replaced C_3H_8 + CH_4 hydrate with CO_2/N_2 gas. (**b**) PXRD patterns of the CH_4 hydrate, initial C_3H_8 + CH_4 hydrate, and replaced C_3H_8 + CH_4 hydrate with CO_2/N_2 gas. Asterisks indicate hexagonal ice. (**c**) Guest composition and standard deviation of the initial C_3H_8 + CH_4 hydrate and replaced C_3H_8 + CH_4 hydrate with CO_2/N_2 gas [82].

Overall, although a higher proportion of N_2 helps improve the displacement efficiency and CO_2 sequestration rate, it also introduces new challenges, such as the formation of multiple hydrates and the encapsulation effect, which may adversely affect the methane displacement recovery rate. Therefore, when using mixed gas displacement methods for natural gas hydrate extraction, it is necessary to comprehensively consider factors such as gas ratios, displacement efficiency, and the formation and encapsulation effects of hydrates to achieve optimal recovery and CO_2 sequestration outcomes.

4.3. The Displacement of CH_4 Recovery Is Improved by Using CO_2 Mixed with N_2 and H_2

When using a CO_2 and N_2 mixed gas for natural gas hydrate displacement, it can promote the dissociation of hydrates under lower pressure conditions, theoretically improving gas displacement efficiency. However, in practical operations, the displacement effect of the mixed gas has not fully achieved the expected recovery rate, indicating that some factors during the actual dissociation process may have affected the enhancement of displacement efficiency.

The main reason for this phenomenon is that the formation of CO_2 hydrates and N_2 hydrates adversely affects the recovery of internal natural gas hydrates, primarily methane hydrates. Specifically, CO_2 and N_2 interact with the hydrate structure during the displacement process, forming CO_2 hydrates and N_2 hydrates. These newly formed hydrate structures can encapsulate the original methane hydrate particles, hindering their dissociation and release, thus impacting the methane recovery efficiency [29].

Therefore, under different pressure and temperature conditions, Chaturvedi et al. [83] studied the effects of CO_2, N_2, and H_2 combinations on hydrate formation and methane recovery. At higher N_2 concentrations, even with the introduction of a small amount of H_2, the methane recovery rate in natural gas hydrates could still be increased to 67–69 mol% (Figure 14a–c). When the H_2 concentration increased from 0.1 mol% to 1 mol%, the methane recovery rate further improved to 71 mol%, and the presence of hydrogen at high pressure promoted natural gas hydrate formation. Previous studies have paid less attention to the synergistic effect of H_2 in the presence of N_2, and a possible explanation has been proposed: small gas molecules like N_2 occupy smaller hydrate cages, preventing CO_2 from entering larger cages and thus hindering hydrate formation. However, H_2 competes with N_2 for the smaller cages, allowing CO_2 to fill the larger cages for a longer time until the hydrate channels close. This theory, however, still requires further verification. Although the hydrogen promoter increased the hydrate yield and expanded the hydrate formation region, its effect remained limited at high N_2 concentrations. Further research indicated that when the H_2 mol% exceeded 1 mol%, the methane recovery rate no longer increased significantly (Figure 14d–g), possibly because H_2 occupies hydrate cages, hindering further natural gas hydrate formation.

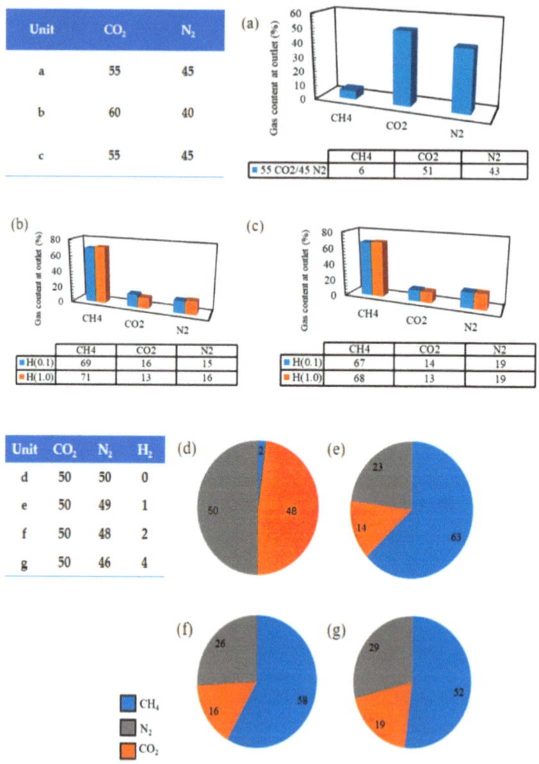

Figure 14. (**a–c**) Role of H_2 dosing in the flue gas of high N_2 concentration on the recovery of CH_4 from gas hydrates via CO_2-CH_4 recovery. (**d–g**) Role of high volume H_2 dosing in flue gas of high N_2 concentration on the recovery of CH_4 from gas hydrates via CO_2-CH_4 recovery [83].

The mechanism can be further illustrated in Figure 15. The addition of hydrogen helps optimize the hydrate formation process of the CO_2 + N_2 mixed gas, and H_2 helps increase the rate of hydrate formation, thereby accelerating CO_2 storage and conversion and optimizing the gas displacement effect for methane. However, if the hydrogen concentration is excessively high, it could disrupt the stability of CO_2 hydrates. Elevated hydrogen levels may compete with CO_2 for hydrate formation, thereby hindering the formation of CO_2 hydrates and diminishing its ability to replace methane effectively. Moreover, hydrogen itself does not have a strong ability to form hydrates, and its hydrate formation rate is much lower than that of CO_2, and excessively high hydrogen concentrations may lead to adverse changes in the structure and thermodynamic properties of the hydrates, affecting the overall stability of the reservoir. Therefore, to ensure reservoir stability and maximize methane recovery, it is essential to strictly control the hydrogen proportion when injecting CO_2 + N_2 + H_2 mixed gas [27].

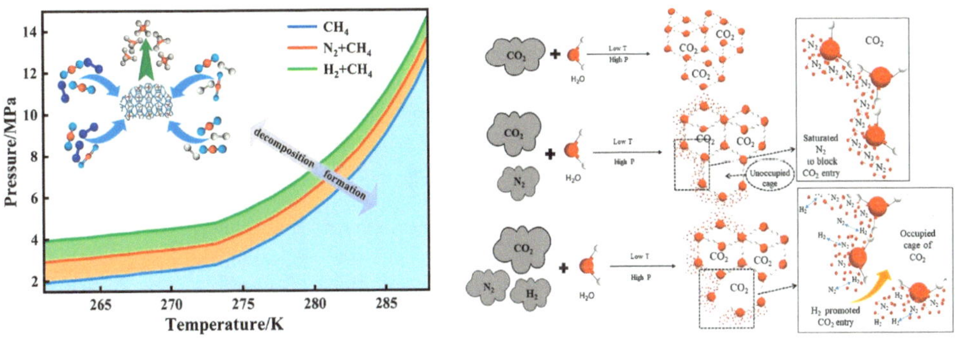

(a) (b)

Figure 15. (**a**) Schematic diagram of N_2 and H_2 boosting the hydrate phase equilibrium pressure to promote CH_4 hydrate decomposition [72]. (**b**) Proposed schematic showing H_2's role in increasing CO_2 hydrate formation in the presence of N_2 [29].

5. Conclusions

Displacement reactions, as an important method for the production and storage of multi-component gases from hydrates, have made significant progress in basic research and laboratory stages, demonstrating their broad application potential in areas such as energy storage, gas separation, and environmental protection. Storing gases in hydrates not only significantly increases the gas storage density but also enables efficient gas release and reuse under low-temperature and normal-pressure conditions.

However, despite some success in laboratory studies of hydrate displacement reactions, several challenges remain in practical applications. First, the efficiency and reaction rate of the displacement reaction are still limited, and the thermodynamic issues and optimization of reaction conditions need further research. Second, the selective displacement and stability of multi-component gases need to be addressed through additional theoretical and experimental work.

To further improve the technical efficiency of methane displacement with mixed gases and achieve sustainable development, the following research directions need to be explored:

Improving the accuracy of experimental devices and numerical models: Future work should focus on developing more accurate experimental equipment and numerical models that are closer to real extraction environments, validating the performance and effectiveness of different gas mixtures.

In-depth study of the internal mechanisms of the displacement process: Future research should use high-resolution experiments and theoretical simulations to clarify the thermodynamic and kinetic characteristics of different reaction paths, providing a theoretical basis for optimizing displacement efficiency.

Optimizing the gas mixture ratio and injection strategy: For the injection of CO_2, N_2, and H_2 mixed gases, optimizing the ratio and injection strategy to achieve the best methane recovery rate and CO_2 storage efficiency is a key area for future research. Additionally, the synergistic effects of salt additives and gases should be considered, exploring the combined impact of CO_2+N_2+salt additives on reservoir stability, recovery rate, and CO_2 sequestration efficiency.

Improving CO_2 sequestration and stability: Future research should focus on the long-term safety of CO_2 sequestration, the storage capacity of reservoirs, and potential leakage pathways to ensure the environmental safety of the technology.

Enhancing the dual utilization benefits of CO_2 and methane: Combining the extraction of CO_2 and CH_4 hydrates with long-term CO_2 sequestration through integrated energy recovery and greenhouse gas emission reduction measures can enhance the overall benefits of natural gas hydrate extraction. The secondary sequestration method combining high-temperature CO_2 with deep geothermal energy can improve CO_2 sequestration rates while potentially enhancing methane recovery, promoting the development of this technology in a green and energy-efficient direction.

Author Contributions: Z.Z.: data reduction and writing and editing; X.Z.: methodology and review and editing. S.W.: conceptualization and data analysis; L.J.: literature investigation and summary; H.D.: review and revision. P.L.: methodology, resources, supervision, and editing. All authors have read and agreed to the published version of the manuscript.

Funding: This study was supported by the National Natural Science Foundation of China (Grant No. 52376175, 52106213), Shanxi Scholarship Council of China (2024-115), and Key Talent Team for Science and Technology Innovation in Shanxi Province (202204051002023).

Conflicts of Interest: The authors declare no conflicts of interest.

References

1. Yu, Y.-S.; Zhang, X.; Liu, J.-W.; Lee, Y.; Li, X.S. Natural gas hydrate resources and hydrate technologies: A review and analysis of the associated energy and global warming challenges. *Energy Environ. Sci.* **2021**, *14*, 5611–5668. [CrossRef]
2. Musakaev, N.; Khasanov, M.; Borodin, S.L. The mathematical model of the gas hydrate deposit development in permafrost. *Int. J. Heat Mass Transf.* **2018**, *118*, 455–461. [CrossRef]
3. Wei, N.; Pei, J.; Li, H.; Zhou, S.; Zhao, J.; Kvamme, B.; Coffin, R.B.; Zhang, L.; Zhang, Y.; Xue, J.; et al. Classification of natural gas hydrate resources: Review, application and prospect. *Gas Sci. Eng.* **2024**, *124*, 205269. [CrossRef]
4. Ma, K.; Li, D.; Liang, D. Reservoir stimulation technologies for natural gas hydrate: Research progress, challenges, and perspectives. *Energy Fuels.* **2023**, *37*, 10112–10133. [CrossRef]
5. Makogon, Y.F. Natural gas hydrates—A promising source of energy. *J. Nat. Gas Sci. Eng.* **2010**, *2*, 49–59. [CrossRef]
6. Sun, H.; Chen, B.; Zhu, Z.; Zhang, L.; Yang, M.; Song, Y. Research development in the traditional methods and water flow erosion for natural gas hydrate production: A review. *Energy Technol.* **2023**, *11*, 2201011. [CrossRef]
7. Ji, C.; Ahmadi, G.; Smith, D.H. Natural gas production from hydrate decomposition by depressurization. *Chem. Eng. Sci.* **2001**, *56*, 5801–5814. [CrossRef]
8. Fitzgerald, G.C.; Castaldi, M.J. Thermal Stimulation Based Methane Production from Hydrate Bearing Quartz Sediment. *Ind. Eng. Chem. Res.* **2013**, *52*, 6571–6581. [CrossRef]
9. Farhadian, A.; Shadloo, A.; Zhao, X.; Pavelyev, R.S.; Peyvandi, K.; Qiu, Z.; Varfolomeev, M.A. Challenges and advantages of using environmentally friendly kinetic gas hydrate inhibitors for flow assurance application: A comprehensive review. *Fuel* **2023**, *336*, 127055. [CrossRef]
10. Li, X.-S.; Yang, B.; Zhang, Y.; Li, G.; Duan, L.-P.; Wang, Y.; Chen, Z.-Y.; Huang, N.-S.; Wu, H.-J. Experimental investigation into gas production from methane hydrate in sediment by depressurization in a novel pilot-scale hydrate simulator. *Appl. Energy* **2012**, *93*, 722–732. [CrossRef]

11. Zhang, Y.; Li, X.-S.; Chen, Z.-Y.; Wang, Y.; Ruan, X.-K. Effect of Hydrate Saturation on the Methane Hydrate Dissociation by Depressurization in Sediments in a Cubic Hydrate Simulator. *Ind. Eng. Chem. Res.* **2015**, *54*, 2627–2637. [CrossRef]
12. Wan, K.; Wang, Y.; Li, X.-S.; Zhang, L.-H.; Meng, T. Pilot-scale experimental study on natural gas hydrate decomposition with innovation depressurization modes. *Appl. Energy* **2024**, *373*, 123921. [CrossRef]
13. Cranganu, C. In-situ thermal stimulation of gas hydrates. *J. Pet. Sci. Eng.* **2009**, *65*, 76–80. [CrossRef]
14. Yamada, H.; Chen, L.; Lacaille, G.; Shoji, E.; Okajima, J.; Komiya, A.; Maruyama, S. Experimental Study of Methane Hydrate Dissociation and Gas Production Behaviors under Depressurization. *Int. J. Mech. Eng. Robot. Res.* **2017**, *6*, 140–146. [CrossRef]
15. Zhao, E.; Hou, J.; Du, Q.; Liu, Y.; Ji, Y.; Bai, Y. Numerical modeling of gas production from methane hydrate deposits using low-frequency electrical heating assisted depressurization method. *Fuel* **2021**, *290*, 120075. [CrossRef]
16. Liao, B.; Wang, J.; Sun, J.; Lv, K.; Liu, L.; Wang, Q.; Wang, R.; Lv, X.; Wang, Y.; Chen, Z. Microscopic insights into synergism effect of different hydrate inhibitors on methane hydrate formation: Experiments and molecular dynamics simulations. *Fuel* **2023**, *340*, 127488. [CrossRef]
17. Yagasaki, T.; Matsumoto, M.; Tanaka, H. Effects of thermodynamic inhibitors on the dissociation of methane hydrate: A molecular dynamics study. *Phys. Chem. Chem. Phys.* **2015**, *17*, 32347–32357. [CrossRef]
18. Wang, P.; Li, Y.; Sun, N.; Han, S.; Wang, X.; Su, Q.; Li, Y.; He, J.; Yu, X.; Du, S. Hydrate technologies for CO_2 capture and sequestration: Status and perspectives. *Chem. Rev.* **2024**, *124*, 10363–10385. [CrossRef]
19. Yan, P.; Luan, H.; Jiang, Y.; Liang, W.; Liu, M.; Chen, H. Influence of depressurization mode on natural gas hydrate production characteristics: One-dimensional experimental study. *Geoenergy Sci. Eng.* **2024**, *234*, 212671. [CrossRef]
20. Lee, H.; Seo, Y.; Seo, Y.T.; Moudrakovski, I.L.; Ripmeester, J.A. Recovering methane from solid methane hydrate with carbon dioxide. *Angew. Chem. Vol.* **2003**, *115*, 5202–5205. [CrossRef]
21. Wang, Y.; Lang, X.; Fan, S.; Wang, S.; Yu, C.; Li, G. Review on Enhanced Technology of Natural Gas Hydrate Recovery by Carbon Dioxide Replacement. *Energy Fuels* **2021**, *35*, 3659–3674. [CrossRef]
22. Ohgaki, K.; Takano, K.; Sangawa, H.; Matsubara, T.; Nakano, S. Methane exploitation by carbon dioxide from gas hydrates—Phase equilibria for CO_2-CH_4 mixed hydrate system. *J. Chem. Eng. Jpn.* **1996**, *29*, 478–483. [CrossRef]
23. Yezdimer, E.M.; Cummings, P.T.; Chialvo, A.A. Determination of the Gibbs free energy of gas replacement in SI clathrate hydrates by molecular simulation. *J. Phys. Chem. A* **2002**, *106*, 7982–7987. [CrossRef]
24. Ota, M.; Abe, Y.; Watanabe, M.; Smith, R.L.; Inomata, H. Methane recovery from methane hydrate using pressurized CO_2. *Fluid Phase Equilibria* **2005**, *228–229*, 553–559. [CrossRef]
25. Park, Y.; Kim, D.-Y.; Lee, J.-W.; Huh, D.-G.; Park, K.-P.; Lee, J.; Lee, H. Sequestering carbon dioxide into complex structures of naturally occurring gas hydrates. *Proc. Natl. Acad. Sci. USA* **2006**, *103*, 12690–12694. [CrossRef]
26. Sloan Jr, E.D.; Koh, C.A. *Clathrate Hydrates of Natural Gases*; CRC Press: Boca Raton, FL, USA, 2007.
27. Ding, Y.-L.; Xu, C.-G.; Yu, Y.-S.; Li, X.-S. Methane recovery from natural gas hydrate with simulated IGCC syngas. *Energy* **2017**, *120*, 192–198. [CrossRef]
28. Wang, X.-H.; Sun, Y.-F.; Wang, Y.-F.; Li, N.; Sun, C.-Y.; Chen, G.-J.; Liu, B.; Yang, L.-Y. Gas production from hydrates by CH_4-CO_2/H_2 replacement. *Appl. Energy* **2017**, *188*, 305–314. [CrossRef]
29. Chaturvedi, K.R.; Sinha, A.; Nair, V.C.; Sharma, T. Enhanced carbon dioxide sequestration by direct injection of flue gas doped with hydrogen into hydrate reservoir: Possibility of natural gas production. *Energy* **2021**, *227*, 120521. [CrossRef]
30. Wei, W.-N.; Li, B.; Gan, Q.; Li, Y.-L. Research progress of natural gas hydrate exploitation with CO2 replacement: A review. *Fuel* **2022**, *312*, 122873. [CrossRef]
31. Arora, A.; Cameotra, S.S.; Balomajumder, C. Techniques for exploitation of gas hydrate (clathrates) an untapped resource of methane gas. *Energy Fuels* **2015**, *7*, 108–111.
32. Lo, H.; Lee, M.-T.; Lin, S.-T. Water vacancy driven diffusion in clathrate hydrates: Molecular dynamics simulation study. *J. Phys. Chem. C* **2017**, *121*, 8280–8289. [CrossRef]
33. Peters, B.; Zimmermann, N.E.; Beckham, G.T.; Tester, J.W.; Trout, B.L. Path sampling calculation of methane diffusivity in natural gas hydrates from a water-vacancy assisted mechanism. *J. Am. Chem. Soc.* **2008**, *130*, 17342–17350. [CrossRef] [PubMed]
34. Schicks, J.M. Thermodynamic properties and phase equilibria characteristics of natural gas hydrates. In *Advances in Natural Gas: Formation, Processing, and Applications. Volume 3: Natural Gas Hydrates*; Elsevier: Amsterdam, The Netherlands, 2024; pp. 65–86.
35. Anderson, G.K. Enthalpy of dissociation and hydration number of carbon dioxide hydrate from the Clapeyron equation. *J. Chem. Thermodyn.* **2003**, *35*, 1171–1183. [CrossRef]
36. Ma, Z.W.; Zhang, P.; Bao, H.S.; Deng, S. Review of fundamental properties of CO_2 hydrates and CO_2 capture and separation using hydration method. *Renew. Sustain. Energy Rev.* **2016**, *53*, 1273–1302. [CrossRef]
37. Hachikubo, A.; Miura, T.; Yamada, K.; Sakagami, H.; Takahashi, N.; Hyakutake, K.; Abe, K.; Shoji, H. Phase equilibrium and comparison of formation speeds of CH_4 and CO_2 hydrate below the ice point. In Proceedings of the 5th International Conference on Gas Hydrates, Trondheim, Norway, 13–16 June 2005.

38. Goel, N. In situ methane hydrate dissociation with carbon dioxide sequestration: Current knowledge and issues. *J. Pet. Sci. Eng.* **2006**, *51*, 169–184. [CrossRef]
39. Circone, S.; Stern, L.A.; Kirby, S.H.; Durham, W.B.; Chakoumakos, B.C.; Rawn, C.J.; Rondinone, A.J.; Ishii, Y. CO_2 hydrate: Synthesis, composition, structure, dissociation behavior, and a comparison to structure I CH_4 hydrate. *J. Phys. Chem. B* **2003**, *107*, 5529–5539. [CrossRef]
40. Xie, W.; Wang, H.; Vandeginste, V.; Chen, S.; Gan, H.; Wang, M.; Yu, Z. Thermodynamic and kinetic affinity of CO_2 relative to CH_4 and their pressure, temperature and pore structure sensitivity in the competitive adsorption system in shale gas reservoirs. *Energy* **2023**, *277*, 127591. [CrossRef]
41. Yin, Z.; Khurana, M.; Tan, H.K.; Linga, P. A review of gas hydrate growth kinetic models. *Chem. Eng. J.* **2018**, *342*, 9–29. [CrossRef]
42. Koh, D.-Y.; Kang, H.; Lee, J.-W.; Park, Y.; Kim, S.-J.; Lee, J.; Lee, J.Y.; Lee, H. Energy-efficient natural gas hydrate production using gas exchange. *Appl. Energy* **2016**, *162*, 114–130. [CrossRef]
43. Sloan, E.; Koh, C.A. Estimation techniques for phase Equilibria of natural gas hydrates. In *Clathrate Hydrates of Natural Gases;* Taylor & Francis Group: London, UK, 2008; pp. 320–523.
44. Zhao, J.; Xu, K.; Song, Y.; Liu, W.; Lam, W.; Liu, Y.; Xue, K.; Zhu, Y.; Yu, X.; Li, Q. A Review on Research on Replacement of CH_4 in Natural Gas Hydrates by Use of CO_2. *Energies* **2012**, *5*, 399–419. [CrossRef]
45. Speight, J.G. *Natural Gas: A Basic Handbook;* Gulf Professional Publishing: Houston, TX, USA, 2018.
46. Liu, T.; Wu, P.; Chen, Z.; Li, Y. Review on carbon dioxide replacement of natural gas hydrate: Research progress and perspectives. *Energy Fuels* **2022**, *36*, 7321–7336. [CrossRef]
47. Ota, M.; Morohashi, K.; Abe, Y.; Watanabe, M.; Smith Jr, R.L.; Inomata, H. Replacement of CH_4 in the hydrate by use of liquid CO_2. *Energy Convers. Manag.* **2005**, *46*, 1680–1691. [CrossRef]
48. Bai, D.; Zhang, X.; Chen, G.; Wang, W. Replacement mechanism of methane hydrate with carbon dioxide from microsecond molecular dynamics simulations. *Energy Environ. Sci.* **2012**, *5*, 7033–7041. [CrossRef]
49. Nakano, S.; Moritoki, M.; Ohgaki, K. High-pressure phase equilibrium and Raman microprobe spectroscopic studies on the CO_2 hydrate system. *J. Chem. Eng. Data* **1998**, *43*, 807–810. [CrossRef]
50. Tung, Y.-T.; Chen, L.-J.; Chen, Y.-P.; Lin, S.-T. Growth of structure I carbon dioxide hydrate from molecular dynamics simulations. *J. Phys. Chem. C* **2011**, *115*, 7504–7515. [CrossRef]
51. Cao, X.; Wang, H.; Yang, K.; Wu, S.; Chen, Q.; Bian, J. Hydrate-based CO_2 sequestration technology: Feasibilities, mechanisms, influencing factors, and applications. *J. Pet. Sci. Eng.* **2022**, *219*, 111121. [CrossRef]
52. Uchida, T.; Ikeda, I.Y.; Takeya, S.; Kamata, Y.; Ohmura, R.; Nagao, J.; Zatsepina, O.Y.; Buffett, B.A. Kinetics and stability of CH4–CO2 mixed gas hydrates during formation and long-term storage. *ChemPhysChem* **2005**, *6*, 646–654. [CrossRef] [PubMed]
53. Kuang, Y.; Zhang, L.; Zheng, Y. Enhanced CO_2 sequestration based on hydrate technology with pressure oscillation in porous medium using NMR. *Energy* **2022**, *252*, 124082. [CrossRef]
54. Farahani, M.V.; Guo, X.; Zhang, L.; Yang, M.; Hassanpouryouzband, A.; Zhao, J.; Yang, J.; Song, Y.; Tohidi, B. Effect of thermal formation/dissociation cycles on the kinetics of formation and pore-scale distribution of methane hydrates in porous media: A magnetic resonance imaging study. *Sust. Energy Fuels* **2021**, *5*, 1567–1583. [CrossRef]
55. Wallmann, K.; Riedel, M.; Hong, W.; Patton, H.; Hubbard, A.; Pape, T.; Hsu, C.; Schmidt, C.; Johnson, J.; Torres, M. Gas hydrate dissociation off Svalbard induced by isostatic rebound rather than global warming. *Nat. Commun.* **2018**, *9*, 83. [CrossRef]
56. Zhao, J.; Chen, X.; Song, Y.; Zhu, Z.; Yang, L.; Tian, Y.; Wang, J.; Yang, M.; Zhang, Y. Experimental study on a novel way of methane hydrates recovery: Combining CO_2 replacement and depressurization. *Energy Procedia* **2014**, *61*, 75–79. [CrossRef]
57. Lee, Y.; Deusner, C.; Kossel, E.; Choi, W.; Seo, Y.; Haeckel, M. Influence of CH_4 hydrate exploitation using depressurization and replacement methods on mechanical strength of hydrate-bearing sediment. *Appl. Energy* **2020**, *277*, 115569. [CrossRef]
58. Chen, Y.; Gao, Y.; Chen, L.; Wang, X.; Liu, K.; Sun, B. Experimental investigation of the behavior of methane gas hydrates during depressurization-assisted CO_2 replacement. *J. Nat. Gas Sci. Eng.* **2019**, *61*, 284–292. [CrossRef]
59. Yang, J.; Okwananke, A.; Tohidi, B.; Chuvilin, E.; Maerle, K.; Istomin, V.; Bukhanov, B.; Cheremisin, A. Flue gas injection into gas hydrate reservoirs for methane recovery and carbon dioxide sequestration. *Energy Convers. Manag.* **2017**, *136*, 431–438. [CrossRef]
60. Sun, Y.; Zhang, G.; Li, S.; Jiang, S. CO_2/N_2 injection into CH_4 + C_3H_8 hydrates for gas recovery and CO_2 sequestration. *Chem. Eng. J.* **2019**, *375*, 121973. [CrossRef]
61. Defu, M.; Yingxia, Q.I. Experimental study on the replacement of methane hydrate by CO_2 with thermal excitation. *Energy* **2017**, *33*, 13–18.
62. Zhang, L.; Yang, L.; Wang, J.; Zhao, J.; Dong, H.; Yang, M.; Liu, Y.; Song, Y. Enhanced CH_4 recovery and CO_2 storage via thermal stimulation in the CH_4/CO_2 replacement of methane hydrate. *Chem. Eng. J.* **2017**, *308*, 40–49. [CrossRef]
63. Tupsakhare, S.S.; Castaldi, M.J. Efficiency enhancements in methane recovery from natural gas hydrates using injection of CO_2/N_2 gas mixture simulating in-situ combustion. *Appl. Energy* **2019**, *236*, 825–836. [CrossRef]

64. Fan, S.; Yu, W.; Yu, C.; Wang, Y.; Lang, X.; Wang, S.; Li, G.; Huang, H. Investigation of enhanced exploitation of natural gas hydrate and CO_2 sequestration combined gradual heat stimulation with CO_2 replacement in sediments. *J. Nat. Gas Sci. Eng.* **2022**, *104*, 104686. [CrossRef]
65. Mohammadi, A.H.; Eslamimanesh, A.; Richon, D. Semi-clathrate hydrate phase equilibrium measurements for the CO_2+H_2/CH_4+tetra-n-butylammonium bromide aqueous solution system. *Chem. Eng. Sci.* **2013**, *94*, 284–290. [CrossRef]
66. Khlebnikov, V.; Antonov, S.; Mishin, A.; Bakulin, D.; Khamidullina, I.; Liang, M.; Vinokurov, V.; Gushchin, P.A. A new method for the replacement of CH_4 with CO_2 in natural gas hydrate production. *Nat. Gas Ind. B* **2016**, *3*, 445–451. [CrossRef]
67. Heydari, A.; Peyvandi, K. Study of biosurfactant effects on methane recovery from gas hydrate by CO_2 replacement and depressurization. *Fuel* **2020**, *272*, 117681. [CrossRef]
68. Choi, W.; Mok, J.; Lee, J.; Lee, Y.; Lee, J.; Sum, A.K.; Seo, Y. Effective CH_4 production and novel CO_2 storage through depressurization-assisted replacement in natural gas hydrate-bearing sediment. *Appl. Energy* **2022**, *326*, 119971. [CrossRef]
69. Adibi, N.; Mohammadi, M.; Ehsani, M.R.; Khanmohammadian, E. Experimental investigation of using combined CH_4/CO_2 replacement and thermal stimulation methods for methane production from gas hydrate in the presence of SiO_2 and ZnO nanoparticles. *J. Nat. Gas Sci. Eng.* **2020**, *84*, 103690. [CrossRef]
70. Lee, S.; Lee, Y.; Lee, J.; Lee, H.; Seo, Y. Experimental verification of methane–carbon dioxide replacement in natural gas hydrates using a differential scanning calorimeter. *Environ. Sci. Technol.* **2013**, *47*, 13184–13190. [CrossRef]
71. Sun, Y.-F.; Wang, Y.-F.; Zhong, J.-R.; Li, W.-Z.; Li, R.; Cao, B.-J.; Kan, J.-Y.; Sun, C.-Y.; Chen, G.-J. Gas hydrate exploitation using CO2/H2 mixture gas by semi-continuous injection-production mode. *Appl. Energy* **2019**, *240*, 215–225. [CrossRef]
72. Zhang, X.; Zhang, S.; Yin, S.; Guanyu, H.; Li, J.; Wu, Q. Research progress of the kinetics on natural gas hydrate replacement by CO_2-containing mixed gas: A review. *J. Nat. Gas Sci. Eng.* **2022**, *108*, 104837. [CrossRef]
73. Sun, Y.-F.; Zhong, J.-R.; Li, R.; Zhu, T.; Cao, X.-Y.; Chen, G.-J.; Wang, X.-H.; Yang, L.-Y.; Sun, C.-Y. Natural gas hydrate exploitation by CO_2/H_2 continuous Injection-Production mode. *Appl. Energy* **2018**, *226*, 10–21. [CrossRef]
74. Wang, M.; Wang, X.; Deng, C.; Liu, B.; Sun, C.; Chen, G.; El-Halwagi, M. Process modeling and energy efficiency analysis of natural gas hydrate production by CH_4-CO_2/H_2 replacement coupling steam methane reforming. In *Computer Aided Chemical Engineering*; Elsevier: Amsterdam, The Netherlands, 2019; Volume 47, pp. 131–136. [CrossRef]
75. Xu, C.-G.; Cai, J.; Yu, Y.-S.; Chen, Z.-Y.; Li, X.-S. Research on micro-mechanism and efficiency of CH_4 exploitation via CH_4-CO_2 replacement from natural gas hydrates. *Fuel* **2018**, *216*, 255–265. [CrossRef]
76. Xie, Y.; Zheng, T.; Zhu, Y.; Sun, C.; Chen, G.; Feng, J. H_2 promotes the premature replacement of CH_4–CO_2 hydrate even when the CH_4 gas-phase pressure exceeds the phase equilibrium pressure of CH_4 hydrate. *Renew. Sustain. Energy Rev.* **2024**, *200*, 114582. [CrossRef]
77. Sun, S.; Hao, Y.; Zhao, J. Analysis of gas source for the replacement of CH_4 with CO_2 in gas hydrate production from the perspective of dissociation enthalpy. *J. Chem. Eng. Data* **2018**, *63*, 684–690. [CrossRef]
78. Pandey, J.S.; Solms, N. Hydrate stability and methane recovery from gas hydrate through CH4–CO2 replacement in different mass transfer scenarios. *Energies* **2019**, *12*, 2309. [CrossRef]
79. Lee, Y.; Kim, Y.; Lee, J.; Lee, H.; Seo, Y. CH_4 recovery and CO_2 sequestration using flue gas in natural gas hydrates as revealed by a micro-differential scanning calorimeter. *Appl. Energy* **2015**, *150*, 120–127. [CrossRef]
80. Niu, M.; Wu, G.; Yin, Z.; Sun, Y.; Liu, K.; Chen, D. Effectiveness of CO_2-N_2 injection for synergistic CH_4 recovery and CO_2 sequestration at marine gas hydrates condition. *Chem. Eng. J.* **2021**, *420*, 129615. [CrossRef]
81. Zhou, X.; Fan, S.; Liang, D.; Du, J. Replacement of methane from quartz sand-bearing hydrate with carbon dioxide-in-water emulsion. *Energy Fuels* **2008**, *22*, 1759–1764. [CrossRef]
82. Seo, Y.-j.; Park, S.; Kang, H.; Ahn, Y.-H.; Lim, D.; Kim, S.-J.; Lee, J.; Lee, J.Y.; Ahn, T.; Seo, Y. Isostructural and cage-specific replacement occurring in sII hydrate with external CO_2/N_2 gas and its implications for natural gas production and CO_2 storage. *Appl. Energy* **2016**, *178*, 579–586. [CrossRef]
83. Chaturvedi, K.R.; Sharma, T.; Kumar, G.S. Experimental investigation for comparative effectiveness of CO_2+N_2 and CO_2+N_2+H_2 on integrated methane production and carbon storage from natural hydrate media. *J. Environ. Chem. Eng.* **2023**, *11*, 109388. [CrossRef]

MDPI AG
Grosspeteranlage 5
4052 Basel
Switzerland
Tel.: +41 61 683 77 34

MDPI Books Editorial Office
E-mail: books@mdpi.com
www.mdpi.com/books

www.ingramcontent.com/pod-product-compliance
Lightning Source LLC
Chambersburg PA
CBHW041406210926
51850CB00013B/3191
* 9 7 8 3 7 2 5 8 3 7 1 3 7 *